CAMBRIDGE LIBRARY COLLECTION

Books of enduring scholarly value

Life Sciences

Until the nineteenth century, the various subjects now known as the life sciences were regarded either as arcane studies which had little impact on ordinary daily life, or as a genteel hobby for the leisured classes. The increasing academic rigour and systematisation brought to the study of botany, zoology and other disciplines, and their adoption in university curricula, are reflected in the books reissued in this series.

The Life, Letters and Labours of Francis Galton

A controversial figure, Sir Francis Galton (1822–1911), biostatistician, human geneticist, eugenicist, and first cousin of Charles Darwin, is famed as the father of eugenics. Believing that selective breeding was the only hope for the human race, Galton undertook many investigations of human abilities and devoted the last few years of his life to promoting eugenics. Although he intended his studies to work positively, for eradicating hereditary diseases, his research had a hugely negative impact on the world which subsequently bestowed on Galton a rather sinister reputation. Written by Galton's colleague, eugenicist and statistician Karl Pearson (1857–1936), this four-volume biography pieces together a fascinating life. First published in 1930, this second part of Volume 3 includes records of Galton's travels and occupations, and recollections of friends and family. Pearson himself was later appointed the first Galton professor of eugenics at University College London.

The Life, Letters and Labours of Francis Galton

Volume 3 – Part B: Characterisation, especially by Letters

Karl Pearson

Cambridge University Press

CAMBRIDGE UNIVERSITY PRESS

Cambridge, New York, Melbourne, Madrid, Cape Town,
Singapore, São Paolo, Delhi, Tokyo, Mexico City

Published in the United States of America by Cambridge University Press, New York

www.cambridge.org
Information on this title: www.cambridge.org/9781108072434

This edition first published 1930
This digitally printed version 2011

ISBN 978-1-108-07243-4 Paperback

THE

LIFE, LETTERS AND LABOURS

OF

FRANCIS GALTON

Francis Galton, when about 75 years of age.

THE

LIFE, LETTERS AND LABOURS

OF

FRANCIS GALTON

BY

KARL PEARSON

GALTON PROFESSOR, UNIVERSITY OF LONDON

VOLUME III[B]

CHARACTERISATION, ESPECIALLY BY LETTERS

INDEX

CAMBRIDGE

AT THE UNIVERSITY PRESS

1930

PRINTED IN GREAT BRITAIN

ILLUSTRATIONS TO VOLUME III[B]

Frontispiece. Francis Galton, when about 75 years of age

PLATE		*to face page*
XLIII.	Francis Galton's Room at Trinity College, Cambridge	453
XLIV.	From Galton's Egyptian Sketchbook. Bob, aged 10, left in charge; Ibrahim	454
XLV.	From Galton's Syrian Sketchbook. Ali, Galton's servant in Palestine and Syria	454
XLVI.	d'Arnaud Bey, a sketch from a photograph	455
XLVII.	Water-colour Sketch of Francis Galton in the Fallow Years . . .	456
XLVIII.	Francis Galton's Niece Milly—Mrs J. C. Baron Lethbridge . . .	471
XLIX.	Francis Galton, the Meteorologist, when about 65 years of age . .	472
L.	The Medallion of Erasmus Darwin by Fassie	473
LI.	Facsimile of one of Francis Galton's weekly letters to his Sister Emma, 1893	494
LII.	Francis Galton, aged 75. Royat, July, 1897	502
LIII.	Francis Galton's Great-Niece Evelyne Biggs—Mrs Guy Ellis . .	507
LIV.	Experiment of Francis Galton on the exhibition of all aspects of a bust on a single negative	520
LV.	(i) "Sister Emma"; (ii) Francis Galton as Geographer; (iii) Francis Galton's Mother	531
LVI.	Galtonia (*Hyacinthus candicans*) from tropical South Africa . .	534
LVII.	Sample of the conventionalised Finger-Print Ornamentation on the Stones at Gavr'inis, from the series of photographs in the *Galtoniana* . .	543
LVIII.	Francis Galton and his Great-Niece, Evelyne Biggs, at Bridge End, Ockham, in August, 1906	576
LIX.	The Rock on the Lake at Otchikoto, Ovampoland, where Galton's name can still be read, inscribed in 1851	617

Specimen of Galton's Annual Medallions commemorating an event of the year. This marks the year in which he obtained composite photographs of the tuberculous patients at Guy's Hospital.

Bookplate of Samuel Tertius Galton.

CHAPTER XVII

CHARACTERISATION, ESPECIALLY BY LETTERS

"His interest was nearly as promptly and vehemently kindled in one subject as in another; he was always boldly tentative, always fresh and vigorous in suggestion, always instant in search." Morley, *Diderot*, Vol. II, p. 37.

In this final chapter the reader will find printed a selection from the innumerable letters which Francis Galton, during a long life, wrote to his family circle and friends. My aim in the earlier chapters of this biography, after describing Galton's ancestry and childhood, has been to give a full account of his contributions to science, and to reproduce portions of his scientific correspondence. I have not wholly excluded from this chapter letters to scientific friends—failing a complete publication of Galton's letters it must be an "omnibus" appendix—but its main purpose is to paint a side of Galton's nature which I fear has not been adequately emphasised in the earlier chapters. Francis Galton to his scientific colleagues was courteous, generous and marvellously humble. To his relatives and close friends he was sympathetic, helpful and always full of fun. His wonderful patience with an invalid wife, and after her death his splendid loyalty to her memory, can only be lightly touched on here. His limitations were as well known to himself as to others; he had no musical sense, and art, whether in colour, form or verbal expression, was not for him an essential need of his being. I do not think he was fond of animals, nor had he a keen comprehension and love of young things; perhaps they suggested too bitterly what was lacking in his own life; he seemed to be alarmed by children, and did not find the right words to say to them. I doubt whether he could have placed a child on his knee and told it a tale. With young people beginning to think and to take an interest in life's problems he was wholly sympathetic; he respected their views, however callow, and entered with jest and anecdote into all their fun. At "biometric teas" his presence was never over-aweing, indeed it was he who generally started and led the mirth.

The letters which follow will show clearly how deeply he could sympathise with those who failed to appreciate the contributions he was making to the great revolution in human thought which marked the last quarter of the nineteenth century. I have said that he did not understand children, yet he did understand and sympathise with those simple childlike natures which still found comfort, and a crutch for the conduct of life, in the faiths of mankind's infancy. He would endeavour to interpret their conceptions in terms of his own wider aspirations. To those who stood nearer to his own standpoint he made no pretence of reconciling the old with the new—"It aids them, but it would be of no service to you and me." Thus he would explain to his biographer that sympathy in expression and action which might not unreasonably appear irreconcilable with his own faith. Without being deeply interested in history he had, as every man of culture, an understanding for the past; he realised that each worn-out phase of mankind's mental evolution is not a ruin to serve as a quarry for to-day's uses, but rather a monument to be preserved, even fenced about to protect it from the ravages of the profit-seeker, or indeed from the sacrilege of the scoffer. It is in this spirit that the reader must weigh some of the letters of Galton and some of the statements about him in the following pages. Galton was as strong an agnostic as Darwin or Huxley, but he was not, like the latter, an iconoclast; as I will venture to put it, the stirps of Galton and Darwin had a more generous historical background than that of Huxley, and this even more so in Galton's case. He has spoken in several places of the unconscious working of the mind. There is a conscious family tradition, and again an unconscious one; our mentality is what it is in accordance with the tradition of our stirp, and works unwittingly in the track of the past. The Galton stirp—witness its quakers and its devout catholics—had a deep religious sense—not unbroken by a tendency to wander at times from the current phases of morals and of religion, but it had also a kingly spirit in the best sense of the words—an understanding of the nature and the needs of those dependent upon it. Roll into one the characteristics of the Plantagenet, the Stewart, the Savile, the Sedley and the Darwin stirps, and we can thus, and only thus, fully appreciate the complex nature of Galton's mind. We can trace therein his impulse towards travel, his fallow years, his inventive genius, his sympathy with deeply religious natures, his zeal for knowledge, and his mirthfulness. Width of mind in any individual usually takes its origin in the happy combination of several stirps of strong but diverse intellectual character. A danger arises when intimates, especially relatives, appraise a great and wide-minded man; they are apt to emphasise that side of his character which has appealed most strongly to them, and of which for that very reason he may have sounded the note. In the case of blood relatives that note may be the characteristic of the part of their stirp common to both, or indeed, if they are of the full blood, as brothers, the one may be dominated exclusively by one factor of their common stirp*.

In reading family letters written originally for no other eyes than those of the recipients, we must ever bear this in mind. When a man soars above his fellows to altitudes they have not yet attained, it is only natural that his intercourse with them should remain largely on the old plane familiar to all of them. The letters of Galton show him as son, as brother, as uncle, and as great-uncle—those which might have limned him in his courtship and marriage failed to reach his biographer. Yet the letters which I have seen, apart from their bearing on Galton's own history, cover upwards of a century of family life, and are in themselves witness to the great changes

* Thus in the children of Charles Darwin one marks in isolation factors which were combined in their father and great-grandfathers.

which took place in our national life, both economically and psychically, during the nineteenth century. We see children, born in an age of canal-building and stage-coaches, dying as motor-boats and airplanes come into being. We note men, great and mediocre, passing from the vigour of youth to the weakness of old age and leaving behind them records of actions which will survive for generations, or which have already perished. And we ask why, with a common environment, does one man achieve and another fail to do so? The answer can only be: "Such is the law of inheritance," and that was Galton's answer. But even with his work, supplemented by that of Mendel and the followers of both, we yet fail to solve the riddle of family history—why one man here and there is so markedly differentiated from his stock, noteworthy as that may be.

It will probably aid the reader to have a short account of the environment of Galton during the successive years of his life, taken from the hasty notes written down by his wife or himself on different occasions.

Brief Record of Galton's Travels and Visits, 1853–1883.

Francis Galton was married on August 1, 1853, and toured in Switzerland and Italy, spending the winter in Florence and Rome.

1853 Rome. '54 Chambord. '55 Farnborough, Paris Exhibition. '56 Innsbruck, Vienna, Dresden. '57 Courmayeur, Gressoney, Corniche. '58 English Lakes. '59 Lakes, Bonn.

1860 Richmond, Pyrenees. '61 Zermatt, Monte Moro. '62 Glarus, Pilatus, Champéry, Chamounix. '63 Glion, Stresa, Corniche. '64 St Gall, Haudères, Sepey. '65 Spa, Holland, Birmingham. '66 Cannes, Mentone, English Lakes, Nottingham. '67 Mentone, Sorrento, St Moritz. '68 Auvergne. '69 Heidelberg, Berchtesgaden. '70 Grindelwald, Folkestone. '71 Scarborough, Whitby. '72 Brighton. '73 Ilmenau, Moselle. '74 Lynton, Chettle. '75 Fontainebleau, Mürren. '76 Bavarian Lakes, Venice. '77 Tunbridge Wells, Bournemouth. '78 Vichy, Mont d'Or, Paris Exhibition. '79 Vichy, Black Forest. '80 Dinant. '81 Bournemouth, York, Vichy. '82 Baden, Constance. '83 Devonshire.

Notes on Galton's Visits, Friends and Occupations, 1875–1883.

1875 with Emma and Lucy to Fontainebleau; in June to Seelisberg, to Paris, met Emma. F. G. went to Bristol, British Association and returned to Paris. Russell Gurneys went to Egypt. Twins. Sweet peas. *Phil. Mag.* Law of Error and Ogives.

1876 Sweet peas. Miss Christie. Loan Collection. S. Kensington; Whistler. Jenkinsons at Fawley. Groves at Syston. With Emma and Brodrick to Bavaria, Venice and Italian Lakes. Arthur engaged to be married.

1877 Louisa's bad illness. Tunbridge Wells. Bournemouth. Plymouth British Association. Arthur married. Mr Holland died.

1878 Rheumatic gout in my knee, March. Composite photos. R. Gurney died. Silver wedding year. Vichy, Mont d'Or, Tours, Blois, Paris Exhibition. Folkestone, F. G. back to Paris.

1879 Lady Grove died. Bournemouth (Tower). Vichy with (Roes and Martindales). Gérardmer. Black Forest. Judge Grove at Wakehurst in Sussex. Generic Images.

1880 Vichy. Boulogne (May and Spencer). Swansea, Brit. Assoc. Liphook. Meuse (Dinant). Spa. Brussels Exhibition. Photographed at home.

1881 Home early in October. Easter Leamington. Guy's Hosp., composite photography. Bournemouth. York, British Assoc. Vichy late. Paris. Douglas broke his ankle. Composite phthisis photos.

1882 Easter Claverdon and Leamington. Rhine. Black Forest. Baden Baden. Constance. Axenfels. C. Darwin died. Graef picture. Guy's Hospital photography (Bethlehem and Hanwell). Life Histories. George, Canon of Winchester. Anthrop. Album.

1883. Georgina died. H[enry] Smith died. Bournemouth. New Forest. Spottiswoode died. Jenkinsons. Newton Abbot. Torquay. Southport B.A. Adèle died. Leamington. Home Oct. 1.

The above records, written in a scarcely legible hand, were found in one of Galton's note-books. I give a continuation from Louisa Galton's diaries to 1897, and have then filled in the remaining years to 1911 from my own knowledge. The series will serve as a convenient reference for the later of the following letters.

1884. Early February Brighton one week. Easter Ventnor April 1–17, Rede Lecture at Cambridge on May 27, 4 days. No British Assoc. but were prevented going to S. France by outbreak of cholera. Lakes, Windermere on July 15, Keswick 17th for about a month, Patterdale for a week. At end of August home for a week, then Claverdon and Leamington, home September 24. October 7–11 to Stanmore.

1885. February 7–9 Leamington. Easter family gathering at Harrow, a week at Tunbridge Wells and a week at Ramsgate. End of July Holmbury St Mary for 6 weeks. British Assoc. Aberdeen. Owing to cholera in Italy had a dull week at St Leonards and a visit to Oxford.

1886. February 25 Paris, Hyères, Mentone, Rome, Sorrento, Amalfi, Castellammare, Florence. Easter Vevey. Home May 5. Late in July to Contrexéville 3 weeks, Bussang in Vosges returning by Nancy, Paris and Boulogne. September 1st visited Marianne North at Alderley for 2 weeks, Hills at Newbury, Montagu [Butler] at Gloucester, Claverdon, Leamington; home early in October.

1887. Early March 10 days with Montagu at Trinity. Whitsuntide at Tunbridge Wells. Mid-July Homburg 3 weeks, Freiburg, Constance, Zürich, 6 days' tour to Rhone Glacier, Eggishorn, Rieder [Furka] and Belalp and Brieg to Domo d'Ossola, San Rosso [? M. Rosso, N. of Pallanza] and the Falls of Toce, back to Zürich over Gries. Engelberg a week, by St Gotthard to Locarno, Lugano 9 days, then Basle, Calais and home. Leamington.

1888. Easter Leamington 1 week. In July visited Lethbridges at Clifton and M[arianne] North at Alderley, returning for Montagu's wedding on August 9th. August 10 Brighton 1 week, returned to Alderley. September 1st Vichy 3 weeks, Paris 3 days, home end of September.

Visits: Arthur [Butler], Mr Brodrick, Stanmore (the John Hollonds) and Trinity.

1889. Early March–May 3rd: Cannes, Mentone, Genoa, Milan, Baveno, Cadenabbia, St Gotthard, Paris. July nine days with M. North. August Paris one week (Congress on Heredity), August 12 Carlisle to Corby Castle (Hills) 10 days, and 2 weeks at Wetheral. Alnwick with E. Wheler for 3 days, Newcastle B.A. meeting, staying at Gateshead Sept. 10–18. Saw Durham and York. Leamington for Lucy Wheler's wedding on Sept. 26. Nov. few days at Trinity.

1890. Easter fortnight at Alderley, ending with 2 days in Forest of Dean. July, Switzerland, St Beatenberg, Stoos [? Stoss, near Altstetten], Freudenstadt, Strasburg, Nancy, Paris, home September 8th. Leamington, home end of September.

1891. Easter Leamington 10 days. In May 2 days at Cambridge with the George Darwins. Tunbridge Wells mid-July for a fortnight, returning to London August 10–15 for Congress of Hygiene. Vichy to September 9th, Châtelguyon, Royat, the Lozère, the Gorges of the Tarn (long F. G.'s ambition), Montpellier-le-Vieux, Nimes, Avignon, home October 9th.

1892. First week of the year with Emma, Leamington. End of March Biarritz 2 weeks, Cambo, home early in May. August 8th to Corby for 10 days, Edinburgh for 2 weeks, Callander. September 9th to the E. Whelers at Alnwick, 23rd–27th to Lady Welby near Grantham; a little time at Peterborough, Leamington. Home early in October.

1893. March 13 Riviera: Hyères, Valescure, Cannes. Home April 20. May 11–15 Trinity. End of July Holyhead, 3 weeks (the Institute under Miss Adeane). Short visit to Lady Stanley at Penrhos, Bettws-y-Coed 3 weeks, tour by coach to Capel Curig, Beddgelert, Aberglaslyn to Port Madoc, toy railway to Festiniog, and next morning through Bettws-y-Coed to Conway and Chester 2 nights, Shrewsbury and Leamington for 2 weeks. Home September 30.

1894. February 27–April 7 Cimiez. In May to Edmond Hills at Darland House. May 12–14 at Merton. June 1 Trinity. June 19 Arthur [Butler] at Oxford. Towards end of July Spa for 2 weeks. Köln, Heidelberg, Bodensee, Landeck, Innsbruck, Berchtesgaden nearly 2 weeks, Salzburg, Ischl, Gmunden Sept. 17–19, Regensburg, Nürnberg, Würzburg, Köln. Home on September 27. Leamington 1 week.

1895. April 9 Leamington for 9 days, then to Mrs McLennan at Hayes for a short visit. May 13-16 Cambridge for D.Sc. In June to Mrs Hodgson at Tanhurst (Leith Hill). July 3 Nauheim, Garmisch for more than a month. August 12 Munich, Nürnberg, Rothenburg 2 nights, Frankfurt, Bonn, home August 27. Fortnight at Tunbridge Wells. September 21–30 Leamington, and then home.

1896. March 13 Hastings. March 25th Eastbourne. April 2 for Easter, Leamington. July 10 Wildbad (treatment for L. G.), Oberstdorf (treatment for F. G.), Lindau, Davos 10 days, Zürich, Freiburg, Strasburg, Rheims, Amiens. Home September 7. Sept. 25 to Oct. 2 Leamington. Oct. 13 Trinity a few days.

1897. March 24 Bournemouth. April 20–23 Leamington. June 5–8 Oxford (Arthur). July 14 Boulogne, 15 Royat. (August 13 Louisa died.) F. G. left Royat Tuesday 17th, home on 18th. Leamington 1 week. Himbleton nearly a week. Corby till Sept. 13. Sept. 25 Boscombe. Weymouth (to Emma). Oct. 16 to G. Brodrick 3 days. Oct. 23 Leamington, to Emma. Dec. 4 to Bessy, and then to the Studdys.

1898. I have but few records of this year. Galton was at Rutland Gate in January, July, and November, and travelled with his wife's nephew, Frank Butler, to Royat and to the Riviera, and then to Italy (Castellammare).

1899. March 21 at sea; 22 Gibraltar; 23 Ronda; 26–April 3 Seville; April 3–5 Cadiz; 5 Tangiers; Morocco, Malaga; 13 Granada; 18–20 Toledo; 20–26 Madrid; Barcelona, Carcassonne, Nîmes, Clermont-Ferrand. July and August Royat for 3 weeks. Switzerland for 2 weeks. Home middle of August. December 15 Luxor; 18–22 Assouan.

1900. January 1 Luxor; to the Petries for a week. January 22–February 9 Luxor; February 15–March 4 (?) Cairo.

1902. November 12 at home; November 28 Valescure.

1903. January 8– Rome, Naples, Ischia; April –8 Siena, Bologna, Milan (for Easter), Cologne, Brussels; April 20 London. June 10–12 Loxton; 14–16 Cambridge; July 22 Norwich for 2 or 3 days. August–September (?) Norfolk Broads; December Italy—Sicily.

1904. February –2 Girgenti; February 2– Taormina. August –26–Sept. 6– Bibury, Fairford; 15–19 Bovey Tracey; 19–20 London; 20–24 Claverdon. Home early October. November 4–7 Leamington; December –10 Bournemouth.

1905. January at home; February 18–20 Calais; February 20–May 1st Bordighera; May 1st home. July 29–August 1 Claverdon; August 1–5 Lakes; 5–7 or 8 Highhead Castle, near Carlisle; August 17–September 27 The Rectory, Ockham; 27–30 Hindhead; 30 home. November 1 (?) started for Pau; November 10 (?)–December 1 (?) Pau; December 4– Biarritz.

1906. January –10 Biarritz; 10–February 1 (?) St Jean de Luz; February 1–27 Ascain; 27–March 9 Biarritz; March 9–26 San Sebastian. Home April 6. April 21–26 Claverdon; May 15 Cambridge (Trinity); June 3 days at Oxford (Arthur Butler's); 20 (?) Trinity Fellowship. July at home. August 2–29 Bridge End, Ockham; September 1–10 Bovey Tracey; September 16–24 Malthouse, Bibury, then home; October (1 week) Sidmouth; October 19 Edymead, Bovey Tracey; November 7– The Hoe, Plymouth.

1907. January –30 The Hoe, Plymouth; January 30–March (?) Hoe Park Terrace, Plymouth. July 21 (?)–26 Helmingham Hall, Stowmarket; August 7–September 12 Yaffles, Haslemere; September 12– Quedley, Haslemere.

1908. Quedley, Haslemere to February. August 30 Shirrell House; September 16–23 Claverdon; October 26– Meadow Cottage, Brockham Green, Betchworth.

1909. February 26 Meadow Cottage, Brockham Green, Betchworth; February 26 or 27–April 3 Crown Hotel, Lyndhurst; April 3–21 Forest Park Hotel, Brockenhurst; then home till August 10 (?)–October 4 Fox Holm, Cobham, Surrey; October 4– The Rectory, Haslemere.

1910. March 21 The Rectory, Haslemere; then home till August 16 or 17, then The Court, Grayshott; moved November 15 to Grayshott House, Haslemere.

I shall commence this chapter of Galton's *Life and Letters* with an appreciation of her uncle by his niece, Mrs Lethbridge—Millicent Bunbury—the child of Galton's beloved sister and instructress Adèle. She most kindly prepared it for me on Galton's death, 19 years ago, and it seems best to me to publish it now just as it was written. She is the "Milly" of many of the letters printed below. It is very characteristic of Galton that when his "home" letters ceased on the death in 1904 of his Sister Emma, aged 93, he felt the need of continuing the family correspondence, and selected his niece, Mrs Lethbridge, to exchange letters with him.

Recollections of Francis Galton by Millicent Lethbridge.

I will begin these short "recollections" of my dear uncle, Francis Galton, by repeating the child-stories my mother has told me, but first you must allow me a digression that I may explain the share she had in his early life. My mother spent a dreary childhood and girlhood, seldom leaving the sofa to which she was condemned owing to curvature of the spine. She had little to amuse or interest her in those weary years, until, when she was eleven, my uncle Francis was born. My grandfather took the baby to her, saying: "Here, Adèle, is a baby brother come as a present for you! How do you like him?"—"Like him!" A new life began then and there for my mother. She set feverishly to work, teaching herself Latin, Greek, German, Italian, and I know not what besides, to fit herself for the task of educating the baby. All her interests, thoughts and ambitions were wrapped up in the little creature. It lay by her side on the sofa, and with the enthusiasm and impatience of a child, she lost no time in cramming it with all her miscellaneous, self-acquired knowledge. I believe the baby could read at two, and what it had learnt by the age of four, I do not venture to report! Strange to say, the baby throve on the system, and delighted as much in learning as his sister in teaching. The two were devoted to each other, and it was a bitter wrench to my mother, when, at eight years old, her darling was sent to a school at Boulogne.

I recollect two or three anecdotes my mother told me of his very early years. My grandfather, anxious to render his boys self-reliant, sent Francis, then about seven years old, to pay a visit to a relative at some distance. The child was to ride his pony, spend the night at a certain inn, and finish the journey next day. A servant was instructed to follow (unknown to the boy) two or three miles behind in case of accidents. When Francis was questioned about his adventures, he related how, on reaching the inn, he had ordered supper and a bedroom, and had then proceeded to empty his purse and hide a shilling under a pillow, a sixpence under a chair and so on, "because then, if a robber came, he might take some of my money, but not all, so that I could still pay my bill!" I am sorry, however, to say that I cannot verify this story, my uncle having entirely forgotten the occurrence.

He had a remarkably sweet temper, and it used to be a joke between his brothers to see if they could not make him angry. Do what they would, they hardly ever succeeded. My mother once said: "Frank, how *can* you keep your temper as you do?" "I don't," he answered, "but I've found out a capital plan. I go to my room as soon as I can get away, and I beat and kick my pillow till I'm tired out, and by the time I've finished, my temper's all gone." In later life my uncle's self-control was really wonderful. I have seen him, on more than one occasion, "keep himself in hand" under the greatest provocation, although I presume the "pillow-recipe" had long been abandoned.

Another child-story is that of his falling off his pony into a ditch, and being dragged out by the legs by his elder brother, the seven or eight year old boy, half-choked with mud, spluttering out Hudibras,

"I am not now in Fortune's power,
He that is down can fall no lower!"

One more story and I have done. A lion had escaped from a menagerie and the child was in terror lest it should suddenly pounce down upon him. His father found him trembling in bed, and said: "Why, Frank, you know the lion has no pocket-money to pay the turnpike, so of course he can't come through!" "I never thought of that, papa," said the child, as with an immense sigh of relief he turned over and went to sleep.

If my uncle derived his genius through his Darwin mother, it is nevertheless certain that the Galton father was most in sympathy with the boy's character. His devotion to his father's memory was most touching, and only a few weeks ago, when the Copley Medal was offered him, he wrote: "People are always very kind to me, but I wish my father were alive. It would have given him real pleasure." His father had then been dead 66 years!

I looked upon my uncle Francis as my special uncle, ever since I was quite a little child, but it was not until after the death of his wife (whom I also loved dearly) in 1897, that I was admitted to a closer friendship, and that I ventured to discuss many things—religious matters especially—with him.

Later still, in 1904, when his beloved Sister, Emma Galton, died, he asked me to correspond regularly with him, just as she had done for many years, so that the custom became established from that time forward until his death, for me to write every Friday, and he every Monday or Tuesday.

I have an amusing recollection of a little trip to Auvergne which he and I took together in the summer of 1904 only a few weeks before he sustained the great sorrow consequent on his Sister, Emma Galton's, death. The heat was terrific, and I felt utterly exhausted, but seeing him perfectly brisk and full of energy in spite of his 82 years, dared not, for very shame, confess to my miserable condition. I recollect one terrible train-journey, when, smothered with dust and panting with heat, I had to bear his reproachful looks for drawing a curtain forward to ward off a little of the blazing sun in which he was revelling. He drew out a small thermometer which registered 94°, observing: "Yes, only 94°. Are you aware that when the temperature of the air exceeds that of blood-heat, it is apt to be trying?" I could quite believe it!—By and by he asked me whether it would not be pleasant to wash our face and hands? I certainly thought so, but did not see how it was to be done. Then, with perfect simplicity and sublime disregard of appearances and of the astounded looks of the other occupants of our compartment, a very much "got-up" Frenchman and two fashionably dressed Frenchwomen, he proceeded to twist his newspaper into the shape of a washhand-basin, produced an infinitesimally small bit of soap, and poured some water out of a medicine bottle, and we performed our ablutions—I fear I was too self-conscious to enjoy the proceeding, but it never seemed to occur to him that he was doing anything unusual!

He had ordered rooms at Royat, insisting that they should have a southern aspect. On arriving at the Hotel it was found that they looked due north. Then, for the first and only time since I had known him, he was guilty of a very forcible and by no means parliamentary ejaculation. A minute or two later he turned round and saw me. He appeared exceedingly uncomfortable, and at last could stand it no longer: "Er—er—did you hear what—er—I said just now?" I could not resist the temptation of declaring myself extremely pained and shocked, but he was so genuinely distressed I had to hasten and assure him I was only talking nonsense.

He half-killed me by his energy at Royat. We used to sally forth at 4 a.m. and take a walk before the heat of the day. That was really enjoyable, but I felt by no means enthusiastic when we started off again when the sun was at its highest, and walked and trammed wheresoever it was hottest. He always chose the sunny side of the road, but occasionally I rebelled and left him to his sun whilst I walked in the shade. He really was a salamander! I can see him now, sitting at his work-table in the window at Royat, with the broiling sun streaming down upon his bald head. Even to think of it is almost enough to give one a sunstroke.

But it was not long after our Royat visit (where he had gone to visit his wife's grave) that his strength gradually began to fail. His sister's death, soon after our return, was a terrible blow to him. I do not know what he would have done, but for his great-niece Eva Biggs, who devoted herself to him as if she had been his daughter. The few remaining years of his life brought him much sorrow—the death of his eldest sister at the age of nearly 98 and of his brother, aged 94, leaving him the only survivor of his family. My Mother—his Sister Adèle—had died many years before. However, with the exception of his deafness, he retained all his faculties to a wonderful extent. His eyesight was extraordinarily good, and he could read the smallest print up to the last. The diaries he kept for many years were not, I suppose, more than 2 or at most 2½ inches square, and his writing in them was necessarily so minute that I could not see to read it. His sense of smell was also singularly acute, and I imagine that of taste likewise. He enjoyed his food as keenly as a child, although he was a very small eater and most abstemious in every way. He delighted in after-dinner coffee, of which he allowed himself two teaspoonfuls, and that only when I, or some other coffee-drinker, was staying with him to set a bad example!

The *joie de vivre* remained strong in him even after he had lost the power of walking, and when he could not rise from his chair without help, and then only with pain. Still he was as keen and full of zest as ever, and I believe that if a ten, or even a twenty years' extension of life had been offered him he would gladly have accepted it, for his heart was bound up in his beloved "Eugenics" and he would have loved to watch its progress, even at the cost of prolonged pain, weariness and suffering.

Whilst he was still able to move about a little, his indomitable energy prompted him to do extraordinary things. For instance, at a time when he could hardly stand alone, I have known him (by holding on to things) climb out of the staircase window on to a sort of lead roof, where he would spend an hour or so in the open air. It was a perilous proceeding, and on one occasion he had the narrowest possible escape from an accident which, if it had actually occurred, would certainly have killed him.

He was touchingly "grateful for small mercies." I remember his telling me one day that he had had a "glorious time" that afternoon. The "glorious time" was just sitting in a bath-chair, helpless and unable to move, in a garden-shelter watching the trees and sunshine. Any little ingenious contrivance was an absolute delight to him, and I have known him amuse himself for quite a long time with some penny toy such as those hawked about the London streets. I do not think he *could* "do nothing." His brain was always busy even when his hands were idle. It is true that sometimes when I asked him what he had been doing, he would quote from *Punch*: "Sometimes I sits and thinks, and sometimes I only sits"—but I never believed it.

He was fond of reading aloud, and he read better than almost anyone I ever heard. He enjoyed reading Tennyson and Shakespeare to me, but I think he excelled himself in reading the Bible. On one occasion he read the Book of Esther right through, and although I had imagined I knew it well enough already, he convinced me I had never known it at all before then. The whole scene started into life, I was transported into the Oriental surroundings, thousands of years back—the dramatis personae lived and moved, and I felt as if I had dived into another world. If it had all been acted before me, the impression could not have been more vivid.

I think he cared little for fiction unless he was tired or poorly. On those occasions I found *Don Quixote* was oftenest in requisition. In novels he evidently preferred fun to sentiment and last year (1910) he delighted in Countess von Arnim's *Princess Priscilla's Fortnight*, *The Caravanners*, etc. Art to a certain extent, and Music entirely, seem to have been omitted in his composition—an inheritance perhaps (or rather non-inheritance) from his Quaker ancestry. He delighted, however, in the artistic nature of his great-niece, Eva Biggs, as much as she in her turn prided herself in his science. Music, I think, he positively disliked, although he only confessed to "not caring for it." His brothers and sisters were also, one and all, absolutely unmusical. Certainly he was a living refutation of Shakespeare's "The man that hath no music in himself etc."—for never was any man further from "Treasons, stratagems and spoils!"

It would be hard to find anyone with so high an ideal of duty as his, and I do not hesitate to affirm that nothing—not self-interest, praise, blame, or anything else, would have made him swerve a hair's breadth from what he conceived to be right. To that which he believed to be true, he felt bound to give utterance, even though it cost him the disapprobation and even the deep sorrow of some whose love and sympathy he most valued. This was especially the case when his work on *Human Faculty* came out in 1883, with a chapter on prayer, which I rejoice to find is suppressed in a recent edition*. Although the chapter in question only attacked the crudest and most materialistic notion of prayer, and was obviously written under a complete misapprehension of the real Christian position with regard to it, nevertheless a storm of indignation was raised, and some whom he most loved, and whose good opinion was dearest to him, were distressed and scandalised. I always felt that his attitude with regard to Religion was absolutely misunderstood. I have heard him called hard names—"Atheist," "Unbeliever" and so on. My own description of his creed would be that of a Religious Agnostic. Faith was denied him, and, as he has often told me, all *intuitive witness* to the

* At the urgent request of the publishers.

Divine. The question of the reality of this "intuitive witness" in others, however, interested him deeply, and he would have given much to convince himself whether it was real or imaginary, subjective or objective. James's *Varieties of Religious Experience* was a book that occupied his thoughts a great deal, and I have a copy that he gave me. That he had the *will to believe* I am sure, but it was the power that was denied him. If he could not believe however, he could seek, and a more earnest truth-seeker could surely not be found. He has told me that at one time of his life (I imagine when he was very young), the asceticism of the Roman Church appealed to him very strongly. His admiration for the uncompromising monotheism of Mohammedanism was recurrent. I imagine that he was latterly much attracted by Spinoza. But early love and sentiment were all on the side of Quakerism. He would sometimes ask me where such or such a parable or discourse of Our Lord was to be found, and on finding it for him in the New Testament, he would read it aloud, saying, half to himself, as he shut the book "Perfect—*very* perfect."

He was scrupulously careful not to say anything on religious topics that could possibly distress or injure the faith of anyone—especially the young—and I never knew him say anything that was not absolutely reverent. He was apparently incapable of accepting anything he considered unproven. Thus, although a devoted admirer of Tennyson, I never heard him allude to *The Two Voices* without a stamp and "Pshaw!" of impatience. The philosophic discussion being concluded by nothing more convincing than the emotional "Sabbath Morn" and "Church-bells" irritated him beyond endurance. In spite of his much-abused chapter on Prayer in *Human Faculty* I know he used to pray himself, indeed in one of his letters to me he wrote (May 12, 1907): "Did I ever tell you that I have always made it a habit to *pray* before writing anything for publication, that there may be no self-seeking in it, and perfect candour together with respect for the feelings of others."—And in another letter (April 9, 1907): "I think in earnest prayer of you and poor F. for I *can* pray, and *do* pray, conscientiously and fervently, though probably in a different form to that you yourself employ. God help you*."

There were many beautiful traits in my Uncle's character upon which I cannot now expatiate. His old-world courtesy, displayed not only in society, but still more at home, to those with whom he was in daily intercourse, and to his servants (falsifying the saying "that no man is a hero to his valet-de-chambre")—his almost exaggerated dread of appropriating any laurels due to others, which feeling led him to the opposite extreme of magnifying the achievements of others whilst minimising his own—his horror of self-advertisement, coupled though it was with a naïve delight in unsought appreciation—all this is familiar to those who had the privilege of knowing him.

I do not think I have more to say. His patience and cheerfulness during the helplessness of the two or three last years was very wonderful, even when his sufferings were aggravated by the constantly recurring attacks of asthma which made every breath a struggle. His devoted nurse and great-niece, Eva Biggs, told me that the last thing he said, when the breathing became very painful and she asked him if he suffered much, was: "One must learn to suffer and not complain."

I could not be with him in his last illness, being ill myself at the time, but he was surrounded with love and affection. Eva Biggs, his valued nephew, Edward Wheler, and his devoted Swiss servant, Gifi, who had been in his service for 40 years, being with him at the last.

Thus ended the earth-phase of a great life.

Selection from the Galton family Letters.

Letter from Adèle Galton to her Sisters, Emma and Bessie.

[1830?]

My dear Sisters, We have just received a letter from sweet Francis, and I cannot help thinking (at least hoping) from its contents that he still preserves his taste for study but here is the copy:

"My dear Papa, I hope you have been pretty well lately. It is now the Easter Holidays and I was asked out last Monday and then I saw a review of the National Guards which I liked very much—It has been very warm for some months and I think we shall soon begin to

* See my footnote, pp. 271–2 of Vol. iii^A^.

bathe in the sea—Thank you for buying me those five shillings of flower seeds—please to thank Emma for taking such care of my Garden and Bessy for my carnations when they return Home—I suppose that almost all the flowers at Home are beginning to blow—I hope that little Herman is better of the Croup—Please to tell me if the Alderney Cow has calved. I can now speak french pretty well. In your next letter please to tell me if Adèle has any german master or mistress as Miss Abick is married.—It will not be more than three Months to the Midsummer Holidays—I have been learning a great deal of Conchology lately—I hope that all at Home are quite well—Have you had any letters from Darwin or Erasmus*. When do you think Erasmus will come home? for it is a long time since he left us—I suppose that Mrs French has a great deal of land to herself as Mr Millington is dead—I have neither begun dancing or fencing. Good bye and believe me your most affectionate Son, F. GALTON."

Is it not a nice letter? dear little Fellow, I am sure he is not aware what pleasure it gives us all to hear from him else he would write oftener. We all enjoy Leamington much and were it not from a calculation that I have made viz. that I shall lose *thirty-six* hours of practising that is allowing three hours for each day, I should wish *never* to leave it. Lucy has told you almost every thing except that we have seen Mr Jones the Surgeon who alas! did not recognise us. We have just received a letter from Darwin, who still seems to be anxious to enter the Army but has not yet received Papa's letter about advising him to enter the Infantry, instead of the Cavalry. The letter is written in very good French, and he tells us that Uncle Howard is going to make a Tour in the south of France and that Little Robert is growing a Beauty. Uncle Darwin† has also sent us a *very kind* letter, saying that he has had a personal interview with Lord Hill, who has been *most* gracious and condescending with regard to Darwin, and assures him that *his* Nephew requires *no* introduction, and that he will send my Uncle in writing his opinion about what steps ought to be taken, so now I think Darwin is in a fair road for entering the Army. Really I begin to like Uncle Bob after all.—Thank you, dear Bessy, for your letter. How happy you both seem to be. What a kind (or what Francis would call *kindissimo*) Aunt Mrs Gurney is. From your affectionate Sister, M. A. GALTON.

Letters of Adèle and Emma Galton to their Sister Bessie at Duddeston on the death of their Grandfather, Samuel Galton (see Vol. I, p. 40 *et seq.*).

[1832.]

DEAREST BESSY, I cannot tell you how often I have thought of you and my dear Aunts during this great trial, more especially of you as this being the first time you have witnessed death it must have made such a deep impression on you. I also wish with you that I had been able to see my dear Grandfather's remains. I never shall forget the last time he shook hands with me; I felt as he walked out of the Dining room that I might never see him again and so it has happened. Thank you for telling me he mentioned my name among those of the other members of the family for it did indeed make me very glad to think that he had so kindly remembered me. We all, dear Bessy, feel *very much obliged* to you for writing such nice long letters to us, what a deal of writing you must have had to do and how happy you must feel in being of use to my Aunt. I must own I felt very sorry to hear that Aunt Sophia‡ has fixed

* Francis Galton's elder brothers.

† Dr Robert Darwin, Charles Darwin's father.

‡ Aunt Sophia: see Vol. I, Plate XXXV. The following lines of Tertius Galton on his sister Sophia may be cited here:

A description of Miss Galton of Dudson by S. T. G. 1831 to her amusement.

"My head wears a cap that makes all the world stare,
My face sports a nose of dimensions most rare,
My eyes like two saucers that roll in their sphere,
My waist thin as a lath, my back straight as a spear,
My manners precise, yet my looks full of fun
And tho' rather coquettish, yet grave as a nun;
A very neat seamstress, I make my own frocks;
A very good housewife, knit stockings and socks,
If one farthing is missing I make a great fuss.
My age, upwards of forty—my name it is 'Puss.'"

to live near Birmingham as I could not help hoping that she might live near here; we could see her so much oftener and there are such a many pretty houses of all sizes and descriptions; however wherever she may be I *do* hope we may often be able to be with her for I do love her most affectionately. From your letter I fear Aunt Booth has suffered much, will you give my love to her as well as to my Aunt Sophia. Lucy comes to us to-morrow, we shall as you may suppose be delighted to see her and James. Only think sweet Francis sets off from Boulogne to-day week. Dearest Child, how rejoiced we shall be to kiss again his dear freckled face. A card was left at our door to say that a Mr White from Cambridge is anxious to give lessons to private pupils in Greek, Latin and Mathematics. I have kept the card as it may hereafter be useful either to Francis or myself—Good bye dear Bessy.

Ever believe me your very affectionate Sister, ADELE GALTON.

[1832.]

MY DEAR BESSY, I wish you would tell Aunt Sophia and Aunt Adèle how much I feel for them, and I should have written to tell them so, but knowing how much they have to do and think about, felt that it would only be a trouble to them. Tell Aunt Sophia I wish I could have made myself useful to her, but Papa requires so much attention that I really think it is quite necessary I should be here, as Mama and Adèle are neither of them strong enough to walk up and down stairs much. It will give us such pleasure to see Aunt Sophia here, and I am sure nothing shall be wanting on my part to make her as comfortable as I can, for I can never forget how very kind she and Aunt Adèle have always been to us, indeed they have been more like sisters than aunts. Tell Aunt Booth she *has* promised to come and see us soon and that I am looking forward with such pleasure at the thought of seeing her, and that I think she will be pleased with the pretty views about this place. What a consolation it must be to Aunt Sophia, to consider how materially she has conduced to my poor Grandfather's comfort during his life time and how she has given up the enjoyments of her friends' society that she might be always with him. I am so glad you have been at Dudson, for I think you may be useful to Aunt Sophia. Believe me ever, Your very affectionate Sister, E. S. GALTON.

At Dr Jeune's School.*

[*February* 12, 1838.]

MY DEAR BESSY, I would have written before, only I have had so little time and that time was spent in writing Valentines as I have bought a Valentine Book and I also am so happy at thinking that the Glorious Conquest of ST VINCENT was fought on the 11th, and by the by please send me a list of the days of the month that the principle [*sic*] battles were fought, like your card. To-day we had a poor fellow handcuffed in my presence for trying to commit a grievous assault by means of his fist on the person of......whereby the said......was put in extreme bodily fear for he would have been hurt without the said......had luckily sprung back and avoided the blow (this is his indictment only the worst is I cannot put in—for I have forgotten them) well the fellow kicked at......and knocked off another boy; we and the person who was with us chased, when he veering to the larboard up Bennet's Hill with about two hundred small craft nearly all Free School boys after in chase till at last we came alongside and captured the prise [*sic*] and then towed him aloft in the Free School where he was ordered to gaol until the petty sessions. I am very happy indeed and am glad I am come here I have only seen P. once and have hardly got time to do anything. X. is a radical, says he hates Wellington and, as he says, his country, and likes the French and Italians more; he does not know what ship Nelson was in when he fought Trafalgar nor that he lost an arm. Is it not shameful? Now I wash from top to toe every morning, head and all. I feel as if I know a great deal more than when I first came here. We fag a great deal, for instance we have to learn 50 lines of Homer and to parse any word and also the derivations in 2 hours only, which is very hard work, but now I begin really to like stewing. I would write a longer letter only I have been

* King Edward's Grammar School, Birmingham. This letter is very difficult to decipher, and the spelling and grammar sufficient to send Dr Jeune, had he seen them, into hysterics!

allowed to sit up after bed time, only for a little time, as I have not 20 minutes to myself at any other time. Good bye and believe me ever your brother, F. GALTON. P.S. I have lately got another hand in writing as I find I can do it much quicker. Give my love to all. I am very well, indeed much better. Mrs Ridges is an odd craft. I like Dr J. very much, we always touch our hats to him. They say that the P.S. is the most important part of a letter, at least it is here for I just want somebody to remember that Monday 16th is my birthday, a little grubbing is very accepttible [*sic*] here.

42, RUTLAND GATE, S.W. *June* 2, 1906.

MY DEAR LUCY*, The letter, which I return as you wish, is an amusing reminder. I see it was written according to the post mark in *1838*, and according to its contents on *Feb. 12th*, and from Dr Jeune's School. His house, where I and a dozen other boarders lived, was at the Five-Ways, Edgbaston. Its garden (a rather large one) is wedge-shaped and its wall forms the angle of one of the 5 blocks. It was a daily walk of one mile, to and fro from the Free School as it was then called. It is now commonly called King Edward's School. The present buildings did not then exist but the school was held where the Theatre is, in a big room just opposite to Bennet's Hill and in New Street.

I had quite forgotten the incident about X. He *was* such an ass, and was a butt of ridicule. But he improved as he grew older and when he had married.

The letter testifies to the influence of your Mother over my social and historical creeds. I wonder where I got the nautical language from. Mrs *Ridges* was the housekeeper, a good and kind-hearted old soul with peculiarities, some of which struck us as funny.

I am getting well of a sharp sort of feverish attack which kept me from going to Claverden last Monday, and at the last moment. The Doctor was peremptory and I felt myself fit for nothing but bed. So we telegraphed, and to bed I went, and was rather bad for a time. We look forward much to seeing you. Affectionately yours, FRANCIS GALTON.

This pen scratches abominably. Heartiest congratulations on your *embroideries*.

BIRMINGHAM HOSPITAL. *August* 16, 1839.

DEAR BESSY, Beg pardon, full of 'trition—May go of course—took that for granted before asking—thought you did ditto. Ready on Monday, the 2nd (I think). Hodgson was most amiable—thought it would do me good. Uncommon clever man to have found that 'ere out. Tell the Governor that I have got to be examined before I can enter at Cambridge by a Trinity M.A., but Lea is out, so that won't do. However as there is nothing like two strings to one's bow I have got both Mr Gedge and Dr J. Johnstone, separately and individually, to promise to get me examined at the time of the Association. I should have written before, only I thought that I could have been examined here by any Cambridge M.A., but I was dished. Poor old Mr Corrie died suddenly this morning—Hodgson told me so.

Now then about our travelling. First comes the *Tin*. I propose that you carry the fund, and give me some sum, say £2, every morning, and *every evening* balance accounts, thus making you the banker and me the paymaster. The route I leave to your "superior judgement." I will come to Leamington on the Saturday before our departure and go with you to Coventry on Monday morning. Please write to me pretty soon about your arrangements.

Bye the bye how often on an average daily are we expected to cry over the different affecting places, because it will make a considerable difference in the number of my pocket handkerchiefs. I must however give warning that when I come to Leamington neither Mamma, Emma, nor Stone are in any way to help, alias incommode, me in my packing up—neither are they to inspect nor give their judgement thereupon without, of course, my sovereign will and pleasure. I have got nothing else to say. Bye Bye, loves right and left, FRAS. GALTON.

* Lucy was Mrs Studdy, the daughter of Mrs Wheler (Sister Bessie). She obtained prizes for her very beautiful embroidery.

PLATE XLIII

Dear Emma. This is a hurried sketch
of my room. I will send you a vignette
in each of my letters to pay you "in kind"
for your two Water colour Paintings — Cambridge
has some beautiful college views greatly
set off by rows of Elm trees. I must try
& sketch you some. Fra. Galton

Francis Galton's Room at Trinity College, Cambridge, before the "improvement," i.e. the transfer of the sofa to face the fire-place. Cf. Vol. I, Plate LI.

17, NEW STREET, SPRING GARDENS, LONDON. Wednesday, [1839].

MY DEAR BESSY, Bless your innocence! with regard to the postage and envelopes—I enclose this in one of a new pattern which may perhaps be explanatory of the allegorical design of the original; and in case that your understanding should still continue obfuscated, I enclose also some newspaper lines for your edification—The reason why I asked about your painting mishaps was that, in case that you did so, I might coax you to illuminate some letters in my prize folio *Seneca*, but as you have got other things to amuse yourself with, I wouldn't ask you on any account to do it, as first of all it would make you stoop—2ndly when you are at the sea—why, what's the good of going there if you don't make the most of it?—3dly I should like to have a previous consultation with you on the correct colours etc. and 4thly there is no hurry. Therefore I don't want you to do it now. Q.E.D. As to the inelegance of the word "splodge" I confess it to be very great but the fact was, that after the previous night's "excitement" (as Mrs Wititterly would say) all my wits were tending to fly sky-high, and indeed to leave me altogether—As an opposing force I therefore used all the most matter-of-fact expressions, and commonplaces, possible, which two forces, acting conjointly, produced a happy medium in the current of my mind. Indeed the night previous I was so completely "knocked off my legs" by Persiani, Erni, Rubini and Tamburini in Don Giovanni that I awoke up bawling away the air "Là ci darem la mano etc." and only got to sleep again by means of a perpetual singing in my ears of "Batti, batti, bel Masetto etc."

I am sorry poor Lucy* is so poorly and also the children, loves to them, please. I enclose you a bottle of Gold Size and hope it won't break. N.B. I write an answer by return of post, in order to keep up the new character which you have given me of letter-writing. I find the newspaper which contained the verses is burnt, so I write from memory and therefore excuse mistakes.

FRAS. GALTON.

[LONDON, 1839?]

MY DEAR BESSY, As to my last letter, it must have been the very same postman that was in fault before when you sent me the missal, who has lost mine now. I am quite glad that you are so much better but "better" won't do in your accounts of yourself—they must be more medical. Such as: *Jan. 27.* Little sleep at night. Slight shivering on getting up. p.m. Headache and nausea. Acid taste in mouth. Occasional numbness in feet with tendency to be hysterical. After dinner—listless and a good deal of gaping. Appetite better.—Now you ought to send me such an account as this—as much longer as you like—of about one day in a week or so.

I called on the Horners the other day. Was shown up—only the Miss Horners there—in extra deep mourning—crape enough on them to furnish Woodhouse and Haddon's shop. I thought that one of the old birds had hopped the twig and was just going to tell them that I was very sorry that I had inadvertently intruded at such a time etc. etc. when one of them said that they had all been at a ball the night before till 2 o'clock. I don't know if mourning is considered here as a sort of fancy costume in character or not. Well, on leaving them I tooled to Charles Darwin, when!!! round the knocker, with the utmost care and precision a grey kid glove was wound—I couldn't make out the knot—he could only have tied it so well by long practice†. This proceeding à la Kenwigs very much astonished me. Of course, I did not call. What am I to do? I have not heard anything about his having a little Perpetuation, but there can be only one way of interpreting the kid glove. Please write me word in your next letter, what you know, and all you know, of the Myners' Family of Weatheroak, near Birmingham, especially as relates to *Miss* Myners. After all neither purse, money nor door key were pick-pocketed, but the Governor has I suppose told you the circumstances.

Give my love to Lucy*, James and Animalculae. How is Mrs Howell? You would very likely have been just as bad as she was if you had not gone to Moor Hall.

I hope you have learnt the tune of "Nix my dolly pals—fake away etc." Darwin would like it of all things. It is quite necessary to know it to get on in the fashionable world. Good Bye etc. FRAS. GALTON.

* Galton's sister Lucy, who married James Moilliet.

† Probably the birth of Charles Darwin's eldest son, William Erasmus Darwin, is referred to.

Letter of Dr Robert Waring Darwin to his sister Violetta Galton.

SALOP. Saturday, *August* 1, 1840.

MY DEAR SISTER, Susan, who is the only one of my family with me, joins in congratulations on the intended marriage of her cousin* with Miss Phillips. it is a fortunate attachment being so agreeable both to you and to Mr Galton. I trust they will be as happy as you wish them.

You kindly mention my sons; they are both far from well tho' from what I hear better than they have been and improving. Susan while on a visit to her brother Charles had the pleasure of seeing your son Francis of whom I hear a most satisfactory account in every respect. I have not heard of Sir Francis† since he wrote to communicate the marriage of his daughter and as you observe, it is only on such occasions we write. I trust I have the prospect of more letters from him. He did hold out some hope of their coming this summer to see us.

My daughters Marianne and Caroline are both well. Catharine is gone on a visit to some cousins in Pembrokeshire. With our kindest regards to your family circle, ever dear Mrs Galton,

Your affectionate brother, RBT DARWIN.

BEYROUT. *9th*, [1846].

MY DEAR GALTON, I was much shocked to hear to-day from Mr Heald of the death of poor Ali‡, and also that you yourself are suffering from this infernal climate. I have been at this place about 8 days; the day we arrived both Delahaut and Fontinillia another Frenchman with whom I am travelling fell ill of the fever, Delahaut very seriously. He is now recovered, and intends to go to France as soon as he can. I expect to be at Damascus in about 10 days or sooner, and from there my movements are quite uncertain. I have been offered a passage in a French brig to Aleppo, (which) it is likely I shall accept after having been at Damascus. We came from Cairo by the short desert, very slowly, for my friends feel the heat very severely. As yet I have been all right, and have not much felt the climate, but all that I have seen of the country coming from Jerusalem here has disgusted me much, for there is really nothing worth seeing. This place is the only pretty thing I have seen.

I hope, my dear fellow, that when I arrive at Damascus I shall find you set up again, and able to continue your travels; what with nursing poor Ali and the climate of Damascus, I am sure you must have had a sorry time of it. I remain, Yours very truly, —— ——

DAMASCUS§. *September* 30, 1846.

MY DEAR GALTON, What an unfortunate fellow you are to get laid up in such a serious manner for, as you say, a few moments' amusement. I had been told you were unwell at Beyrout, but I had no idea you had been suffering so much. I trust this letter will find you getting stout and well again. I do not start for Bagdad for some little time as yet, and am in doubts whether I shall not prefer going alone to the caravan. Meantime, I am flourishing, installed in my old house and leading an exemplary life studying the "Alf Leyla, we Leyla"|| which I have bought here in two large old volumes. The town is very full and gay by reason of the pilgrims who will start on Sunday, and the departure is to be unusually grand. There is a very nice fellow staying here by name Stobart, who is a great acquisition. I have been in treaty for the purchase of a slave, and have had several Abyssinians brought for show, but none as yet sufficiently pretty. Dr Thompson desires his love and remembrances, and regrets as much as myself your not returning to cheer the solitude of El Sham el Kebira. I am thinking of buying his grey mare. The Han Houris are looking lovelier than ever, the divorced one has been critically examined

* Darwin Galton. I have inserted this letter from Dr Robert Darwin as it refers to the health of his son, Charles Darwin.

† Sir Francis Darwin, son of Dr Erasmus Darwin and his second wife, Mrs Chandos-Pole. He lived at Breadsall Priory, his father's old house: see Vol. I, pp. 22–25, and Plates XVIII, XLIII and Vol. II, Plates XV, XVI.

‡ As to the death of Galton's devoted servant Ali, see *Memories of my Life*, pp. 89, 103.

§ These letters throw some light on the doings of young Englishmen in the near East in those days. The writer was a College friend of Galton, and they may be taken to illustrate the "Fallow Years," for which see *Memories of my Life*, p. 85.

|| *Arabian Nights* in the Arabic.

PLATE XLIV

From Galton's Egyptian Sketchbook. "Bob," aged 10, the lad who was left in charge of the boat on the Nile during the visit in 1846 of Galton and his friends to Khartoum, and Ibrahim their servant. The originals are in water-colour. See Vol. I, p. 201.

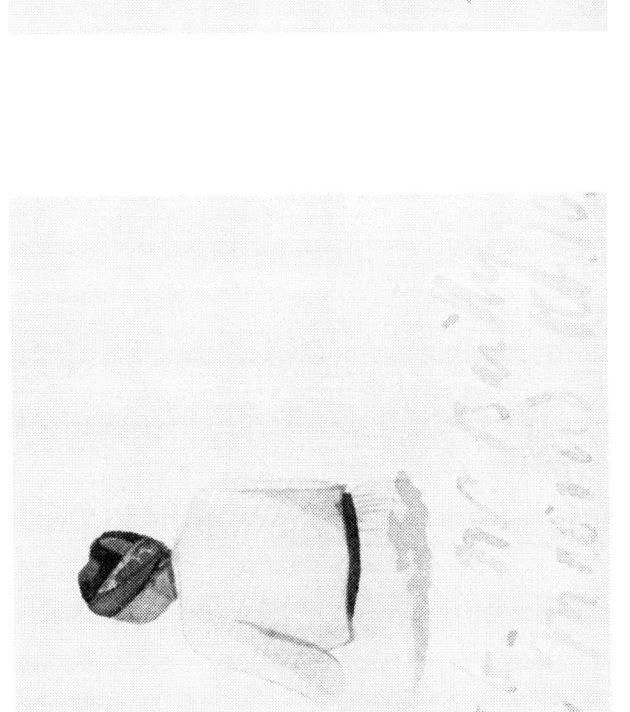

From Galton's Syrian Sketchbook. Galton's faithful servant Ali. Original in water-colour.
See Vol. I, p. 203.

PLATE XLVI

d'Arnaud Bey, a sketch from a photograph.

and pronounced a virgin. We are expecting a large fête and rejoicing in one of the Jewish families, and I hope I shall get invited. I have made acquaintance with Mlle Hanoum of singing celebrity. Don't tell the Colonel. There has been one of the regular Damascus pieces of tittle tattle about the French doctor, who was living here and whom I take to be no great things. I don't know anything about the rights and wrongs of the accusations against him, but you know how the people here must have talked about a little bit of scandal, which is their greatest treat. I staid at Eden some days, and made great friends with all the people there especially your acquaintances the well-whipped Welleds, one of whom offered for sale a cake of chrome yellow that I suspect has been in your paint box before this. I staid too a day or two at the Cedars. I must bring this scrawl to an end, but before so doing I must assure you, my dear fellow, of my best and hearty wishes for all possible health and good luck in all your future peregrinations, whether homeward or outward bound. If you return to England pray give my remembrances to any of my friends you may happen to come across. With the same once more to yourself, believe me to remain, Ever sincerely yours, —— ——

There are so few records of the "Fallow Years" that I gladly note here what contributed to their termination.

The d'Arnaud Bey Incident.

Life, Vol. I, p. 200.

"However, we got on very well and made him talk of his travels and tell us of the country ahead, we had then no map and knew nothing hardly. He said: 'Why do you follow the English routine of just going to the 2nd cataract and returning? Cross the desert and go to Khartoum.' *That sentence was a division of the ways in my subsequent life.*"

Memories of my Life, p. 96.

"That chance meeting with Arnaud Bey had important after-results for me by suggesting scientific objects for my future wanderings. I often thought of writing to him in order to bring myself to his remembrance, and to sincerely thank him, but no sufficiently appropriate occasion arose, and it is now too late.

"In the winter of 1900–1901 I visited Egypt again and, calling at the Geographical Society there, learnt how important and honoured a place d'Arnaud Bey had occupied in its history. He had died not many months previously, and I looked at his portrait with regret and kindly remembrance. Being asked to communicate a brief memoir to the Society at its approaching meeting, I selected for my subject a comparison between Egypt then and fifty years previously. I took that opportunity to express my heartfelt gratitude to Arnaud, which posthumous tribute was all I had the power to pay."

March 20, 1900.

DEAR SIR, I am happy to present you hereby M. Bouda's letter with the interesting photograph. Mr Somers Clarke will come with us to-morrow. I send the carriage at 8½ to your Hôtel. Yours sincerely, G. SCHWEINFURTH.

A photograph of d'Arnaud was enclosed and from this our sketch was made.

"Il habitait dans une petite hutte de terre, entourée par une légère palissade de roseaux et que son génie avait transformée en quelque sorte en un sanctuaire de philosophe au milieu d'un pays presque barbare. La distribution artistique de sa petite collection d'instruments scientifiques, livres et curiosités, donnait un air singulièrement recueilli et studieux à cette modeste demeure. J'étais enchanté de sa conversation et des nombreuses anecdotes qu'il nous racontait des pays du sud."

....

"Je n'ai plus eu l'occasion, depuis lors, de voir ou d'être encore en rapport avec d'Arnaud Bey, mais je l'ai toujours considéré dans mon cœur comme un bienfaiteur, car c'est lui qui m'a donné l'idée des voyages sérieux que j'ai faits plus tard en explorant la terre des Damara dans le sud-ouest de l'Afrique; c'est lui qui a dirigé mon goût et mon énergie vers la géographie et les sciences; j'ai su employer mon temps utilement au lieu de le consacrer à de frivoles amusements."

From Galton's "Souvenirs d'Égypte." *Bulletin de la Société Khédiviale de Géographie, No.* 7, *Mai,* 1900.

COLRAIN, BONAR BRIDGE, SUTHERLANDSHIRE. *August* 28, 1848.

DEAREST MOTHER, The grouse won't come just now. All those killed as yet Fazakerly of course disposed of. He may give us some few to send soon; if he does I trust they will arrive safely at Claverdon. He is a capital fellow. I enjoy myself more than I have for a year and a half—everything is so free and open. I stay with him till he goes if I like. His son, Col. Wallington and Paddy Johnson are the party. Louis Napoleon was to have come too but was prevented. We have every variety of field sport,—pitch tents and hack ponies.—Johnson stuck a dog into me to buy (Emma will explain, if I am not intelligible), it has proved a beauty and I with my one dog see more game than my fellow guests with two each. Shouldn't you like to buy him at the end of the grouse season? He would find you hares so well in Paul's Piece and I would part with him to my Mother (mind, only to my Mother) for the money I gave for him (£10). Write me an occasional note and tell me how Lucy is, for I am most anxious to hear.

Ever your affectionate Son, FRANK GALTON.

The heather is beautifully out on the moors—I pulled a piece to send to you as a memento but I have lost it. Lots of fishing and really of everything. It is a most civil thing of Fazakerly asking me.

Letters and note-books are very scant during the years that followed Galton's marriage. Probably like other husbands, he left correspondence to his wife.

Tea Making, My Experiments [1859].

There are among Galton's papers and note-books accounts of various experiments made by him, scarcely with a view to publication, but rather with the purpose of amusing himself and gratifying his insatiable desire to observe and measure. One especially characteristic series of experimental measurements dates from early in 1859, and deals with the

"Flavour, Freshness, Body and Softness" of Tea.

The experiments were made morning and evening, and must have tried severely the patience of Mrs Galton, and not unlikely of the household. Galton begins with the following preliminaries:

"The teapot holds 26 ounces $= 3\frac{1}{4}$ breakfast cups. One breakfast cup holds 8 ounces. The teapot requires 3 minutes to become warmed through. It radiates heat at the rate of 2° per minute."

Then we have the categories to be used:

$G=$ good, $B=$ bad, $D=$ decocted, $W=$ weak, $F=$ flavour,
$C=$ body, $a=$ best, $b=$ 2nd best, $c=$ 3rd best.

We next proceed:

"To find the capacity for heat of the teapot.

$n=$ number of ounces of water used,

$e=$ excess of its temperature above that of the teapot,

$t=$ additional temperature attained by the pot after the water has been poured in,

$C=$ required capacity,

$$C+ne=(C+n)\,t, \quad C=n\,(e-t)/(t-1)."$$

I will give a few illustrations from the notes which extend through February and March.

Water-colour Sketch of Francis Galton in the Fallow Years.

Experiments.

Original temp. of pot A	Temp. attained after mixture B	No. of ounces of boiling water n	$e=212^\circ-A$	t=additional temp. $=B-A$	$e-t$	$t-1$	C
75°	170°	6	137°	95°	42°	94°	2·6
91°	183°	9	121°	92°	29°	91°	2·8
119°	187°	10	93°	68°	25°	67°	3·7
58°	158°	6	154°	100°	54°	99°	7*
64°	148°	6½	148°	84°	64°	83°	5
57°	151°	6½	155°	94°	61°	93°	4·3
54°	161°	7	158°	107°	51°	106°	3·4

* N.B. "Carter [presumably the maid] 'frothed up' the water."

"Feb. 13, 1859. Sunday Evening. [Pot] heated to 140°. Put in tea at VI *h.* 35 *m.* and water $\dfrac{192^\circ}{180^\circ}$ at VI *h.* 46 *m.* Tea good, but a little too much of a decoction. $\Big\|\ \dfrac{184^\circ}{174^\circ}$ at VI *h.* 64 *m.* Tea weaker but decocted somewhat.

Feb. 15. Tuesday Morning. [Pot] heated to 174°. $\dfrac{178^\circ}{169^\circ}$ [in] 4 minutes $+\dfrac{194^\circ}{186^\circ}$ [in] 7 minutes hot and decocted. 2nd cup $\dfrac{194^\circ}{174^\circ}$ [in] 11 minutes hot and weak.

Feb. 16. Morning. L. G. fecit $\dfrac{172^\circ}{[?]}\,2^{m}+\dfrac{192^\circ}{187^\circ}\,5^{m}+\dfrac{187^\circ}{184^\circ}\,3^{m}$ Black not decoct., fullish body, fresh, hot. 2nd cup $\dfrac{188^\circ}{170^\circ}\,14^{m}$ fairly good. If more tea had been placed in the pot (4 spoonfuls) I think the brew would have been successful.

5 *ozs. of tea to be henceforward used in the morning.*

Feb. 16. Evening. $\dfrac{40^{m}\,|\,191^\circ}{48^{m}\,|\,178^\circ}$. Decoct. slight. Louisa says fresh and little body (I have a cold). $\Big\|\ \dfrac{48^{m}\,|\,?}{58^{m}\,|\,160^\circ}$ very good. $\Big\|$ 3rd cup 7^{m}, 142°. No water had been added, good, a little bitter.

Feb. 20. Sunday Morning. $\dfrac{40^{m}\,|\,180^\circ}{44^{m}\,|\,170^\circ}$ (filled up to make it so) $+\dfrac{44^{m}\,|\,186^\circ}{45^{m}\,|\,178^\circ}$ $+\dfrac{46^{m}\,|\,184^\circ}{48^{m}\,|\,?}$ (filled up); very excellent (it is true it had cream). $\Big\|\ \dfrac{48^{m}\,|\,189^\circ}{61^{m}\,|\,168^\circ}$ good, a little flat. 3½ cups altogether."

It is difficult now to determine exactly what the temperatures signify; presumably the fall in temperature of the teapot in the intervening number of minutes; a plus sign seems to denote a filling up or repeated experiment, while || signifies second cup. By the middle of March the record is systematised, but more cabbalistic:

	1st cup				2nd cup			
March 13	192° 180°	186°	8	*Gb . Db*	184° 174°	179°	8	*Gc . Dc*
March 16	172° 2m + 190° 5m + 185° 3m	185°	10	*Gb . Cb*	?	179°	14	—

About March 27 the experiments appear to have been discontinued, but were started afresh in November of the same year. Presumably the same tea was used throughout. But no conclusions are drawn, and we are left in doubt as to the meaning of the values recorded. We are not left in doubt as to Galton's taste for a very strong cup. "Quite good, I think it would bear strengthening. L. G. says not." "Admirable, strong and fresh and pure (there was plenty of tea put in), excellent." There is evidence that visitors were occasionally present during these experiments, and the mistress of the house must have had some difficulties when the tea was weighed out and the thermometer popped in and out of the teapot.

I have not cited these experiments for any result that flowed, or indeed was likely to flow, from them, but solely to indicate how strong was Galton's passion for measurement, and that, already in 1859, he was giving full play to his statistical tastes. These teapot data are indeed the "Puffing Billy" stage of Galton's statistical career!

CAMBRIDGE. *August* 3, 1863.

MY DEAR GALTON, In consequence of Phillips's* retirement from the Office of General Secretary, which he has held temporarily for the last year, a Committee was recently appointed by the Council of the Brit. Assoc. to recommend a successor to the Office. According to the general rule of the Association there ought, as you are probably aware, to be two General Secretaries, and one paid Assistant Secretary. Now Mr Griffith of Jesus Coll. Oxford, has succeeded Phillips in that office, and during the past year Phillips has nominally held the office of one of the General Secretaries in consequence of my illness last autumn. The purpose at present is to elect a second permanent General Secretary as my coadjutor, Griffith taking the *labouring oar* as Phillips had done before him. Now comes the question—Will you accept the office if offered to you? The Committee are Sir R. Murchison, Sabine, Vernon-Harcourt, Phillips and myself, and I think I may venture to say that in proposing you there will be no dissentient voice. The Office is a very pleasant and gentlemanly one, requiring of course attention and courtesy, without much time or trouble. On account of my absence last year Phillips will act with me at Newcastle *this* year. After that he will retire entirely, but I am now getting pretty *au fait* at the work, and should of course take it as much as might be necessary on myself till my future coadjutor should have gained the requisite experience. I need scarcely say, I hope, how much I should rejoice if you could be installed as my *partner*.

Believe me, Yours very truly, W. HOPKINS†.

* Professor John Phillips, the geologist. He was Assistant Secretary of the British Association, 1832–1859.

† Galton's instructor in mathematics, the famous Cambridge Coach.

The Château in the Heart of the Ardennes.

Poem found among Galton's papers. I do not know when the visit to the Château in the heart of the Ardennes came off, or who wrote the skit; if it was Galton, it was at his own expense. The visit may have been as late as 1879 or 1880, but probably much earlier.

They told me of a château in the heart of the Ardennes,
A pension kept charmingly by two young châtelaines;
They told me of some English people who had summered there,
On next to nothing for the best and most abundant fare;
They could not tell me where it was, or who the châtelaines,
But they knew it was a château in the heart of the Ardennes.

The heart of the Ardennes is large, if it be somewhat cold,
And châteaux are in plenty there, the homes of barons bold;
The ruins that were homes in ages past, that is to say,
And not at all like pensions where English people stay;
But all the information that they really could obtain
Was this,—it was a château, in the heart of the Ardennes.

They both were very anxious to be able to make out
The way to reach the château they had heard so much about;
'Twould be so charming, after all the towns where they had been
And after all the gay and noisy places they had seen,
To go and live for nothing, far from all the haunts of men,
At a veritable château in the heart of the Ardennes.

They left. No more they déjeuner'd at healthy Souvenière,
No more they meant to déjeuner at distant Géronstère;
No more abused the Ninth for all the tuneless things they played,
No more encored "La ronde qui passe" in Leopold's arcade.
They left; and I was lonely for a day or two, and then
I went to find the château in the heart of the Ardennes.

There met me, on the way to join the luggage at the gare,
About the most experienced of travellers that are,
The Art himself of Travel; and, though not born yesterday,
I listened to the guileful tale he told me by the way;
For he told me with descriptive tongue, as clever as his pen,
Of what sounded like the château in the heart of the Ardennes.

He told me of the demoiselles who kept a charming place;
Of English people, how they praised its cleanliness and space;
He told me of a brother, too, who helped his sisters dear,
And how for almost nothing they gave most delightful cheer.
It was not called a château by his friend, he said, but then
It really was a pension in the heart of the Ardennes.

A Belgian lady staying in the Britannique hôtel
Had told him. That was where and how my ladies learned as well.
It clearly was the very place. I took the train at one;
Then drove across the bitter moors, and when the day was done
We pulled up in a dirty town amid the drenching rain,
And o'er the door was painted H.—not château—des Ardennes.

A hugger-mugger maid appears, with pail and brush in hand,
And makes a sound or two which she perhaps may understand;
And then there comes another, with a wart upon her nose,
And she must be, as I at length unwillingly suppose,
At least the mother of the pair of blooming châtelaines
Who kept the charming château in the heart of the Ardennes.

But if a pair they ever were, the other's not alive,
And this one is the only one, and she is fifty-five;
The brother is a page in blouse, who won't do what he's bid,
The people call him Jacquot, but with Madame he's "stupide"—
We've thus disposed of brother and of blooming châtelaines;
But what about the château in the heart of the Ardennes.

I'm ushered in, and there, I find, are fellow victims three,
Prepared to eat their souper, fixed for sept heures et demie;
A monsieur with a napkin tucked beneath his double chin,
A mother, and a giggling girl for ever on the grin.
Then knives begin to shovel in the meat and beans, and then
I feel I'm in a pension in the heart of the Ardennes.

The mother tells of glories which have quite possessed her brains,
The salons of a wealthy fabricant of counterpanes;
Discusses is it proper for a Vérificateur
To ask to dance the daughter of a public Inspecteur.
It sounds perhaps a little insignificant, but then
We're very near a château in the heart of the Ardennes.

Monsieur gets purple over "non!" and shouts it six times o'er;
And when he feels affirmative, a dozen "si's" or more;
Élisa nips her mother when I don't take haricots,
Which smell so strong of onion I'm glad to see them go;
And this within a yard or two, not more than eight or ten,
Of a most undoubted château in the heart of the Ardennes.

The morning breaks in beauty, and romantic dreams take flight,
As through the open window streams the sun's fast growing light,
Romantic dreams of sylvan courts, and eke of banished dukes,
And pensive Jaqueses who meditate by sweet meandering brooks.
I rise and seek the window, feeling sure that there and then
I shall realise the château in the heart of the Ardennes.

The noises that the pigs are making really pass belief;
The cocks are louder still,—to shut the window's no relief;
And ah! for dreams of sylvan glades so sweet and fresh and pure,
At every door are soaking heaps of excellent manure.
But what are trifles such as these, when close within my ken
There stands at last the château in the heart of the Ardennes.

The guide book says ninth century, but carved in stone the date
Of this remaining morsel is but sixteen twenty eight;
It's now a shop for carpet-slippers, sweets and boots and wool,
And Madame takes a room in it when her "hôtel" is full;
The rest was all "fait sauter," not by Revolution men,
But to build a new Hôtel de Ville in the heart of the Ardennes.

The meats are very tender, and the bedrooms very good;
Madame is very pleasant, and there's quite sufficient food;
The coffee's sometimes perfect, and there seem to be no fleas,
And it costs you very little by the day at Houffalize;
But yet I'm not at all inclined to go and see again
That smelly—not a château, in the heart of the Ardennes.

Letter of Charles Darwin to his Aunt, Violetta Galton (née Darwin), Francis Galton's Mother.

DOWN, BECKENHAM, KENT. *July* 12, 1871.

MY DEAR AUNT, I am very much obliged to you for your great kindness in writing to me in your own hand. My sons were no doubt deceived and the picture-seller affixed the name of a celebrated man to the picture for the sake of getting his price.

Your note is a wonderful proof how well some few people in this world can write and express themselves at an advanced age. It is enough to make one not fear so much the advance of age, as I often do, though you must think me quite a youth!

With my best thanks, pray believe me with much respect,

Your affectionate nephew, CHARLES DARWIN.

*Letters to George Darwin, Esq.**

BRITISH ASSOCIATION, BRADFORD. Wednesday, *Sept.* 24, 1873.

The paper came off yesterday and, as an amusing fact, Carpenter had afterwards to speak about some "current" questions and found the mercator's map of the north parts so inscrutable that he left it and *went to your globe* to point out to the audience what he meant.

The application that most commends itself at present to me, is to have the hexagon-pentagon map on the scale of about a 9 ft. globe, to mount the map on screens, stoutly made (? with *projecting mouldings* to represent the mountain chains, made by pasting a few successive contours

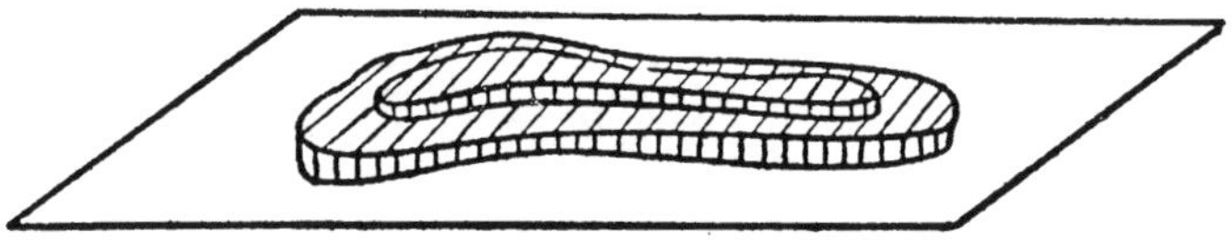

upon it), and to have a couple of stout frames to hang them on, one having a hexagon and the other a pentagon as its middle compartment.

I will take care, and I am sure Strachey will too, that the plan gets properly discussed at the Geographical. Here, in a room full of ladies and no one to understand, it is impossible to do so.

I have often thought of procuring a really artistically made and coloured globe and once had much correspondence about it. Ruskin wrote a very good letter. It seems to me that one might set to work by making a spherical shell, then cutting it up *into convenient parts* like a puzzle-map, and mounting the parts that were temporarily wanted on a convex table for consultation. These could be multiplied by casts, also by electro-type.

With my kindest remembrances to all your party. Ever yours, FRANCIS GALTON.

(I return off and on to London.) 5, BERTIE TERRACE, LEAMINGTON. *Oct.* 3, 1873.

MY DEAR DARWIN, Mr Geach forwarded your note. I extracted the enclosure and sent it to him. Also I sent for the *Contemporary*, but instead of the August they sent the number of this month with quite another subject of yours; but I will get the August one.

I am most grieved to hear of your Father's recent illness, but I firmly believe in his powerful underlying constitutional powers as sure to assert themselves whenever there is real need.

Do you know or has Dr Clark † heard of that half incredible but uncontradicted assertion made in a long paper at Bradford before a room crowded with physiologists, that albumen mixed with water in a short time becomes undistinguishable from the contents of the lacteals, white corpuscles, etc.!!! (so that you could assimilate it without any stomach at all!) and the very practical conclusion was drawn that if an egg be broken into cold water (just as it is broken into hot water for poaching) and left to stand 12 hours, it becomes opaque,—then if you boil the whole affair slightly, the result is a food that the author asserted to be digestible when nothing else could be digested!

It seems worth trying.

I enclose a printed solution of a problem which I received yesterday and which I think (and hope) may interest you. I sent the question to the *Educational Times* some months ago, when a Mr Carr of Woolwich gave an answer making a frightful mull of it,—a total misconception. Then I asked Watson who got the enclosed very elegant result, but still it is not one of practical applicability. Is it really hopeless to obtain a more manageable solution?

Would you please send me back the paper in a few days as I want to have it put in the *Statistical Society Journal* and I have no other copy.

* Later Sir George Darwin, second son of Charles Darwin.

† Afterwards the well-known consultant, Sir Andrew Clark.

Send me some "cousin" circulars that I may distribute. I heard of them at Bradford. My antecedents of scientific men is fairly in hand. Out of the 186 asked, between 120 and 130 have either sent or promised. I have about 80 in hand now.

Are you quite sure *Hadley* of St John's *is* a relation. Miss Parker's* eldest daughter married Mr Hadley (there was one other daughter who died unmarried) and had one son, Dr Henry Hadley, and one daughter who died unmarried. Ever very sincerely, FRANCIS GALTON.

To GEORGE DARWIN, ESQ.

Copy of Genealogical Tree enclosed in letter of October 3 to George Darwin.

I got this from my Sister Emma.

Miss Parker and Dr Erasmus Darwin

Mr Hadley m. Daughter
Surgeon

Daughter
(d. unmarried without issue)

Son
Henry Hadley
Surgeon

Children[1]

Daughter
(d.
without issue)

[1] The Hadleys of St John's are not descendants of this Henry Hadley (?) †.

Miss Parker ultimately married a Mr Day and had two or three children, of whom one daughter turned out very ill.

42, RUTLAND GATE, S.W. Xmas Day, 1874.

MY DEAR GEORGE, I also quite forgot about your maps till just after you went; but Gen. Strachey is the man. He has the thing in his hands and I am only an occasional assessor. It is the framework that gives the difficulty. He had, at last, two great machines constructed down in the Isle of Dogs, by an Engineer who makes bridges and the like for his department. They were both heavy and crooked. I went down with him and we suggested a much amended plan of which he sent me the working drawings but my illness has prevented my seeing him since. The immediate object is to produce two frames, 10 ft. diameter, that can go in a cab or be sent by luggage train and yet be easily mounted in the lecture room. The great point is to have them as *the* regular maps at the British Association at Bristol next year, when Strachey may be counted on as being elected President of the Geography Section. You had better write to him and keep at him periodically, and whenever I see him I also will "nag."

Thanks greatly for your bits of criticism, they are all valuable to me and helpful. I am gratified to hear that your Father is interested in the book.

Henry Parker ‡ is not wholly my fault; the entry in your Father's schedule is "distinguished classic, and good artist and chemist." I quite see now that the last half of the sentence was intended to be amplificatory, merely for my own information, but it happened to chime in with some vague recollection I had of his having occupied himself much with chemistry and I did not inquire further but put in the "chemist" (or whatever the exact phrase was—my book is not at hand).

* Miss Parker was the mother of two natural daughters of Erasmus Darwin: see Vol. I of this *Life*, p. 17 and Plate X. The surgeon with the spurs is Hadley.

† The Hadleys of St John's College, Cambridge were distinguished mathematicians, and the problem was, and remains, whether they were related to the Darwins. Sister Emma's diaries continually refer to the Hadleys. But the pedigree of the Derby Hadleys has not yet been ascertained.

‡ See *English Men of Science*, p. 48.

I was grieved beyond measure at reading of your brother's ill-luck in New Zealand with Venus.

As regards that ogive* of which we were talking, I was stupid and explained myself ill, and boggled. In the ordinary way x is the magnitude and y the frequency. In my plan y is the magnitude and x is the *sum* of the frequencies, the frequencies being taken from the $e^{-\frac{x^2}{c^2}}$ tables and the sum of the frequencies from the tables of the integration of it, viz. Tables I and II respectively of the usual publications (? II and III in the *Encycl. Metropolitana*).

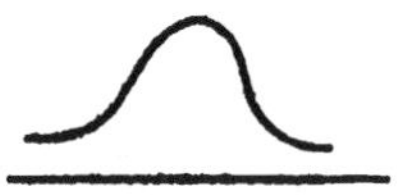

What a pleasant man Dr Andrew Clark is! He examined me most thoroughly, pronounced it a concurrence of irregular gout and influenza and that my heart was weak. I mend, but not over-fast. Best Xmas greetings to you all. Ever yours, FRANCIS GALTON.

Extract from a Review by Francis Galton in "The Academy," Jan. 30, 1875.

"Heredity; a Psychological Study of its Phenomena, Laws, Causes and Consequences." From the French of Th. Ribot, author of "Contemporary English Psychology." (Henry S. King & Co., 1875.)

"It may be affirmed with much truth that if we wish to learn what pursuit ranks highest in public opinion, we shall find it in the career of those men to whom statues are erected by public subscriptions. It happened that the writer of these lines not long since revisited Cambridge, where, as he walked admiringly among the many new improvements, his eyes fell upon a recently erected bronze statue. It was the only out of door statue in the whole town; it occupied a commanding position in the market-place, hard by the University Church, and only a few steps from being in full sight of the Senate House. He walked reverently up to it, pondering as he went as to the manner of the man whose memory it so proudly perpetuated, and lo! it was Mr Jonas Webb of Babraham, the famous breeder of Southdown sheep. The erection of this statue by the agriculturists of a county in whose capital a great university happens to be located, is worthy of note. It expresses their genuine appreciation of the practical application of the laws of heredity to all descriptions of farm produce, and it may be accepted as an omen that the time is near when the study of those laws and of their logical consequences shall permeate the philosophy of the university. It must do so, because there is no branch of science which refers to bodily structure or to mental aptitudes, neither is there any theological doctrine in which the theory of heredity, either directly or as one of the principal agents in evolution, can hereafter be left out of consideration.

"In the course of formation of every science there has always been an embryonic or pre-scientific period. Nothing then existed but detached pieces of evidence, of an unsatisfactory kind, laxly discussed and explained by wild hypotheses. But, at length, the methods of science succeeded in catching with a firm grip some of the loose materials, then more were seized, and so, with an ever-increasing rapidity of conquest, the whole of them became gathered together within the pale of law. Heredity has, at the present time, developed into a science; much is definitely established, and many questions seem to require for their solution little more than direct experiment or the simple but careful collection of statistical facts. There is consequently some need of a work that shall concisely and clearly set forth what is already known, and what are the undecided questions which most urgently call for solution and might at the same time be solved by any person who chose to devote a fair amount of intelligent and steady work to the purpose."

The remainder of the Review deals with Ribot's book, emphasising its inadequacy.

* Galton's "Ogive Curves," giving the deviations at the percentiles, etc. See our Vol. II, pp. 387–390.

June 4, 1875.

DEAR MRS HERTZ*, Fechner's *Elemente der Psychophysik*, Leipsic, 1860 (Breitkopf und Härtel) is a 2 vol. 8° containing in the aggregate 1000 pages, not very closely printed. It is a thoroughly standard work and lays the foundations of a new science which is beginning to attract serious attention in Belgium, France, America and England. In Belgium, Delboeuf's memoir upon it in the *Acad. Roy.* last (?) year (reprinted in a separate pamphlet by F. Hayez, Brussels) shows the primary importance of the work, though Delboeuf criticises and pushes the investigation a step further. In France, Ribot has lately been an exponent of Fechner's, or rather of Delboeuf's, views in a slight article in the *Revue Scientifique*. In America, Nipher (or one of his set) has recently been referring to him in *Nature*† and in England Sully in his papers in the *Fortnightly*, recently republished as a separate volume on "Intuition" (? exact title), renders full justice to Fechner. A mass of work by Arago, Herschel, and various astronomers, falls in as a part of the wide generalisations of Fechner, and much criticism and recognition of him will be found in Helmholtz. Therefore though the work dates as far back as 1860, it must rank practically as a new book, and the reading world is only now prepared to recognise its merits. Its object is, in a few words, to show that one fundamental law connects the amount of *sensation* (in the widest sense of the word) with the magnitude of the *exciting cause*. The generalisations are exceedingly curious and the experiments upon which the law is founded are most delicate and ingenious. The very science of such experiments, suitable for other applications, is laid down in the book and is one of the valuable parts of it. Fechner modestly ascribes the discovery of the law to his old master, Wagner, but it is Fechner who, by the admission of all who know about the matter, is practically the founder, exponent and establisher of the law. I should be heartily glad if an English publisher were to bring his work out in translation, believing that it would interest many scientific men and introduce a new and much needed branch of scientific investigation into England.

Very faithfully yours, FRANCIS GALTON.

42, RUTLAND GATE, S.W. *November* 18, 1875.

MY DEAR BESSY, Overleaf is the prescription and description. I heartily hope it may also succeed with you. The merits of this, compared with what I have had before, lie principally in the opium and in the absence of spirits of wine, etc. Those dulled the ear and disagreed with it; this does not, but is bland. After putting it in, of course the hearing becomes more defective as the wax is softened and plugs the ear effectively; but when the time comes for syringing the wax is all driven away quite easily. No forcible syringing is wanted but *you can't do it properly yourself*, you must have a gentle surgeon. Heroic surgeons (like Pritchard) assassinate the ear. Mem. Hamlet's uncle murdered his brother by dropping hellebore into his ear; I protest against being hung, if any ill effects follow my prescribing opium to be dropped into my sister's ear.

What a happy and moist time Edward‡ is having in Devonshire. Many loves.

Affectionately yours, FRANCIS GALTON.

Galton was very fond of prescribing on the basis of his early medical experience.

To GEORGE DARWIN, ESQ., TRINITY COLLEGE, CAMBRIDGE. *May* 2, 1876.

[Post-card] What a very interesting memoir you have sent me. It does one good to read about such large subjects. I wonder if the conditions of a nebula shedding a satellite could be illustrated by a whirling drop spluttering off, as shown and analysed by that curious method by which (in the last but one (?) number of the *Proceedings of the Royal Society*) the successive shapes assumed by a drop of water splashing down on a plate were investigated.

My wife is going on quite comfortably, and gaining strength, but Sir J. Paget, who saw her last Saturday, confirms all that Dr Chepmell has said. Sufficient for the day is the evil thereof! FRANCIS GALTON. 42, RUTLAND GATE.

* Mrs Hertz was a lady, who established a "scientific salon," and it flourished from 1865 onwards. On her death letters to her from Huxley, Galton, Clifford, etc. were sold to booksellers, the above and others being purchased by the Galton Laboratory.

† See May 20, 1875.

‡ Galton's nephew, sister Bessy's son, Edward Wheler.

42, RUTLAND GATE, S.W. *August* 1, 1876.

MY DEAR GEORGE, Mrs Jebb's account of the twins and the way she puts it, is most striking. How one wishes one could have such a case under close examination. A single instance verified in a large number of particulars would carry such immense weight. Thanks very many for sending it to me.

What a pleasant Autumn you have before you. We shall not meet first, as we leave Town to-day week (Aug. 8) to stay with Judge Grove and thence on Aug. 24 we go abroad to the Tyrol.

I am rejoiced at the fair promise of all your earth axis work and especially at the fact that you can do so much without being upset by it. What laborious work it must have been.

I have just left Hooker at the Club, very matrimonial-looking, studying the Bravo case*.

Ever yours, FRANCIS GALTON.

To GEORGE DARWIN, ESQ.

42, RUTLAND GATE, S.W. *January* 5, 1877.

MY DEAR GEORGE, How wonderfully inventive you are. I am most anxious to learn your plan about the curve-drawing.

May I venture to trouble you with a request, not a great one? It is to look through a short, clearly written (orthographically, I mean) memoir on "Typical Laws of Descent" which I propose sending to the Royal Society and which would occupy four to five pages of the *Proceedings*, and tell me if it is sufficiently intelligible.

You did me real good service in burking my memoir of last year. This is certainly very much better than that, but tell me—is it good enough? I will send it at once, if you will have it. Affectionately yours, FRANCIS GALTON.

P.S. Pencil anything you like on it. If possible I want to send it in *soon* to the Royal Society so as to be read before my February 9 lecture.

To GEORGE DARWIN, ESQ.

42, RUTLAND GATE, S.W. *January* 12, 1877.

MY DEAR GEORGE, How can I thank you sufficiently. I am aghast at the trouble my unlucky memoir gives, and at the great pains you have taken to put clearness into it. I will certainly adopt your suggestions generally and rewrite the thing.

Let me mention an illustration of one of the principles (Family Variation), which I think may interest you. You recollect that apparatus of mine with the shot;—well, suppose I want to show by a modification of it, how it comes to pass that when the ordinates of an exponic†

mountain subside, each of them, into an exponic hillock, as in the sketch, the sum of the hillocks is an exponic curve of larger modulus.

In I (see p. 466), I pour shot, and it makes an exponic heap at the bottom. In II, I have cut the apparatus across at *AB*, and have interposed a row of vertical compartments with trap door bottoms that I can pull out and in to form a temporary landing for the shot, when I so desire. If these are open, the shot falls through and of course makes an exponic mountain at the bottom of II, exactly as it did in I. But if they are closed, they intercept the shot and an exponic mountain (of less

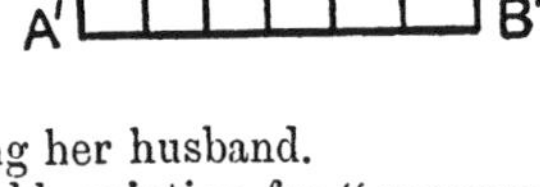

* A famous trial of that day; Mrs Bravo was tried for poisoning her husband.

† I do not remember Galton using this word elsewhere as an abbreviation for "exponential." It seems itself slightly "out of place."

modulus) is formed on $A'B'$. Now I open the trap doors, successively; the shot in each vertical compartment rushes down and forms its own exponic hillock, and we have already seen what

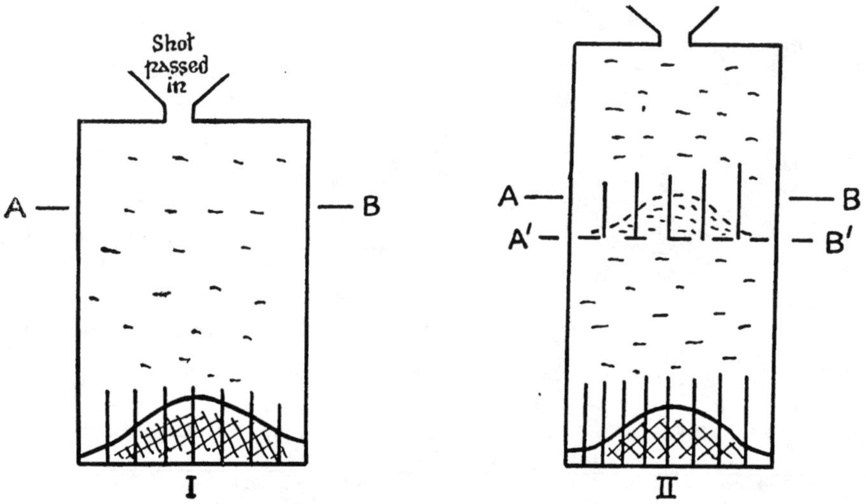

the sum of them will be. The ratio of the moduli of these heaps is self-evident (they vary as the square root of the indices which vary directly as the length of passage of the shot). For

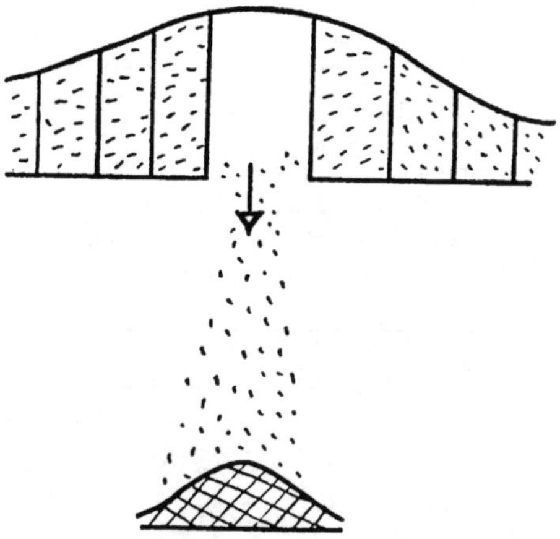

my Royal Institute lecture, I shall simply go into generalities to show what Reversion, etc., mean and how a law is possible, and shall hang up the formulae, but not speak a word about them. Affectionately yours, FRANCIS GALTON.

To GEORGE DARWIN, ESQ.

The substance of this letter appears in Galton's R. I. Lecture of Feb. 9, 1877: see Vol. IIIA, pp. 6–11.

42, RUTLAND GATE, S.W. *July* 14, 1877.

DEAR STOKES*, With reference to our land meteorology, would you kindly consider and advise on the following point (which notwithstanding first appearances really falls within that branch). It is, what form of mechanical indication or registration would best convey "sea-disturbance"? I presume what is wanted the most is some idea of the ship-wrecking or

* Later Sir George G. Stokes.

even sea-sick-making power of the sea. Now what element or elements should be measured in order to show this? Am I right in supposing that the two measurements of maximum height

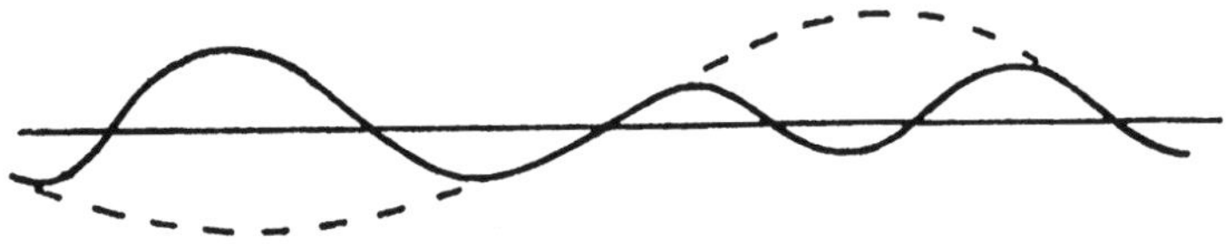

during the past (say) 5 minutes *and* the sum of the heights during the same or some other uniform period would give this? The first, alone, would distinguish between big waves and little waves, the last would make the further distinction between an abrupt tumultuous sea and simple regular waves. Do not trouble about the mechanics part as yet. These and many other elements can easily (I think) be measured and I can readily explain and show drawings. What I merely want to know is *what would* (for the purposes of those who read our weather reports and of ship insurers who dispute claims for wrecks on our coasts, on the ground that the weather was not really bad, and who apply to our office for evidence) be the *best elements to measure.*

Pray look at the July number of the *Philosophical Magazine*, at a paper by George Darwin on interpolation. It may greatly improve our office calculating. I had begged him to examine and investigate the subject, especially with a view of interpolating in three dimensions (latitude, longitude and time), as he has shown how to do in the latter part of the paper. I have asked him to send you a copy of it. Very faithfully yours, FRANCIS GALTON.

42, RUTLAND GATE, S.W. *October* 8, 1877.

MY DEAR PROFESSOR*, We are now not only nearly but quite "in focus," I think. (1) The fiducial marks:—a scale is cheaply cut. We can try the "web" and if it confuses the picture we can ultimately adopt a simpler plan. (2) Weight, or spiral spring?—whichever the instrument maker prefers—(One can't do the equivalent *easily* with a spring, of *lifting up* the counterpoise). (3) Zenith adjustment:—your plan is the simplest and best. (4) Azimuth:—Allah forbid, that I should propose to carry a theodolite about with each instrument, for the sole purpose of laying down rough azimuth. I was merely thinking of Kew and of fixing in the ground there two or more permanent slabs with fiducial marks, and as there is a meridian line laid down, and a theodolite at Kew, I thought it might be just as well to use them—(It is more important to sight one instrument from the other than to get an exact azimuth). (5) Single or double camera:—I quite agree to beginning with a single one, though when the clouds are low and drift rapidly, I doubt whether it would be possible to work with a single one. The expense of the box, single or double, will be trifling. Our first attempt is sure to be not over good, and whether we have fitted one or two lenses to our first camera, they will serve again. Neither need we buy a lens on purpose for trial—we could easily borrow one—I could lend one, but perhaps it would be better to get one of a large angular field of view. I would meet you at the Athenaeum on Wednesday if you are disposed and will send me a post-card to say about when.

Very faithfully, F. GALTON.

42, RUTLAND GATE, S.W. *December* 10, 1877.

MY DEAR PROFESSOR*, I went yesterday to Maudsley's place at Clapham Junction, saw him, and ordered three dozen plates and accompanying gear (solutions and dropping bottles)—total cost £1/8/0 or thereabouts. Also, I have written to Kew and find that Whipple the Superintendent understands our proposed photographic requirements. There will be trouble about the theodolite, I find, as none that they have there will admit, he says, of viewing an object placed vertically below the telescope. (There may prove to be some simple way of lengthening the axis for the occasion, or rather for performing some equivalent process.)

Thanks for your letter. Uniformity of wind velocity and direction at all altitudes can never I fear be expected, as all balloon ascents have shown the contrary. There will probably be some curious effects when the pictures are viewed stereoscopically—as, if the clouds move in opposite directions at different levels, the plates which must be disposed *left* and *right* to

* Professor G. G. Stokes.

give a stereoscopic image of the lower clouds must be disposed *right* and *left* to give one of the upper clouds.

I fear that our real difficulty will relate to time of exposure. I should propose to begin by taking four or five consecutive pictures at somewhat different degrees of exposure, and seeing what can be learnt from them in every way, including various stereoscopic combinations—and will certainly follow your suggestion of making the first trial on a suitable cloudy day, as well as the other suggestion in your letter. Sincerely yours, FRANCIS GALTON.

42, RUTLAND GATE, S.W. *October* 23, 1879.

MY DEAR PROFESSOR*, About MacAlister's paper; it might be well to look at the marked passages in the enclosed letters from him, sent to me a few days back. Do not return them.

The principal people who have used the law of error for vital statistics, since Quételet, are the compilers of the War Department Statistics of the N. American Forces after the war between the N. and S. States. And again, curiously enough, Fechner himself in his *Psychophysik* (I, 108) introduces a long mathematical investigation by his mathematical colleague (I have lent the book and forget his name) wherein a series of law of error tables, "Methoden der richtigen und falschen Fälle," are formed to help him in his own investigations. In short, he ignores his own law! He uses tables on the Arithmetic Mean principle to discuss results of observations on phenomena that have the Geometric Mean condition. So the question treated in the paper is really one of importance to statisticians.

Very sincerely yours, FRANCIS GALTON.

42, RUTLAND GATE, S.W. *October* 14, 1879.

DEAR SPOTTISWOODE†, I venture to enclose some suggestions for increasing the interest—negatively if not positively—of the meetings of the Royal Society. If they seem reasonable to you, perhaps the Council would in due time take them into consideration. The recognition of the fact that very dull papers do not need to be read at all, and that difficult papers should not be discussed after only one simple reading of them, would I think be a boon. I fancy, too, that under the proposed plan the experimental part would gradually develop and the discussions ought certainly to improve. I have talked the matter over with a few persons and thus far with a favourable result, but I leave the matter to your much better judgment.

Ever sincerely yours, FRANCIS GALTON.

Suggested procedure for the Meetings of the Royal Society.

The first publication of a memoir by the Royal Society to be not as at present by reading it to the meeting, but by laying revised copies of it, printed in sheet with paper cover, title and date—in fact, the author's copies—on the table and reading the title only.

The subsequent issue of the memoir in the ordinary publications of the Society to take place exactly as it does at present.

The subjects advertised for such meeting should generally be memoirs that had previously been published. The authors or their deputies should give explanations of them, illustrated as far as may be by experiments and drawings, and followed by discussions. The President to have full power as at present to select the subjects for the meeting and the order of taking them. For the most part they would come before the Society in one or two weeks after their publication. Some however would never be brought forward at all, and others would perhaps be most advantageously discussed on the same day as their publication.

42, RUTLAND GATE, S.W. *October* 23, 1879.

MY DEAR MISS HERTZ‡, Please accept by letter, as you were out when we called and I could not verbally give, our very warmest congratulations and best wishes for many years of future happiness. I can assure you that I think your intended ought to consider himself

* Professor G. G. Stokes. The letter refers to Donald MacAlister's paper on the Law of the Geometric Mean: see our Vol. II, pp. 227–8.

† President of the Royal Society 1878 to 1883.

‡ Daughter of Mrs Hertz (see p. 464 above).

a very fortunate man indeed, and have not the slightest doubt but that is his frame of mind. It will give us great pleasure to make Mr Macdonell's acquaintance, and I hope you will soon give us the opportunity of doing so. What a great deal of new happiness and new life you have before you, and what a break-up of Harley St life will be the result. Once more with our united kindest wishes, believe me, very sincerely yours, FRANCIS GALTON.

Thanks about the Generic Images paper. I have sent to-day a copy to Professor Oscar Liebreich and will gladly send you a few—for friends—in a day or two.

42, RUTLAND GATE, S.W. *October* 28, 1879.

MY DEAR MISS HERTZ, On coming back last night from the country I found your book of readings awaiting me. Thank you so much for it. I have been reading ever so much of it already. What a true idea of yours that is, in the preface, about aesthetic training not being a step by step affair, like that of science, which has to make each foothold sure before venturing another pace. But I suppose the same is true of morale—conduct—and much else besides. Even language; though when this is taught classically it *is* a step by step affair. I find as I write that the subject enlarges and there is evidently much to be said about the two ways of teaching; in fact it seems to open out the whole education question. *Requiescat in pace.*

That "galloping" poem of Browning's is certainly wonderful rhythm. I wonder if a great artist could write a poem in a rhythm that should bore one most insufferably:—a sort of "Ancient Mariner" from whom there was no escape, who bewitched and made one half mad at one and the same time? Very sincerely yours, FRANCIS GALTON.

What a deal of kind, good educational work you must have done by your readings.

42, RUTLAND GATE, S.W. *March* 6, 1880.

MY DEAR GEORGE, About those visualised numerals—of which by the way I have now collected much information—can you easily answer me this question? I want to know whether the graduation belonging to any number, say 10, does really occupy very exactly the same position at all times in reference to the axis of vision and the horizontal plane passing through it. In short—if you look at a ship on the horizon and think of 10, does that number always appear in the same altitude and azimuth reckoning from the ship and horizon?

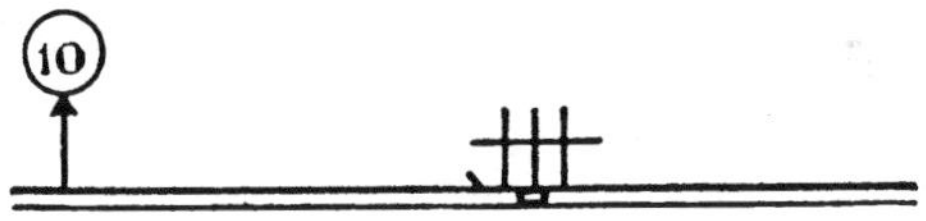

I suspect that in many cases it does so with considerable accuracy, and that these visualised numerals are the strongest case known of "topical" recollection, which implies some system of *division of labour* in the brain elements. In short, that the 10 always occupies a spot corresponding to a speck that would be seen if a certain part of the retina were injured, and also to a spot that would be produced if the part of the brain in physiological connection with that spot on the retina were injured. If this be true of each of the numbers, then it seems to follow that a particular part of the brain is charged with the care, so to speak, of one particular number. It is quite extraordinary how in the great majority of cases (not yours) the want of coincidence between the names and the values of the numbers betrays itself in the numerical forms. There is almost always a hitch and a bother at 12 and at the teens, which repeats itself at 120 and the series hardly ever runs regularly except between 20 and 100, 120 and 200, etc. Children are puzzled and the puzzle continues throughout life as shown by the persistence of the misshapen form.

I wish you had been in Town and that I could have persuaded you to come next Tuesday to a paper of mine about these numerals at the Anthropological. I have got at least 6 "seers" of these things to dine with me and then to go to the Society and stoutly maintain their veracity there: viz. Bidder, G. Henslow, Schuster (wave length), Woodd Smith and Col. Yule—besides Mrs Haweis and (I have no doubt) Roget. My collection exceeds 60 forms, curiously diverse in some respects but almost all alike in fixity, extremely early origin, and in the 12 difficulty—and I have got returns from schools. It seems that about 1 man in 30 has the tendency, and twice as many women. Many other odd things come out. A left-handed twist of the forms is about as common as a right-handed one, etc.

We heard two days ago from Mrs McLennan's* sister, who says that she (Mrs McL.) has been nearly dying but that she is now somewhat better. They are preparing for leaving Davos, and are inquiring for a good place to go to. I am sorry to hear that your visit of charity did little good to your own self. It was very good of you to go. Ever yours, FRANCIS GALTON.

August 12, 1880.

Excuse bad paper, ink, etc., our house is in the plasterers' hands.

DEAR GEORGE, The enclosed was sent to me asking me to read it and forward it to you. The writer, Walter Smith, was a bracketted 2nd Wrangler some few years back and of Trinity College—you would know all about him. I knew his people well, especially his father, Archy Smith.

Did I tell you that during a happy day I spent among the idiots at Earlswood I learnt from the very intelligent medical director, Dr Grahame, that his inquiries about the parents of the idiots quite confirmed your conclusion about cousin-marriages, and that he had said so in print?

I suggested to W. Smith that if he wished to work up the subject *de novo* he should get an old Burke's "Peerage" and "County Families" and pick out the first hundred or so cousin-marriages, also of ordinary marriages that he came across, and partly by the help of more recent editions but chiefly by that of gossips about the aristocracy compare the results. If the difference was not a notable one he might be at rest as to harm done by not forbidding the banns. I wonder if he has a personal interest in the inquiry. What a charming episode in a novel—the conscientious young Scientist collecting laborious statistics before he ventured to propose.

We go to McLennan's to-day, to stay till Saturday afternoon at Hayes Common.

Ever yours, FRANCIS GALTON.

P.S. Thanks for grouse.

42, RUTLAND GATE, S.W. *December* 11, 1881.

MY DEAR GEORGE, Here are the three sets I circulated of Mental Imagery questions. They were usually followed up by correspondence.

What a wonderful application of your earth-history theory is this big tide in early geological times! I want particularly to read your account of the matter when it appears, and to have your own views thereupon. It is a *grand* idea indeed—the grandest since the *Origin of Species*. Have you thought over the corresponding air tidal-wave? Now, in the tropics, the diurnal barometric range is (...? say $\frac{1}{8}$ inch), what will it have been in those times? And what would be the corresponding wind force? I can't understand how any thing could live on dry land under such blasts. Talk of catastrophes, why, that time must have been a continual series of catastrophes. Dante's Hell is nothing to it. But I had rather have the *facts* from you than through the Astronomer Royal of Ireland. Don't of course bother to answer this, but I hope we shall soon read a short article from you in *Nature* or somewhere on this extraordinary revolution in old ideas.

Have you too (I ask not for an answer) talked over or thought about the air flying off from the earth, and notably from the moon, to somewhere else? I mean what we were talking about. Lord Rayleigh seemed to think it worth considering and within range of calculation. Just now I suppose you are busy up to the eyes with Tripos preparations. We look every morning in the column of births in the *Times* for news from Horace†. Ever yours, FRANCIS GALTON.

THE ATHENAEUM. *December* 11, 1881.

What frightful nonsense I have just despatched in a letter to you about air-tides. There was conversation—I had two ideas in my head and they blundered together as in a dream, the letter went and I could not correct it.

In sober sense I should have written: Supposing height of air-tide in an imaginary homogeneous atmosphere to be the same height as water-tide (Herschel says so), say 8 feet, then the corresponding barom. pressure due to air-tide would be 0·008 inch. Under the supposed ancient condition of a 216-fold height of tide this would become $216 \times 0{\cdot}008 = 1{\cdot}728$ inches, so that the barometer would go up and down $1\frac{3}{4}$ inches in every 12 hours, which implies a constant state of hurricane. F. G.

* The wife of Donald McLennan, the writer of *The Patriarchal Theory*.

† Horace Darwin, Charles Darwin's fifth son.

PLATE XLVIII

Francis Galton's Niece Milly—Mrs J. C. Baron Lethbridge.
Compare Plate XIV, Vol. I.

INGLEWOOD, BEDFORD PARK, TURNHAM GREEN. *April* 27, 1882.

MY DEAR MR GALTON, I thank you heartily for your note. And I so fear to trespass upon the profound sorrow that fills the home at Bromley that I cannot venture to obtrude directly even an expression of the gratitude I feel that my name should have been remembered in giving out invitations to the funeral. It was, indeed, with deep satisfaction that I learned that our Minister, Mr Lowell, was to be a pall-bearer, and his countrymen will regard it as a most happy circumstance that they were represented, on such an occasion, by no mere politician but by a man so worthy to bear the pall of Charles Darwin. I see also that the venerable Robert C. Winthrop was present, the President of the Massachusetts Historical Society and in many ways a representative American.

The experience you speak of, in connection with the generalisation worked out by your great relative, corresponds with the experiences of others who were watching by night when the glory of this new star shone around them. A few years ago when, through that considerateness of a heart which could hold a world and at the same time not overlook the smallest opportunity for kindness in it, I was invited to Down, and when I was walking with him in his garden, I felt as if I would fain clasp his feet and try to tell him what he had been to me. At night I well remember lying sleepless for some hours tracking the steps of my pilgrimage which had begun in an Egypt of Darkness and been able to clear Wildernesses by his aid. This spiritual effect of a pure scientific generalisation, as I have known it in myself and in many other minds, is the most significant phenomenon of this age. It is a thing to be pondered on by those who consider what is to be the God-spell or glad tidings of the coming time.

On Sunday last I had a very large audience to attend our memorial service and discourse in honour of Darwin. I am now engaged in preparing a sort of memoir which I shall probably deliver before the American Assoc. for Advancement of Science at their meeting in August. It occurs this year at Montreal, and Steny Hunt has tempted me to cross the ocean merely to remain one month. (I wish I could tempt you to go also.) I shall aim, in what I am writing, to give the facts of Darwin's personal life, so far as I can obtain them; the dates of his works, etc. I shall also try to trace carefully the history of the doctrine of evolution—tracing it from the empirical suggestions of Newton, and then Buffon, to Erasmus Darwin, then to Lamarck, Oken, Goethe, Geoffroy St Hilaire, and Darwin. (And by the way, do you know that more than forty years ago Ralph Waldo Emerson was basing his entire idealistic philosophy on evolution?—in his first book, 1836, writing—

"And striving to be man, the worm
Mounts through all the spires of form."

As for this matter of a memoir concerning Darwin, I should hope to consult you about it at some time.

I send you an American paper with a little Essay of mine written last year. I sent it to Mr Darwin in January. It is not much, but may interest you and Mrs Galton.

Ever yours, MONCURE D. CONWAY.

HARLECH HOUSE, BOURNEMOUTH. *March* 26, 1883.

MY DEAR PROFESSOR*, Thank you much for your pretty cloud problem. I have been on the look out for an opportunity of experimenting with it, but have not hitherto had a chance. It has however suggested to me a plan which I enclose, and which I have tried, that really looks as though it might be regularly employed in many stations where there are cliffs or neighbouring hills, and which might even give good results for clouds up to 2000 or so feet. I experimented by using the Kew Pagoda to serve as the AC in the enclosed. The sea here is bare of ships, but I have tried the method this morning upon one that happened to be passing and it seemed very convenient. Very sincerely yours, FRANCIS GALTON.

5, BERTIE TERRACE, LEAMINGTON. *September* 27, 1883.

MY DEAR MILLY†, From your very liberal standpoint, the arguments in the Chapter on Prayer have necessarily little value. They are directed to those who either (1) like the great

* Professor G. G. Stokes.

† Mrs Millicent Lethbridge, daughter of Galton's Sister Adèle, Mrs Bunbury. Galton is referring to the section on Prayer in his *Inquiries into Human Faculty*: see our Vol. II, pp. 100–101, 115–117, 258–261.

majority of Puritans and theological writers assign a magical—(? right word) power to prayer, or (2) whose ideas are habitually confused as to what they believe, what they doubt about, and what they disbelieve. I fear that everyone belongs in some degree to the last category and that it is most important for reasonable beings to extricate themselves as far as may be out of it. If there is a lingering tendency to believe in the magical (?) objectivity of prayer, which would not be avowed if the question were put in a straightforward way, then I should say try and eradicate that tendency. Let your thoughts and the outward expression of them be conformable. I am sure that the average clerical mind is in hopeless disaccord with its outward expressions, and that was one reason why I wished to discuss a class of views that appear to me (and to most of those who consider them plainly) to be untenable—those which refer to what I call the objective efficacy of prayer.

Your "Einverständniss" view seems to be undoubtedly that which deserves investigation. Is it a reality or is it a fancy? I have endeavoured in the book to show that the solution is by no means so easy as religionists say, because very much of what are commonly taken as evidences of it, innate feelings, aspirations, etc., are demonstrably of very little weight indeed. I want to knock away all fictitious supports, and to get the evidence *pro* and *con* that we possess clearly before us and to look at it fearlessly. Men lead happy, useful and honest lives under so many forms of belief that I cannot suppose the precise form of belief to be of much importance. But it is of course cheering to the heart and ennobling to the mind if the belief be that of being a missionary, as it were, in a high cause affecting humanity. Beyond that I suspect there is little, and that each man puts a great deal of his own self into the ideal that he sets before him.

How infinitely little we know! I like to look at a mongrel cur sitting on the doorstep of the house he belongs to, looking as if he were the master of the situation and as though creation presented no difficulties whatever. He is so like most men in this.

Thank you much for the letter, which I will keep and read again when, if ever, I write on the topic a second time. People are often so crude and unreasonable that I get quite savage and then it does me a world of good to read such letters as yours, which tend to lift the discussion to a higher level.

About the numerals and teaching: have you thought of writing the declensions, etc. not only in different coloured inks but in different *shapes*, even differently shaped borders would be something? If you could somehow associate the shape (or colour) with the matter taught in a *reasonable* or even in a suggestive way, it would be a help. For my part, I think I should recollect best by *gesture* and in

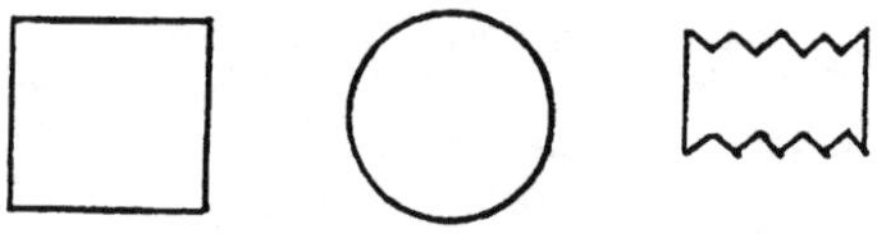

a kindergarten kind of way—thus if I learnt one thing with my right arm waving like a Salvationist's and another while beating a tattoo on the table, I should find the association easy. Some people associate with sound very readily. Thus one declension might be sung to one tune, another to another. Even a high-pitched or low-pitched tone would go some way. But the associations should not be haphazard, they should in some way be natural, whether by a reasonable, a long since acquired, or by a punning connection....

Ever affectionately, FRANCIS GALTON.

42, RUTLAND GATE, S.W. *October* 1, 1883.

MY DEAR PROFESSOR*, I am just back in London and ready at any time for a Council at the Meteorological, and have written to Scott to say so. We must proceed cautiously but firmly in this self-recording observatory matter. One plan would be to draw up a brief and cold-blooded statement of the reasons *pro* and *con* as we understand them, and ask the memorialists whether in their opinion they cover the ground; after receiving their replies, to reconsider and decide. However eminent the men may be, they cannot see the matter in the same light as if they had administered the affairs of the Office and knew details.

I left Southport on Saturday morning and never attended the Committee. Indeed, as I said, it did not seem to me quite the right way of proceeding on the part of the objectors. They might more properly have first sent in a memorial; *then*, if that produced no effect, they might use pressure if they liked; but should not I think have begun with external pressure.

* Professor G. G. Stokes.

PLATE XLIX

Francis Galton, the Meteorologist, when about 65 years of age.

PLATE L

The Medallion of Erasmus Darwin by Fassie, from the copy in the Galton Laboratory.

For my own part, one strong reason for suppressing the observatories and diverting the money saved to more pressing inquiries lies in the belief that hereafter it may become possible to note a *greater variety of data*—such as upper air currents, total humidity of a vertical column of air, some electrical facts, possibly by the captive balloon, and generally, data from the wide field of the now unknown. What we *have* recorded during these past years is such a very little bit of what we want to know before we can understand the weather, that it seems a pity to prolong unnecessarily the present system—we might probably recommence 20 years hence on a much more favourable basis. Very sincerely yours, FRANCIS GALTON.

In 1884 Galton gave the Rede Lecture in the Senate House at Cambridge. Some account of this will be found in our Vol. II, pp. 268–271. The impression formed on the mind of a competent critic is conveyed by the following post-card headed in Galton's handwriting:

"My Rede Lecture. Note by the Rev. G. F. Browne."

You will have heard that you were admirably audible; I only hope I didn't overwork you. It was beyond measure (!) interesting and several of us have vowed that the thing shall be set going for undergraduates. G. F. B.

42, RUTLAND GATE, S.W. *September* 25, 1884.

DEAREST EMMA, The news of the pencil fills my heart with rejoicing. I dreamt an eventful dream last night of which the climax was that it was discovered in the pocket of my dressing-gown, and awoke rejoicing to tell Louisa;—and lo! it was a dream. I must never again wear it together with your door key. The two do not agree in the same pocket. The pencil case is flipped out by its great cuckoo half-brother, which hangs from the end of the watch-chain and is also stowed in the pocket of the waistcoat. This is the second time it has occurred; I have been watchful since the first time, but now I look on the reconciliation of key and pencil case as impossible, and will hereafter carefully separate them lest they quarrel on the sly.

I went to the British Museum to-day with my earthenware god *Bess*. Another, but I am happy to learn a smaller one of the same god, has just been discovered by Flinders Petrie in his excavations in the Delta. I have given mine to the British Museum. They are to give me three casts of it: one to bow down to in my own house as heretofore, the others for the archaeological collections of Oxford and Cambridge respectively. Then I produced the E. Darwin medallion, which was discussed in the medal room just as Lucy's coins were. They say it was by a Scotchman called Fassie, who made many fair medallion portraits about the end of last century in a paste of his own composition. There will not be the least difficulty in making plaster casts of it. They will make a mould and turn out as many as are wanted. I have ordered a batch and you and Bessy shall each have one; also Mrs Oldenshaw (to whom I have sent a line) and Emma Wilmot. When the medallion comes back to me, I will take it both to S. Kensington and to Scharf at the National Portrait Gallery, to see if they also know anything about it. To-day has been a considerable scurry. Louisa will, I am sure, tell you about herself and Chepmell. She discussed six raw Whitstable "Natives" at dinner with considerable gusto (she was told by Chepmell to try oysters), but I fear the pain is not sparing her just at this moment (indeed it is not).

I cannot sufficiently tell you, and it is needless for me to try to express what you know, how much we feel the sense of your affectionate kindness to us both. It comes so much as a matter of course and is received so much at the time in that way, that it looks as though we were not really half as conscious of it as we should be, but we *are*, and I am sure you know it.

Milly is in a way about Eddy's* future, naturally enough. She has written such nice letters in answer to those we sent her; Baron† clings to Edward's being sent to a private tutor and thence to Oxford, while she wants differently. Cyril will have been there by now and I am very curious to learn the result. Best loves to Bessy and the Moilliets‡. I wrote a paragraph at the Meteorological Office to-day about the little inquiry I had made there in reference to Edward's foggy voyage. I dare say it may get quoted in some newspaper in a few days. Ever affectionately, F. GALTON.

* Edward Galton Baron Lethbridge, now of Tregeare.

† Mr J. C. Baron Lethbridge, Millicent Galton Lethbridge's husband.

‡ Francis Galton's second sister, Lucy Harriot Galton, married Mr James Moilliet.

42, RUTLAND GATE, LONDON. *October* 4, 1885.

MY DEAR SIR, Excuse delay in reply, as though I date from town I am still in the country. Let me first cordially thank you for your kind letter and the many interesting remarks it contains.

(1) I have written to the Secretary of the Anthropological to tell you exactly what the annual cost of the journal is, I think it is £1, viz. 4 parts at 5/- each. Also I told him to send for your acceptance from me, a recent number in which there is an exceedingly good paper about the Jews, illustrated by some rather successful "composite" photographs of Jews by myself, which it may amuse you to look at.

(2) I have ordered both the books you speak of: thank you very much for telling me of the latter especially, I mean that about the sex of the child.

(3) You were so kind as to send me some time ago your investigation into the colour of hair, and I feel myself open to blame for not having drawn attention to it already at the Anthropological or elsewhere, but the fact is that I wanted to work up my own data, and to give both results at the same time. My data are now worked up, but there still remains something to be done, so that there will be a little further delay.

Did you ever consider the physiology of *clear green* eyes—bright green I mean, such as Dante says Beatrice had? The common often repeated statement that blue eyes are merely the effect of seeing pigment through a semi-transparent medium, and that there is only one sort of pigment, cannot possibly explain the existence of blue *and* green eyes, both equally translucent. There must be a green pigment somewhere. I have asked all our best physiologists, and have looked through many German and French memoirs, thus far in vain, for a rationale.

I am assured that the pigment particles are not so minute as to affect the light by any *iridescent* effect. In short, that the blue and green cannot be due to such causes as those that make the waters of the Rhone, blue, and that of some of the Tyrolese rivers, green.

Believe me, Very faithfully yours, FRANCIS GALTON.

This letter is a reply to that of Alphonse de Candolle, published in our Vol. II, p. 210.

HOTEL VICTORIA, SORRENTO. *March* 24, 1886.

DEAR GEORGE, At last we are in the promised land, most comfortable, and all most beautiful. It was a disagreeable journey, so far as railway went, to Genoa. Genoa most Italian, and yet quite fresh and full of bustle. Then we tried Nervi but it is cramped. I got a biggish, Ste Agnès*, sort of a walk in the afternoon and we left for Pisa next morning. Pisa *glorious*. I felt there was more in man than I was wont to think looking at the artistic triumphs there. Next day to Rome (Hôtel d'Italie—very recommendable for sunshine, and good generally): Saturday, Sunday and Monday we saw old scenes. We had a very social afternoon with Mrs Grey and Miss Shirreff; also I looked up an Anthropologist (G. Sergi) and saw his studio, and learnt at the Vatican Manufactory much about mosaics, as affording good standards of reference for anthropologists, tints of skin, etc. Left Rome yesterday, Tuesday, morning and got to Sorrento at 8. Slept at another hotel, but rooms not sunny enough so changed here this morning. Vesuvius smokes famously. Yesterday the air was saturated and clouds lay here and there among the hills at all levels. The steam from Vesuvius mixed with the clouds and occasionally showed itself distinctly as *growing in volume* as it left the cone. I strongly suspect the sulphur in it formed centres of deposition for the fresh cloud. The effect was rather striking. We shall, I expect, settle here for a full fortnight.

Tell us how you are going on, and what has taken place at Mentone since we left. Any good excursions? Louisa sends her kindest remembrances,

Ever affectionately yours, FRANCIS GALTON.

To GEORGE DARWIN, ESQ.

April 9, [1886]. In a dull railway carriage, all alone.

MY DEAR GEORGE, You will be in England I suppose now, so I write there and to the Meteorological Office. Both your letters came safely. The first reached me just after I wrote,

* Presumably the well-known excursion from Mentone.

so this is my second letter only. I never have enjoyed a holiday so much and daren't trust myself to look towards its close. We have been three nights at Quisitana*, and I write this in a railway carriage *en route* to Paestum for the day, whither I make a solitary journey of a total of ten hours' travelling and detention in order to get a two hours' view of the ruins, twelve hours

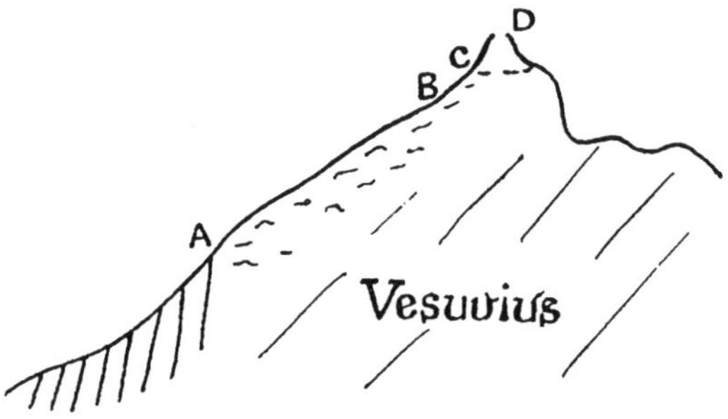

altogether. Yesterday I *did* Vesuvius, with George Butler, a lady-like chaperone, a pretty daughter, and also a Newnham young lady. We got on admirably by being pulled up the ashes on the side *AB* and then up the much more difficult ashes on the cone, *CD*, and I feel this morning as though I had been all night at a ball held on the sandy seashore, dancing reels and not missing one. It gave one a vivid idea of the muscular effort required to fly; that is, to support oneself in a yielding medium. The sulphur colours were glorious, the "lapilli" came up in occasional volleys and fell about us dull-red-hot. It was a grand sight looking into the crater at the steam with its glowing foundation;—then we scuttled down, to get out of the way of the next shower of lapilli. The flames were beautiful last night and reached above the cone at times, to a height equal to the height of the cone. To-night they reach to three times its height. Sorrento was the acme of felicity. Amalfi was a falling off and, to our taste, Quisitana is a further descent. We turn northwards in three days to or towards the Italian lakes, thence to the Lake of Geneva, to stay a few days with my niece Milly Lethbridge, and then home, where I am pledged to be by May 11 at latest and probably a *few* days earlier. My Wife has thoroughly enjoyed herself, but of course is not up to the longer excursions, and is beginning to feel the climate. Josephine Butler † joined her husband at Quisitana last night.—Well, well! one can't talk to her about her favourite topics, holding as I do most diametrically opposite views in nearly every particular of faith, morals, and justifiable courses of action; but for all that she is, or was, very charming and keenly alive and sympathetic.

I was very glad to hear your own satisfactory home news and trust that your American letters show that all is going on well and happily there, and also that your Mother is fairly if not wholly well again. We have not been fortunate in meeting many pleasant people. One agreeable acquaintance was an American, Mr Andrew White, once president of, and now a history professor in Cornell University, and for some years U.S. minister in Berlin ‡. He knows both Oxford and Cambridge pretty well. I dare say you may have met him and his wife.

I wonder whether you have been able to strike out important ideas about our procedure of weather predictions, I am sure you will strike out some new ones, and it is high time that original ones should be struck out.

Will you kindly give the enclosed card, or post it, to Scott? You will see what it says. If it should prove quite convenient to fix the first *May* Meteorological Meeting during the second and not the first week, all the better for me.

Ever affectionately yours, FRANCIS GALTON.

To GEORGE DARWIN, ESQ.

* I have failed to find any such place. It is possibly a very obscurely written Positano, which is S.E. of Sorrento.

† See our Vol. II, p. 130.

‡ During the biographer's student-days there, 1879–1880; he most courteously invited to his house English as well as American postgraduates.

From Louisa Galton: 42, RUTLAND GATE, S.W. *November* 4, [1886]. Private.

(It's not private at all. F. G.)

DEAREST EMMA, See the Telegram just come, quite unexpected but not the less welcome. I am so glad. Frank works on so patiently and quietly, there is less to bring him to the front than with many who do less. He is very pleased but do not talk about it for a few days, as the President of the Royal Society puts "private" till confirmed by the Queen (a mere farce). It is given for his Statistical inquiries and investigations in Biology. You will be pleased, I know, more than anyone next to ourselves. I write in bed having been sick half the night, but hope the attack has passed its worst, still I cannot write much. The encomiums on Montagu* are delightful and not too great. I long to hear we shall soon see him. Friends are so hearty and pleased; we were none of us satisfied with his Deanery from the first and very dissatisfied with Gladstone. I am so thankful he has this rather than a bishopric, but he will sadly need a wife. I am grieved your Cookery has been troubled by Miss Ellis's illness, and that so much has devolved on you. We shall anxiously await tidings, how all goes on, including Miss E.

November 5. DEAREST EMMA, I finish the letter as Louisa is arranging with cook. We are very sorry indeed about Miss Ellis and your troubles in consequence. The full-sized design for the Lichfield memorial is ready to be seen and sent down, and I shall go to the Sculptor this morning, but being rather busy afterwards at distant Committees will not be able to send an account of it to-day. I was so pleased last night about the Royal Medal. Stokes, the President, wrote to me this morning to say it is for my "statistical inquiries into biological phenomena." These things don't get into the papers, certainly not for a couple of weeks, as the names of the Royal medallists have to be submitted to the Queen, because she gives the medals. But this is a pure form. The medal contains some £50 worth of gold. George Darwin got one as you will recollect, two years ago, for his most elaborate researches into the early planetary history. Two Royal medals are given each. year. There are also two others usually given to foreigners, the "Rumford" and the "Davy," but by no means always; beside the great medal of all, the "Copley." Tell Bessy with thanks that the book covers have returned safely.

Ever affectionately, FRANCIS GALTON.

42, RUTLAND GATE, S.W. *November* 15, 1886.

DEAR MRS HERTZ†, Best thanks for "Vain Discourse" which I have thoroughly enjoyed and return.

I stole an "umberella" from your house yesterday, taking it by mistake for my own—really by mistake, and it is a better one, which I know throws doubt on my honesty. If you can tell me who the owner is, I will penitently return it. Yet my own had merits. It was of real silk, very light, and I bought it *new* from an itinerant dealer in Lombard St for 4/6. I could have got one of inferior silk for 4/-. The "umberella" deserved study; it was made up of clippings of silk. It certainly acted and looked handsome and I took it abroad. The one I stole probably cost a guinea or more. Very faithfully, FRANCIS GALTON.

The "Ascidian flippant in an infinite azure" as heraldic bearing for the Darwinians is charming. Can't you get someone to draw it?

Letter to Alphonse de Candolle.

42, RUTLAND GATE, LONDON. *May* 26, 1887.

MY DEAR SIR, It gave me great pleasure to receive the "Extrait" from the *Revue d'Anthropologie* of May 15 containing your article on the relative healthfulness of the brown and blond types. You had told me of the suspicion you then had of the accuracy of the American references and I had long wished to see your article. Their statistics are clearly imperfect

* Henry Montagu Butler, Galton's brother-in-law, appointed Master of Trinity College, Cambridge, in 1886.

† See my footnote p. 464 above. I have a letter of Huxley to her from the year 1870, expressing great sympathy with the Prussians.

from neglecting important data. No doubt, however, you have remarked that the soldiers—the accepted men—of German birth are usually ranked high for their physique in Baxter's Statistics (see Vol. I, p. 169 and again pp. 182, 215, on the one side, and pp. 199, 206, 227, on the other). I cannot find in my English statistics any sign of the dark race supplanting the fair. The persistence of the proportions during four generations between them (see Diagram on p. 405, *Royal Soc. Proceed.*, 1887*—I send the memoir "Hereditary Eye-Colour" for your acceptance) is very remarkable. Neither do my data show that either is more prolific or less healthy than the other. The data are but scanty; still I imagine that the English climate and surroundings may be equally suited to the two types. Moreover the Scandinavian contingent to our population, contributing largely to the blond type in Eastern England and Scotland, seems the most vigorous though least aesthetic of all our stocks. I have failed in obtaining trustworthy results from my data concerning sexual predilection for, or aversion from, concolour marriages; there are too many interfering causes of importance on which I am insufficiently informed. It is, as you most justly say, among the irregular *liaisons* that data are most preferably to be sought. Together with the "Hereditary Eye-Colour" I send "Hereditary Stature" which will I fear hardly interest you being very mathematical in its reasoning, but as the Eye-colour inquiry depends on formulae derived from it I may as well send it also. It also describes my data. Thirdly I send a recent Presidential address, the last part of which beginning at p. 394 may be worth while glancing at.

When I had the great pleasure of making your personal acquaintance a little more than a year ago, you were in domestic anxiety. If you should ever again favour me with a letter, I should be very glad to learn that that anxiety was lessened.

Believe me, my dear Sir, Very faithfully yours, FRANCIS GALTON.

GENÈVE. 20 *avril*, 1888.

MON CHER MONSIEUR, Il y a longtemps que j'aurais dû vous donner signe de vie en reponse à votre lettre obligeante du 26 mai dernier, mais l'âge m'a rendu très lent et m'empêche de faire des recherches d'aucun genre. J'aimerais pourtant bien savoir si vous avancez dans vos utiles publications, auxquelles je porterai toujours de l'intérêt.

Un de mes derniers efforts a été de rédiger pour la *Société de Psychologie physiologique* fondée à Paris, une série de questions à poser sur l'hérédité. J'ai su, par M. Taine, que la Société a reçu des réponses et qu'on s'occupe de les utiliser. Reste à savoir comment les personnes questionnées ont été à la hauteur d'impartialité et de jugement nécessaire. La lecture de la correspondance de votre célèbre cousin Charles Darwin m'a causé beaucoup de plaisir. J'aurais bien aimé connaître les questions qu'il agitait avec Sir Joseph Hooker, en 1852–54, lorsque je m'occupais moi-même de l'origine des espèces au point de vue géographique, ce qui me conduisait en 1855 à constater l'ancienneté géologique des causes de la distribution actuelle. D'un autre côté cela m'aurait retardé dans les recherches et je n'aurais publié ma *Géographie botanique raisonnée* que plus tard, après peut-être l'ouvrage classique de Darwin (1859). Il y a dans ses lettres des phrases caractéristiques et admirables sur les principes et les méthodes dans les sciences d'observation. Je veux relire les trois volumes pour les extraire. Quelle remarquable exposition des idées successives de l'auteur! On ne trouve rien de pareil dans Montaigne, Gibbon, Rousseau et autres qui ont écrit sur eux-mêmes trop tard et avec une impartialité souvent douteuse.

L'hiver a causé des désastres en Suisse comme ailleurs. Ici ce sont surtout des avalanches. Malgré cela nos lacs et nos Alpes auront toujours de l'attrait. Ne viendrez vous pas les visiter de nouveau cet été? Ce serait fort agréable pour votre très dévoué

ALPH. DE CANDOLLE.

Letter to Alphonse de Candolle.

42, RUTLAND GATE, LONDON. *May* 6, 1888.

MY DEAR SIR, It gave me very great pleasure to hear from you about a fortnight ago, and I should have replied at once only I thought the enclosed scrap (which might have been printed a week earlier) would interest you and I delayed till I got it. Dr Venn's memoir will not appear

* Reproduced in Vol. III^{A}, p. 35 of this work.

till November. He is the author of a most thoughtful book called the *Logic of Chance* which young statisticians ought to read, for it explains what statistics cannot as well as can do, in a very masterly way. The third edition is just out. If you happened to think of any logically disposed reviewer it would be worth while suggesting this book to him as well deserving notice.'

I was very pleased to read how much Charles Darwin valued and profited by your labours and views—What an immensity of work in science has been performed in the last 50 years! It must be an endless pleasure to yourself to look back upon your own large contribution to it. It will be very curious to watch the results obtained from your questions circulated by the *Société de Psychologie physiologique*, and the way in which the veracity of the answers may be tested. I have myself lately had a batch of rather disappointing replies to questions circulated among teachers in schools of all grades, concerning the signs and warnings of mental fatigue. There was great absence of skilful self-analysis and of suggestion, and not a few transparent indications of exaggeration here and of suppression there. I was hearing the other day from a particularly trustworthy source, a list of unveracities of one of our own men of science, formerly one of the leaders of science, but whom I must not indicate further. The general facts and many particulars I had long known, but was surprised to learn how much more there was that I had not known. It is strange that a man who had so little care for truth could succeed in science at all. It is a most painful case of psychological interest and made me think how painfully it would have interested you when writing that paragraph on the general veracity of men of science in your *XIXme Siècle*.

I had a pleasant summer last year in Eastern Switzerland, etc., but in the autumn fell suddenly ill with a most severe gastric attack at Lugano and was got home somehow in a wagon-lit. Then I fell ill again in another way with violent catarrh, then again in a third way with inflammation of the cæcum, and lastly in a fourth way with severe bronchitis. In short I had four separate severe illnesses within five months. I suspect there was some microbic poison at the bottom of it. However I am clear of all illness just now.

I was grieved to see the black-edged paper of your letter, and beg of you to accept my sympathy. I shall deem myself very fortunate if the next time that I pass through Geneva I shall have the great pleasure of finding you at home and inclined for a half hour's conversation.

Very faithfully yours, FRANCIS GALTON.

GENÈVE. 28 *mai*, 1888.

MON CHER MONSIEUR, Je regrette d'apprendre par votre lettre du 8 mai que vous avez été si longtemps malade, mais heureusement vous ajoutez que maintenant votre santé est rétablie. Quant à moi les fatigues et le chagrin causés par la maladie et la mort de ma femme ont singulièrement affaibli mes facultés morales pendant que l'ouïe, la vue et la mémoire diminuaient par un effet naturel de l'âge. J'ai perdu mon ancienne activité et ma confiance dans le résultat possible des recherches. Il faut prendre mon parti de la retraite et me souvenir qu'ayant commencé à publier en 1824, ma carrière scientifique n'a pas duré moins de 64 ans. Mon ancien goût pour la statistique persiste encore, au moins lorsqu'il s'agit de suivre de bons travaux faits par d'autres.

C'est donc avec plaisir que j'ai lu votre analyse des recherches du Dr Venn sur la tête des étudiants de Cambridge. Il y a bien des comparaisons probantes à faire sur des jeunes gens de mêmes conditions, âges, etc., qui se conduisent diversement à l'université. Par exemple, comparez les fumeurs intrépides, fumeurs médiocres et non fumeurs, au double point de vue des succès intellectuels et des succès dans les exercices du corps. L'antagonisme entre les aptitudes intellectuelles et corporelles, si bien connu des Anciens, ressortirait sans doute d'une comparaison statistique dans les écoles.

À propos d'exercices, je vous recommande un volume qui vient de paraître dans la collection internationale d'Alcan (autrefois Alglave) à Paris. C'est *Dr F. Lagrange: Physiologie des exercices du corps. 1 vol. in 8°, Paris*, 1888. *Prix* 6 *fr*. L'auteur traite la physiologie des muscles, nerfs etc., d'une manière très savante et vraie, à ce qu'il me paraît, et j'ai remarqué une définition dont on ne parle pas encore, c'est que certains exercices fatiguent à la fois la tête et le corps tandis que d'autres reposent le cerveau tout en employant les muscles. Par conséquent les premiers (escrime par exemple) contribuent au surmenage dont on se plaint dans les écoles, tandis que les autres (la marche par exemple) n'ont aucun inconvénient et offrent beaucoup d'avantages physiques. Il faut recommander les exercices qui exigent une tension d'esprit aux oisifs, et les exercices bêtes aux étudiants qui veuillent travailler aux commis, employés, etc., dont la tête est fatiguée.

J'ai connu deux savants distingués qui n'étaient pas bien véridiques, mais je dois dire qu'ils ne mentaient pas sur des affaires scientifiques, comme leurs expériences ou observations. C'était plutôt pour rendre service à un ami ou pour nuire à quelqu'un qu'ils n'aimaient pas. Tous deux aimaient la vie politique. Les hommes de science manquent parfois de force morale et il en résulte une disposition à cacher certaines opinions plutôt qu'à mentir. En général cependant j'estime que l'habitude des recherches rend véridique.

Si vous passez à Genève cet été je serai très heureux de vous voir. Dans le moment des grandes chaleurs j'irai peut-être dans les montagnes, mais ce ne serait ni loin ni pour longtemps.

Recevez, mon cher Monsieur, l'assurance de mes sentiments les plus dévoués.

ALPH. DE CANDOLLE.

P.S. Je demanderai l'ouvrage de Dr Venn. S'il est trop mathématique pour moi je le communiquerai à quelque calculateur de mes amis.

42, RUTLAND GATE, S.W. *December* 16, 1888.

DEAR DR WARD, Thank you very much indeed for your valuable letter, which will be of considerable guidance in devising and varying the experiments. It shows me that difficulties which I had not seen so clearly before must be evaded. All, for example, that are connected with mistaking an incapacity to make out the "Bulgarian's cat," with imperfection of seeing power. And I think I see how; but will not bother you with details.

By the way, I once made some experiments on the above, intending to bring them into a lecture I had to give at the Royal Institution, but the examples selected seemed rather melodramatic and I had not much time, so I wholly left them out. I used two Magic Lanterns; in one slide was a picture of a number of dots and splashes; in the other slide a selection of these was made that spelt the words Blood and Murder and there was a hand pointing. When the light was faint in slide 2, nothing of the Blood and Murder was seen, but as it increased they began to catch the attention and soon became prominent. On reducing the light again, the level at which the image disappeared was much lower than that at which it first appeared. The curious thing was its *sudden* disappearance. I tested this latter point in many ways with the same result*.

Thank you very much for the reference to "Urbanschicht" which I have already looked up in *Nature*. I had read it at the time, and was greatly struck by it, but had wholly forgotten the name and wanted to refer to it. I shall get the paper in *Pflüger* to-morrow. As regards Kussmaul, I have made less progress in reading his very able and exhaustive work than I had hoped, and must I fear content myself with what little I have already done, which bears on the question in hand, as I am very busy and get through work slowly. The book shall be sent back to-morrow. It should have gone earlier but I delayed a little, partly in hope of hearing from you and partly because I had not your address at hand and did not like to trouble you by sending it to Trinity. Very sincerely yours, FRANCIS GALTON.

GENÈVE. 15 *sept.* 1889.

MON CHER MONSIEUR, Je viens de lire dans le *Times* vos observations sur les examens du Civil Service et j'ai lu également des articles dans le *Nineteenth Century* sur le même sujet, qui m'intéresse particulièrement. Voici pourquoi: J'ai un petit-fils, né en Angleterre, sorti avec honneur du collège de Rugby, et qui se fatigue depuis 18 mois à préparer un examen du Civil Service, qui (hélas!) n'est jamais annoncé et sera peut-être encore renvoyé longtemps. Je n'ignore pas qu'on veut réduire le nombre des places et qu'on ajourne les examens à cause de cela. Mais il serait pourtant équitable d'avoir égard aux jeunes gens qui se préparent. Ce serait *fair play*. En France il y a des examens pour l'entrée dans les écoles polytechnique, militaire, etc., et ces examens ont lieu à des époques fixes chaque année. Ainsi au bout de quelques mois un jeune homme sait s'il est admis ou s'il doit viser à une autre carrière.

* The two slides still exist in the *Galtoniana* and, before seeing this letter, I was much puzzled to discover their purpose. K. P.

Vos réflexions sur les conditions physiques à stipuler dans les examens sont de toute justesse quand il s'agit du service militaire ou du service Indien, et je crois que dans ces cas on a déjà pris des mesures convenables. Quant au Civil Service proprement dit, je remarque un défaut capital. On exige les mêmes connaissances pour des services d'une nature très différente. Ainsi, pour des occupations sédentaires ou actives, pour des services d'ingénieur, de calculateur, etc., qui demandent des mathématiques, ou des services diplomatiques ou littéraires, qui demandent des connaissances de langues vivantes et d'histoire, on oblige les candidats à se bourrer la tête également de grec, latin, mathématiques, histoire, économie politique, etc., etc. La moitié des objets ne servira à rien dans chaque carrière. C'est le système Chinois, dans lequel chaque lettré est supposé apte à tout et utile dans toute carrière. Une première réforme désirable serait d'avoir deux catégories d'examens pour les jeunes gens de la division dite supérieure. Pour les uns on exigerait des connaissances spécialement littéraires et pour les autres spécialement mathématiques. Les conditions physiques ne sont pas du même importance dans tous les cas. Pour inspecteur des travaux publics une certaine force musculaire, une taille élevée, une bonne vue, sont des avantages. Mais pour les métiers sédentaires de la poste, de divers bureaux d'administration et dans la diplomatie, c'est assez indifférent. Les gens un peu faibles et sédentaires par nature valent mieux dans un bureau que ceux passionnés de sports. Le plus habile diplomate du XIXe siècle a été Talleyrand, qui était boiteux et myope. Le vrai principe devrait être d'obtenir des hommes spéciaux: "the right man in the right place." On s'en éloigne dans le système actuel anglais du Civil Service. C'est peut-être la conséquence des idées fausses de la démocratie actuelle qui juge tous les hommes égaux et propres à tout, ce qui conduit à une médiocrité générale.

Je pense avec plaisir à la continuation de vos recherches. Votre persévérance sera récompensée et déjà on sait à quel point elle est méritoire. Recevez, mon cher Monsieur, l'assurance de mes sentiments très dévoués. ALPH. DE CANDOLLE, Cour St Pierre, 3, à Genève.

P.S. Si vous connaissez un emploi dans lequel il soit avantageux de parler et écrire également bien le français et l'anglais, de savoir assez l'allemand et d'avoir des connaissances étendues dans les sciences historiques, je pourrai vous recommander mon petit-fils, âgé de 21 ans. Il rentre dans vos conditions de familles intellectuelles, par trois générations du côté paternel et autant du côté maternel.

Letter to Alphonse de Candolle.

42, RUTLAND GATE, LONDON. *November* 13, 1889.

MY DEAR SIR, The long delay of two months in replying to your very kind letter has been wholly due to the hope that I might have something to say that you would like to hear. The particular scheme about which you wrote, of our introducing marks in our competitive examinations for physical efficiency, has not yet publicly resulted in anything, but from private information I learn officially, though *confidentially*, that the question will almost certainly be examined into by a very favourably disposed committee of one of our great public Departments, among whose officials the need of high physical efficiency is great.

Also several of our public schools are, I believe, making experiments in marking for it, and in seeing how far the examiners agree between themselves and with the general verdict of those masters who know the boys thoroughly in the cricket field, at football, and in other games.

I venture to send you the paper in full that I read at the British Association (of which the last part was published with good illustrations in *Nature*, Oct. 31).

On looking at the second page where I have marked a paragraph, you will see how careful I was not to commit the fault you feared in your letter, of supposing that high bodily efficiency is of universal importance. I only speak of professions in which it *is*.

I was very sorry to hear of the inconvenience to which your grandson has been put, by an absence of an opportunity of competition on which he had reckoned. Probably the expected vacancies did not occur. I do not at all profess to defend the action of our Civil Service Commissioners either in giving a notice of expected examinations which was not fulfilled, or in exacting much the same knowledge from candidates for widely different offices. But they have a very difficult task in fulfilling as far as may be two conflicting wishes. One is not to disturb the regular course of education, so that a youth may be educated at any great school without going to a special "crammer" up to nearly the last moment, and the other is to require a sufficiency of special knowledge.

This is accomplished in some cases by two examinations, the one at a comparatively early age, to qualify for entering; the second one which is special, and not so severe, but that every lad who passed the first might be expected to succeed in the second. Then if he failed in the first, he would be in the same position as other boys who looked forward to any one of a multiplicity of possible careers. No one however seems satisfied with what is now done either in the Government examinations or in the public school teaching; but no one here has yet had the wit to suggest a course that commends itself to the general judgment as an improvement. The question is apparently a *most* involved one; so many interests and prospects being seriously affected by any change of system.

As regards the particular question you put, as to any satisfactory employment for a person having the high qualifications you mention, clearly they must exist in abundance, but personally I have not any one of them distinctly in view at present. I should have thought that a private secretaryship to some political person would be eminently a post to try for, or that to some person in the higher branches of commerce or manufacture, who has varied foreign connections. All such posts give a young man excellent opportunities for afterwards succeeding by his own efforts, and adequately educated candidates for them are hardly equal in number to the demand.

In concluding let me express the great pleasure that it gave me to receive your kind letter, for there are now few persons whose sympathy I prize more than your own on those many subjects in which we feel a common interest. You say nothing of your health but I trust and believe that it is maintained more fully by far than in the great majority of your contemporaries. Believe me, very faithfully yours, FRANCIS GALTON.

42, RUTLAND GATE, S.W. *January* 25, 1890.

DEAREST EMMA, I have hardly anything to tell, owing to being so shut up and seeing and hearing nothing. I *am* glad that at least one of the three brothers, Erasmus, is well again. There is ever so much spasmodic asthma with me, it comes on so oddly and violently and then goes. I wish it would say "good bye" finally.

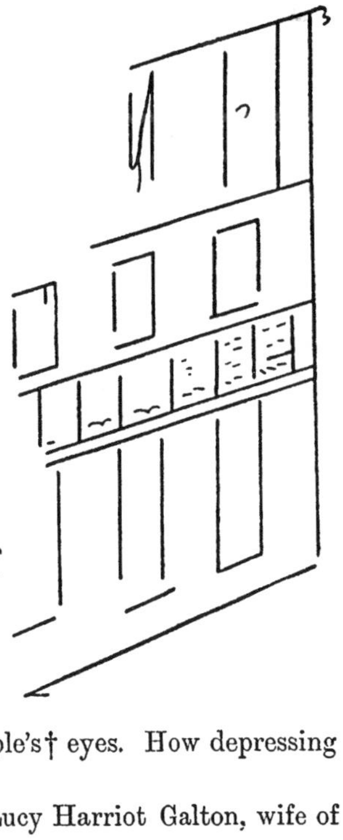

I am trying to get a grand display of weather information stuck on to the balcony of our office in Victoria St. I have long wanted to show, as soon as it arrives, the weather on the coasts near to London. It does not get into the newspapers until five hours after we receive it. My colleagues agree, and it is now a question of detail. I carpentered a board in the proposed way, and painted the lettering thus [see figure below] and we had it up for inspection on Wednesday. Literally on passing the turn to Victoria Station I could see the glimmer of the board all that distance off!!—a good ½ mile. I propose to give the facts for Yarmouth, Dover, the Needles, Scilly, Valencia, Holyhead; all the ways of changing the slips are worked out and feasible, but there are still some details to be fixed and the written permission of the landlord to be obtained. The slips would be changed at 8.30 a.m., 3 p.m., and in summer at 8 p.m. It would make much difference to many persons to know this: for instance, if doubting whether to cross by Dover or Harwich or Newhaven. I am sorry that you think Tertius* not well. I do hope that Bessy and you continue all right. It is grievous about Temple's† eyes. How depressing eye ailments are. Ever affectionately, FRANCIS GALTON.

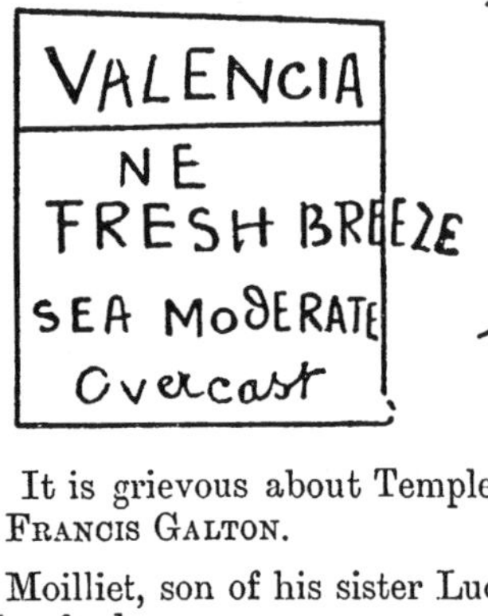

* Galton's nephew, Tertius Galton Moilliet, son of his sister Lucy Harriot Galton, wife of James Moilliet of Cheney Court, co. Hereford.

† A maid of Emma Galton, who had been many years in her service.

Letter to Michael Foster, with a memoir entitled: "Decrease of Mortality by Smallpox, 1838–1887."

42, RUTLAND GATE, S.W. *March* 19, 1890.

DEAR FOSTER, I have gone through the paper, corrected and added. I am ashamed of having sent it in so slovenly a way. Look at the addition to the bottom of p. 9. As the previous part stands by itself it might lead to a misapprehension. I have confiscated the lithographed map (of which doubtless you have plenty of copies) by marking it and attaching it to the paper. It is wanted to explain. If you should find it desirable to put the paper as it now stands into the evidence, I have no objection. Very sincerely yours, FRANCIS GALTON.

The paper to which this letter refers deals with vaccination statistics. The data are divided into three periods: (i) 1838–1853, vaccination optional; (ii) 1854–1871, vaccination obligatory, but not efficiently enforced; (iii) 1872–1887, vaccination obligatory, but more efficiently enforced by the vaccination officers. The treatment of the data is rendered difficult by (*a*) the absence of records for 1843–1846 inclusive, (*b*) by very severe epidemics in 1838 and 1871, and by lesser epidemics in intervening years, i.e. the graph of the mortality rate is very jagged. Galton deals only with *crude* death rates, he had no incidence rates. He could not therefore test the effects of (i) any change in the age distribution of the population, (ii) how far the lower mortality rate was due to better nursing, nor did he (iii) endeavour to allow for any hygienic improvement. The statistical methods he adopts are quite simple, but adequate for his purpose, and his final conclusion is stated in guarded terms: "For the whole period under review the maximum reasonable decrease in the mortality rate is 500 per million and the minimum reasonable decrease is 150 per million." He makes no statement as to what the source or sources of this reduction may be. A discussion of the data brought up to date on Galton's lines would be of interest. I am unaware if the memoir was ever presented to the Royal Commission on Vaccination (1889–1890), or printed elsewhere. Together with the letters of Galton, it was apparently sold by the executors of Sir Michael Foster, and was purchased by me from a Cambridge bookseller on May 9, 1914. This was the first occasion on which I had information that Galton had ever dealt with the statistics of smallpox. He never referred to that topic in conversation with me, although several memoirs dealing with smallpox were issued in *Biometrika*.

The Philosophy of Snoring. Notes found in Francis Galton's Handwriting.

The philosophy of SNORING. Married ladies have remarked that husbands past the age of 50 or 60 are apt to snore. I have enough reason to believe in the correctness of this generalisation to assume it to be true and more generally to ask the reason of it. What is the cause of snoring? I have not found this interesting and domestically important topic treated anywhere in a scientific manner. I write for information. I have only a few ideas and observations of my own of the scantiest, and mention them merely to elicit those of others. First I have been surprised at the silent sleep of men in bivouac. The breathing of some 30 or 40 men, mostly savages, though old men, of whom I had many months' experience in travel, was inaudible. Conversely hot bedrooms stimulate snoring. Again a deep sleep is more accompanied by snoring than a light one. The uvula droops more. It is analogous to the fallen jaw of death.

Lastly the mucus of the throat becomes more tenacious and more copious as life advances, consequently bronchitis then begins to be dangerous.

I suppose that snoring is generally due to the concurrence of two causes, to the drooping of the soft palate and to the presence of much mucus and that of a tenacious kind.

Causes: Mucus in old age; Drooping of the soft palate in hot bedrooms, also in deep sleep.

GENÈVE. 23 *juin*, 1890.

MON CHER MONSIEUR, Je ne vous ai pas encore remercié des informations que vous avez bien voulu me donner au sujet du Civil Service et des examens renvoyés indéfiniment. Mon petit-fils a dû n'y plus penser, après avoir perdu 18 mois à s'y préparer. Il est allé en Allemagne apprendre le droit.

Ne pouvant plus travailler pour la Science, je m'amuse à observer le déclin de mes facultés, et j'ai cru un moment pouvoir ajouter quelque chose à vos recherches sur l'influence relative de "Nature and Nurture." Je me disais: Les facultés qui se maintiennent le mieux chez moi, à 84 ans, sont-elles de naissance (nature) ou le résultat d'un exercice fréquent? Il se trouve qu'elles sont à la fois un effet de naissance et d'un usage continuel. Inversement les facultés devenues très faibles étaient faibles à l'origine et n'ont guère été cultivées pendant ma vie. Ainsi, j'ai conservé la faculté de marcher mieux que beaucoup de vieillards. Or, mon père avait été un grand marcheur dans sa jeunesse et j'ai toujours aimé la marche; j'ai fait autrefois de fortes marches dans les montagnes, c'est le seul exercice que j'ai cultivé.

J'ai hérité de ma mère une mémoire faible. Maintenant elle est très faible. Or, j'ai eu toujours de la répugnance à apprendre par cœur et j'ai cherché toujours à remplacer la mémoire par des notes.

Ma conclusion est que la plupart des hommes font les choses auxquelles ils se sentent naturellement propres, et négligent celles pour lesquelles ils ne sont pas bien doués. L'usage résulte d'une disposition naturelle et le non-usage d'une faiblesse aussi de nature. Voilà qui est bien contraire à ce que pensent les instituteurs, les professeurs et beaucoup de parents. Ils veulent forcer les jeunes gens et les jeunes filles à faire ce qu'ils n'aiment pas, tandis que la jeunesse aimerait faire çà pour quoi chacun se sent bien doué! Il y a ainsi beaucoup de temps et de force perdus; mais la jeunesse échappe bientôt à la contrainte, et alors on voit les jeunes gens qui ne sont pas calculateurs abandonner les mathématiques, les jeunes demoiselles qui ne sont pas naturellement musiciennes fermer leurs pianos, etc. Les pédagogues veulent faire tous les individus semblables et les individus voudraient être dissemblables, ce qui serait un grand avantage pour la société en général.

Avez-vous été informé que les naturalistes ont fait depuis deux ou trois ans de grands progrès sur le procédé de la fécondation dans les deux règnes? Ce n'est plus le protoplasme qui joue le principal rôle mais les noyaux (nuclei) mâles et femelles. Ces noyaux s'accouplent. Ils renferment des filaments, en nombre déterminé, dont les positions changent d'une manière curieuse. Vous pourriez juger de ces découvertes en regardant les planches d'un mémoire de M. Guignard, dans la *Bulletin de la Société Botanique de France de* 1899, qui se trouve aussi dans les *Actes du Congrès de Botanique* à Paris en 1889. Les zoologistes ont observé les mêmes faits.

Recevez, je vous prie, mon cher Monsieur, l'assurance de mes sentiments les plus dévoués.

ALPH. DE CANDOLLE.

Letters of W. F. R. Weldon dealing for the first time with the Correlation of Characters in Organisms other than Man.

1, HOE VILLAS, ELLIOT STREET, PLYMOUTH. *May* 14, 1890.

DEAR MR GALTON, Forgive the long delay in the preparation of the correlation curves. The measures are practically finished—1000 Plymouth, 400 Southport, 380 Sheerness. Unless you feel anxious to see the results very quickly I will not do arithmetic by daylight, because I want very much to do some anatomical work. By the help of Crelle* the arithmetic will not take many evenings.

* Crelle's Multiplication Tables, 1000 × 1000. At that date Weldon did not possess any arithmometer.

I want to ask you about a possible experiment with the *Illustraria*—of which I am allowed to rear another set. You remember that though the ratio between the races is the same whether the creatures are reared by Miss Pridham, by Mr Merrifield, or by myself, yet the *absolute size* of each race varies. Call the mean size of Mr Merrifield's A race $= A_m$. I receive eggs whose inherited tendency should be to vary about A_m as a Median. The resultant moths vary about something else $= A_w$ as a Median. It appears that the offspring of my moths, reared at Brighton, vary again about A_m. Therefore the increase of size causing the median value of the race to rise from A_m to A_w was not inherited. This seems a very typical instance of an "acquired" character. Would it be worth while to devote a few spare pairs of one set to the foundation of a race which should live for *several* generations here in Plymouth, and should then be returned to Brighton—in order to see through how many generations the external conditions can act without producing an inherited change?

I shall be here till the end of this year, = 2 generations, and I can easily find someone at the Laboratory who will deal with the following generations.

Yours very truly, W. F. R. WELDON.

I believe I have not yet thanked you for your kind congratulations on the action of the Royal Society.

30A, WIMPOLE STREET, W. *October* 29, 1891.

DEAR MR GALTON, I hope to send you, in a little while, detailed tables of the correlations of which I spoke to you this morning in the Senate House. The organs are the four which I had previously used in the Shrimp: and the rough figures for the relation *a* to *b* or *b* to *a* are at present

Plymouth race	(1000)	0·84
Roscoff (Finistère)	(500)	0·88
Southport	(400)	0·83
Sheerness	(380)	0·85

The values obtained for each deviation clustered about the line $r = 0·85$ so well that I thought it worth while to determine the second place of decimals by taking the arithmetic mean of all observed ratios in each case. Between character *b* and "*telson length*" the ratio is

Plymouth	0·25
Roscoff	0·29
Southport	0·30

All the other values are, in the Plymouth race, so small that I have not thought it worth while to determine them in the other races at present, because of the small number of individuals in each sample. But I have just obtained, and nearly measured, 400 additional Shrimps from Southport: so that I hope soon to have a set of 800 measures of this race, which will give a fair basis of comparison with the 1000 from Plymouth. When these preliminary determinations are finished, I hope to determine a reasonably numerous set of constants for homologous organs in one or two species. An enthusiastic student, to whom I have preached you, has already undertaken to measure 20 organs in each of 1000 Prawns.

Yours very truly, W. F. R. WELDON.

*Selection of Galton's Letters to Mr Howard Collins dealing with Finger-Print Data**.

HÔTEL CHERBOURG, VICHY, ALLIER, FRANCE. *August* 19, 1891.

DEAR MR COLLINS, You must have thought me very forgetful of your most kind offer to help in some of the matters over which I bother myself, and in which I am making far too slow progress. But in truth I have been very far from forgetful, and have delayed only through difficulty in seeing the direction in which I could reasonably ask your help. And the difficulty is not yet overcome, because as a rule my work is in no respects straightforward, but I have to plan as I proceed, and am consequently much bewildered between theory and detail. There is

* These letters, with a considerable amount of Galton's unpublished material on finger-prints, were purchased by his biographer from a Birmingham bookseller.

however one matter which it is just possible you might care for, that does not fall quite into this category, and which if you cared to undertake it for publication as a joint work with myself, would I think repay the trouble well, both from a scientific and a popular point of view. It is to undertake the analysis of a large and growing collection of finger-prints from the *racial* and the *hereditary* point of view. Thus, I have the impression of the three first fingers of the right hand of rather more than 1000 Jewish children, and those of more than 1000 ordinary English ones. My assistant is at this moment engaged with purely Welsh children. Orders are sent by Sir G. Goldie, with the needful materials, to the Niger regions, to procure me the prints of at least five distinct races of Africans, in abundance. Professor Haddon has taken steps to procure me those of natives of N. Australia and on to the Solomon Island groups, and when I come back in the autumn I propose to set much more agoing. My impediment has been to find someone with a genius for classification and power of work. I myself can do but little. As regards families, my collection as yet is small, but I propose to make an effort, and a sustained one, in that direction. The classification is, of course, laborious on account of the numbers, but it is not at all difficult *after* the right way of setting to work is well explained, and those specimens have been examined which are to be accepted as transitional cases between the classes. There would be great difficulty in doing this satisfactorily by written or printed description. Nearly but not quite as much as I can do in this way appears in the last number of the *Royal Society Proceedings* and is hinted at in an article by me in the *Nineteenth Century* of this month. I am sure the inquiry is a promising one. I find, for example, a distinct statistical difference between the finger markings of the Hebrew and the Anglo-Saxon*. I also find them to be as strongly hereditary as anything else. As they are independent of age, and cannot be falsified, they form a solid basis for work. Should you be inclined, when I come back in October, to work at these conjointly with me, you doing the analysis and I advising, but doing little more? The object would be to produce joint papers (1) on racial differences; (2) on the measure of hereditary tendency; I should add a third, or a previous one perhaps, based on other material that is already in hand, viz. (3) on the measure of the tendency to symmetry. I shall be at the above address for nearly three weeks, and a letter to 42, Rutland Gate will always reach me in time. Very faithfully yours, FRANCIS GALTON.

42, RUTLAND GATE, S.W. *October* 22, 1891.

DEAR MR COLLINS, The beautifully neat packet and roll reached me three hours ago with your letter, since which I have carefully gone through the first 59 and purposely cease there, that my pencilling may not interfere with your revision of the rest. I return them both. You clearly are *on the threshold* of doing it quite right, but the threshold is just the place at which people are apt to stumble when entering a house. Your chief difficulty is with the Whorls, not taking a bold enough view of them. You will see what I mean, by looking at my pencillings. Another minor common fault is interpreting an ordinary loop as though it had an eye in it thus

. These Jewesses are deficient in eyes of this kind (however well they may be endowed with real ones). In the Primaries† it is better not to make outlines thus , but thus .

I think it will be a useful guidance if I send you, as I do herewith, a packet of thumb-prints (Nos. 3000–3164) which have been carefully outlined and measured; these are all *rolled* prints, so the nature of the patterns, especially of the Whorls, is much more easily understood than in the finger-prints. They will teach confidence in outlining by inference. (Please let me ultimately have these back again.)

Will you then again go over the Jewesses, and finish the 60–100 by the light of what is now sent, and let me see them when complete and before you take the trouble of making a fresh table? There is a little difficulty about some few imperfect prints. It would not do to

* Galton was much less certain about this later: see Vol. III, pp. 193–4.

† Galton's original name for Arches.

replace them by new ones *taken at hazard*, because these imperfect prints are all Whorls and owe their imperfection to their bigness. They must be made the best of.

If anything in this letter, etc., is insufficiently explained, pray write to me at once.

Very faithfully yours, FRANCIS GALTON.

A Mr T. V. Hodgson, a microscopist of Mason Science College, who writes from 52, Francis Road, Edgbaston, has sent me *beautiful* finger-prints to see, and offers to take further prints. I suggested that he should make your acquaintance and show you what I had written to him as an introduction.

42, RUTLAND GATE, S.W. *November* 12, 1891.

DEAR MR COLLINS, I am sure that I appreciate the general principles that easy writing makes hard reading and that what has to be said ought to be logically put. Alas, for one's incompetence to do what is right! But I can assure you that I will *well* go over all your suggestions and will re-write the chapter.

Let me wait a while, before speaking of the next chapters, as I have had quite a bother about the best plan of the index which has rendered much of what was written nugatory and introduces much modification in already drawn-up tables, so I am behindhand.

Did I tell you that I have another batch of negro prints?

Very faithfully yours, FRANCIS GALTON.

Letters of Galton to Dr W. F. Sheppard.*

"CARDRONA," BERKHAMSTED, HERTS. *May* 17, 1926.

DEAR PEARSON, Would you care to look at the letters I had from Galton between 1891 and 1907? There is not much of interest to the general public in them: they are rather of interest in showing the amount of trouble he was prepared to take in helping other people on. Indeed, looking at the letters now, I seem to have caused him an unjustifiable amount of trouble! His criticisms of my successive efforts were of great value to me. On the whole I think my mathematical work has been fairly lucid; what there is of lucidity in it I really owe to Galton's criticisms. Yours sincerely, W. F. SHEPPARD.

42, RUTLAND GATE, S.W. *December* 3, 1891.

DEAR SIR, Hearty thanks for your full and very interesting account of your Number-forms. They have clearly grown in your case, together with the years, and seem to have done so automatically with possibly a little conscious assistance on your part. The wonder is *why* a particular "form" is so congenial to each several mind. What is the relation between the form and the peculiarities of *association*, in the working of the mind? If you can trace any such relation in your own modes of thought I should be exceedingly glad to hear of it. I fancy I have some slight clue to a relation, but it is very slight, and when I last thought on the matter, I did not find out any good way of putting the notion to the test. With renewed thanks,

Faithfully yours, FRANCIS GALTON.

42, RUTLAND GATE, S.W. *October* 24, 1892.

DEAR MR SHEPPARD, Am I right in supposing that it was you who were Senior Wrangler in 1884? It is needless to say how highly, under those conditions, I value your mathematical remarks. They shall be carefully considered. In the meantime I have read them somewhat cursorily. I wholly agree with you that the book would have been made much better, by giving a brief *résumé* of the mathematical results. It is obscure and confused as it stands, largely owing to misgivings as to how far the basis of the whole would be accepted as established. I think now this might be assumed. What is greatly wanted is a clean elegant *résumé* of all the theoretical work concerned in the social and biographical problems to which the exponential law has been applied. I believe the time is ripe for any competent mathematician to do this with much credit to himself. I am *not* competent and know it. Edgeworth has his own work and interests, and fails in sustained clearness of expression. He is moreover somewhat over fond of using higher and more mathematics than is always necessary. Watson is over busy

* For further letters to Howard Collins, see pp. 488 *et seq.* I hesitated to break too seriously the chronological sequence of letters and papers in this Chapter.

and I think too fastidious and timid. I have often considered what seems wanted and been very desirous of discovering someone who was disposed to throw himself into so useful and such high-class work. He might practically *found* a science, the material for which is now too chaotic.

Faithfully yours, FRANCIS GALTON.

42, RUTLAND GATE, S.W. *June* 9, 1895.

DEAR SIR, Would you give me the pleasure of your company at dinner at the Royal Society Club next Thursday. The enclosed card gives all needful particulars except that it is *not* the custom to dress. You can get away easily by 8½. There are many topics I should like to have the opportunity of talking over. Might I venture in the interim to send you a brief MS. on a new point of very wide application? I propose to send it to the Royal Society if I can persuade some mathematician to communicate a brief supplement to it, much as MacAlister did to one of my papers, H. Watson to another, and Dickson to a third. I can work out the problem in definite cases but it wants generalising. If your occupations preclude the chance of your being able to do this, of course you will tell me; otherwise I fancy that a pretty little stroke of work might be the result. Faithfully yours, FRANCIS GALTON.

Karl Pearson and Burbury are I know both full of "law of frequency" work, so I do not like to trouble either of them with the problem.

42, RUTLAND GATE, S.W. *June* 29, 1895.

DEAR MR SHEPPARD, I am rejoiced at your success in arriving at such wide generalisation of the problem. It will be far better that you should write the paper wholly by yourself, and I feel no doubt that it would be a very acceptable one to the Royal Society. After the recess we shall I hope discuss this. For the present, there is no need. During the vacation you may find time to do what you propose about the table. I am quite indifferent as to the fate of my preamble, the real object with me being to get the problem properly solved. The passage on my p. 12 was indeed most bunglingly as well as inaccurately expressed. What I meant is written in the enclosed (to which 12*a* is put for the page). I should like to keep your MS. for a few days longer, being extremely busy just now. Then, before going abroad, I will return all the papers,—mine, partly for possible convenience to you in future reference and more especially with some curiosity to learn hereafter how far my little tables prove correct.

Very faithfully yours, FRANCIS GALTON.

We leave town on Wednesday.

42, RUTLAND GATE, S.W. *July* 6, 1896.

DEAR MR SHEPPARD, So far as I can judge, you seem to have boiled down the original very judiciously, but I am much below par and not able to read it carefully, only to look through it. The question is,—whether in its present form it is suitable for publication? I should say decidedly so, in the pages of any mathematical serial other than the Royal Society. Whether or no it be suitable for the *Transactions* of the Royal Society, Forsyth would be the judge; but, for the *Proceedings* of the Royal Society, I think decidedly that under the new Regulations, the part you have sent me is not suitable. On the other hand, the Introduction of which you speak ought to be the very thing for the *Proceedings*, and would serve as the "Abstract" if the complete paper were offered for the *Transactions*. I would therefore urge that particular pains should be taken with the Introduction, the business of which is to explain to members of the Royal Society generally, *what the paper is about*, and *wherein its novelty consists*. Imagine that it has been just read to any small representative body of those men,—such as John Venn, Frank Darwin, Inglis Palgrave, who are all statisticians but not especially mathematical. The test would be that they should severally be able afterwards to give a lucid and consistent account, though probably a very imperfect one, of what you desired to show. It was to that end that I suggested the introduction of a few interesting types of problems that your methods enable statisticians to deal with, which otherwise would be very difficult problems. Of course the Introduction would contain your tables, or adequate samples of them. If the Introduction fulfilled the end proposed, it would certainly be translated into French and German, and reprinted in America, and your labours would become widely known and set many persons thinking. It ought to be a work of art—simple, clear of unnecessary detail and readable. I think you have a great opportunity of becoming an exponent of modern theories of statistics and should be delighted if you would rise to the occasion. Very faithfully yours, FRANCIS GALTON.

42, Rutland Gate, S.W. *February* 13, 1892.

My dear Bessy, We will write to Emma after the wedding on Wednesday to tell news of it. I saw Douglas* on Thursday, he had rather a bad boil on his face which looked painful but as though it had reached its worst. There were others by, and I had no family talk. He did not look over well. Yesterday I went to see George Darwin receive his gold medal from the Astronomical Society. The President read an Address of no less than 40 minutes of quick reading on his merits. It is a considerable honour to him, but one that he has more than deserved.

It is such a pleasure to be able to think of Emma in her drawing-room and not in bed. I get strong rather by fits and starts than regularly, and still want a good sleep some time during the day. I think I have now no *illness* left in me, but was not so sure of that five days ago. Poor Reginald†. I often think of old times when he was a sort of glorious Bob Sawyer, as medical student in London. It is pleasant getting back to work again. They want to nominate me as President of the British Association for 1893, but I have definitely declined, as I did for 1891, being out of my element in dining out day after day, and making speeches, which I detest. Besides, I am too deaf to do the ordinary Presidential duties well.

This is of course intended as a letter to Emma also. Dear Mother, I often thought of her yesterday. To think it was as much as 18 years since she died.

We expect to be cut off from London proper this afternoon by the Salvationists, who are to disport themselves in Hyde Park round General Booth, so as I have things to do in London proper, I must start earlier and lunch out. I was very glad to hear that William Eccles had had a favourable crisis. I suppose a big gall-stone cleared itself out. With both our very best loves to you all. Ever affectionately, Francis Galton.

42, Rutland Gate, S.W. *February* 27, 1892.

Dear Mr Collins, I am utterly humiliated. The registered letter was laid quietly on the table, while I was sorting and tying up MSS. books. When I ultimately saw it, I mistook it for the Introductory chapter, which you took such pains about, and which ever since has been distinguished by being wrapped in the same envelope in which it arrived, and which was precisely like the envelope in which you sent the last. So I heedlessly tied this up along with the rest and never opened it. On receiving your telegram I made thorough hunt and found the missing MS. I don't ask you to forgive me, only to *try* to forgive me for causing all this trouble, which I greatly regret.

Thank you very much for your emendations and suggestions to this last chapter, which I have read through and will adopt, except perhaps the transposition, believing still that it is best to show first that the proposed *principle* of indexing is feasible, and secondly to consider the best of many alternative ways of applying the principle. As I said on my post-card, the corrections you made to the previous chapter have vastly improved it.

I have a set of 50 Welsh which you ought to have had, and which I now enclose—they may be acceptable.

I am trying heredity with some success, partly to test the convertibility or relationship of the *patterns* (not classes of patterns). There is no possibility of doubting the tendency here to hereditary transmission. Can you let me have back the "Album" which contains specimens of relations? I want next to revise the set of standard patterns and have already something useful and hope before long to send a revised plan for consideration.

Very faithfully yours, Francis Galton.

42, Rutland Gate, S.W. *March* 1, 1892.

Dear Mr Collins, I send some more Primaries to cut up, for the purpose of defining your frontier. Don't throw away those you don't use, as I should like the opportunity of giving in the book photolithographs of transitional cases, Primary-loops, Primary-whorls and Loop-whorls,

* Francis Galton's cousin, Sir Douglas Galton, the engineer.

† Reginald Darwin, son of Sir Francis S. Darwin, Galton's maternal uncle. He died on Feb. 7 of this year.

and they would be useful for that. It takes many failures before a *neat* collection can be made, equal in depth of tint and all clear. What shall I tell Randall* to collect for you next?

Enclosed I also send back your own family prints that, if you would kindly do so, you might *add* to them a more complete print of your sister's right forefinger. This one has been printed too much on the tip, as in

Fig. e

Fig. *e*, instead of thus  consequently, a very interesting part of it is left out. Please let me have them all back. I shall be able to tell you more about hereditary matters in a few days. I get *much* better results even than those of your family. They take time to work up.

Yesterday evening's post brought me the enclosed from you on the 14–21 patterns, dated (? by mistake) February 22. So I treated it as subsequent to the one received yesterday morning, Feb. 29, though the latter contained "certains" and "doubtfuls"; was I right? There seems *no* racial difference, and it also appears that the English group (at least) is very discordant *inter se*. Please let me have it back.

If all your work should end by showing that race goes for nothing, and if (as I am sure it will) the other work testifies to hereditary transmission, we shall have got, not what was hoped for, but something quite different and of great interest of its own, namely a perfect instance of the effects of "panmixia." This will be charming. There is none other that I know of that approaches it in completeness. The whole subject becomes more and more curious. About the Bar-lock†: I made good progress, but domestic arrangements interfere with its use, at least at present. I write in three different rooms and the click of the thing in the drawing-room after dinner is voted a nuisance. So I sent it back. However my back study is being now fitted up with extra shelves and will be turned into a liveable room, and I may perhaps before long revert to the Bar-lock.

Very sincerely yours, FRANCIS GALTON.

Just before I finished the above, the packet of relations' cards arrived by post.

42, RUTLAND GATE, S.W. *May* 13, 1892.

DEAR MR COLLINS, The results are certainly very curious of the 31 F.'s and M.'s and of the 44 sons and 83 daughters. The comparison between a parent influencing the patterns of the offspring of the same or of opposite sexes shows approximate equality, thus:

Opposite Sex		Same Sex		
F. D.	M. S.	F. S.	M. D.	
52	31	18	69	on same fingers
83		87		
111	72	44	166	on opposite fingers
183		210		

* Sergeant Randall, the officer in charge of Galton's Anthropometric Laboratory.

† At the suggestion of Howard Collins Galton started a typewriter, but typewriting never became customary with him.

But when the comparison is made between the paternal and maternal influences it certainly does seem that the father's influence on the son is uncommonly smaller than the mother's influence on the daughter:

Paternal Influence		Maternal Influence		
F. S.	F. D.	M. S.	M. D.	
18	52	31	69	on same fingers
70		100		
44	111	72	166	on opposite fingers
155		238		

It is 18 against 69. It is true that there are only half as many sons as daughters: therefore the corrected proportions are as 36 to 69, but this is an enormous difference; too great for mere chance, apparently. Also, the other figures give 88 against 166. The influence of mother on son seems also equal to that of mother on daughter: the figures uncorrected being 31 : 69, 72 : 166, or corrected by doubling the sons, 62 : 69, 144 : 166. On comparing paternal with maternal influences, the results are not sufficiently congruent, for 70 : 100 (14·3) is a different ratio from 155 : 238 (13·5). One must not be too much impressed by the lesser magnitude of the latter number. It would be fairer to compare the number of the 31 *families* in which the maternal influence prevailed, than to compare the individuals in those families. Taking the last paragraph into account, I should not dare to ascribe to the results more than a *suspicion* that the mother's influence is stronger than the father's. This really ought to be worked out and placed beyond doubt*. I will see what evidence I can collect for you. In haste,

Very faithfully yours, FRANCIS GALTON.

A Bar-lock is busy in my room, copying the MS. at a great rate.

42, RUTLAND GATE, S.W. *May* 29, 1892.

DEAR MR COLLINS, Here is all that seems to come out of the fraternal heredity. (I will leave the maternal, just for the present.) It would be satisfactory were it not for the curious anomaly of the loops, referred to in the last paragraph. Also, I cannot succeed in bringing these data within the grip of the formula in *Natural Inheritance*, or even to make a proper comparison between the two. It is too puzzling for me at present, the problem being a peculiar one. These data give one much to think about. I will go again at *Race* now. *If* a Royal Society paper can be made out of the Heredity and Race, it will have at this late season of the year to be merely *nominally* read. There is only one meeting after the next, towards the third week in June, and that is technically called the "Massacre of the Innocents." The papers are not read except hurriedly, or only their titles; but they get printed all the same. I am pegging away steadily but the work is slow.

I am truly glad that you really like the book thus far. The chapter on Identification will be greatly improved. The first Introductory Chapter will of course now be written the last. Hardly any of that which was done will do now.

What a glorious day Saturday was. I rushed off after luncheon finding there was just time to catch a special Saturday train to Hampton Court. The boats on the river were most pretty and numerous, and full of nice, merry-looking people.

Very faithfully yours, FRANCIS GALTON.

* I have had to correct Galton's figures in the above schemes, and, corrected, they modify to some extent the results as they stood in his letter. I have cancelled the *very strong* he put before *suspicion*. On the point in question: see Vol. IIIA, p. 192.

42, Rutland Gate, S.W. *June* 1, 1892.

Dear Mr Collins, Bravo thus far. Your figures tally well and conviction is at a measurable distance. Still a total of 44 families is not large. The figures show that the ratio of paternal to maternal influence is as 7 or 8 to 10. Now if 44 families are divided in that ratio they would split into 18 and 26 ($18 \times 10 = 180$; $27 \times 7 = 189$). What your figures really show is that 18 of the families have preponderating paternal influence and 26 have preponderating maternal. If the influences were really equal, the chance against the figures coming out as yours do, would be as that of tossing 44 pence and only 18 of them coming down heads instead of 22. This is not highly improbable by any means. I don't see that the number of children mends matters. They tell the variation in the degree of preponderance of one parent or the other in the various families, but that is all. It strikes me as hazardous in exactly the above proportion to publish the results. You say, I have tossed up 44 pence, 18 only have come down heads, therefore those pence were weighted, like false dice. The judge would say—the evidence is strong but not cogent enough to convict. I should certainly advise your working hard to get more cases, the inquiry being very hopeful; and when you have enough, it might be well to try the effect of eliminating loops, because they *seem* to be less hereditary than arches or whorls, so their inclusion may dilute the results. So glad that you are safe back, well refreshed. All is well with us.

Very faithfully yours, Francis Galton.

42, Rutland Gate, S.W. *June* 3, 1892.

Dear Mr Collins, By all means, in respect to Miss H. I should be delighted to give her £5 towards a holiday with a roving commission for families and Quakers. For my own part, I doubt the Quakers worth inquiry in the face of the fact that racial differences are so minute. Moreover on going carefully into Race, I feel sure that nothing can be substantiated without dealing with large numbers. The Jews are all right being upwards of 1000, but 50 is inadequate, as shown by the non-conformity of 50 groups with the larger ones. I should certainly say "concentrate upon families" never minding their origin, but settle (1) the comparative paternal and maternal influence, (2) the parento-filial, (3) the fraternal. I think (3) is already good enough to justify publication, but it would be improved by more cases. As Quakers drop in you might if disposed work them up apart, as a luxury rather than a necessity. I would write to Miss H. and enclose the cheque, but it would be nicer through your hands. When you have fixed with her, please tell me what I otherwise owe. Very faithfully yours, Francis Galton.

42, Rutland Gate, S.W. *June* 5, 1892.

My dear Mr Collins, Miss H. has written to me enjoying the prospect of her holiday and asking for the papers of lists of relatives, which I have returned to her. I said that I had sent the cheque to you, which I do herewith. There were two reasons for not sending it to her, one was that you might have already given her the money and the other is that *quaere* would it not be better to give her more tether, another two or three pounds will do for twice as many extra days. So I send the cheque unfilled but with "not exceeding £10" written on it for you to fill up. (Besides I am in your debt for much else which this will not and is not intended to cover.) Might it not be well that I should anyhow pay her as usual for the families she gets, *in addition* to her holiday fund. It would prompt her to work all the harder. You will understand better than I do what is right and reasonable, without fear of spoiling her.

I am going through the chapters you corrected at Ryde. Your criticisms are most just and the corrections most welcome. I am sincerely obliged for them. This Whitsuntide breaks into postal arrangements so I don't know when you will get this.

Very faithfully yours, Francis Galton.

September 20, 1892.

Dear Collins, I welcome the briefer and more cordial mode of address. Your paper reached Callander yesterday in time for getting *everything* done before post time; the contents being re-written in the train and posted *en route*. I shall get proofs of these, of Index, etc., here, and then that too will be closed.

We stay here (Alnwick) for three days—then go to Lincolnshire for three or so more, and thence to Leamington, 5, Bertie Terrace, on this day week, or the next day to that. I hope it will be fine and may tempt you to come as you have proposed after all.

You are indeed most good to offer continued help and I appreciate its value. We will have a good talk. I have had an inquiry in view for long and must now begin to beat the bushes as it were, to see if the covert I want to shoot over holds a fair quantity of game. If it promises well, I shall think of taking it up steadily. It would be a growth out of one chapter or rather paragraph of *Human Faculty*, which must in the meantime be disposed of in a 2nd edition. I am very glad you like the job just completed. I wonder if it will produce results. A good deal of routine work will now be set going, in continuation of what has been done, at the Laboratory. Among other things, I must form a good standard collection of enlarged prints. I must start a photographer to attend some hours, say three days a week, at the Laboratory. I must talk over with you my plans about all this.

What grand mountaineering you have had. But it will be difficult to change into a sedentary life. Till we meet. Very faithfully yours, FRANCIS GALTON.

Address to: 42, RUTLAND GATE, S.W. *September* 22, 1892.

DEAR COLLINS, It would take long to explain, as the idea is at present nebulous and capable of concentrating round one or other of many alternative centres, but briefly it is on the measure of motives. For example, that interesting little book of Leffingwell showed how little influence race, religion, etc., etc., severally had on illegitimacy. What is their relative influence and what is it that governs the variety of result? Men coolly face death under many conditions—what *are* these conditions (this is the substance of the paragraph to which I referred)? Bribery can do so and so—*what* can it do? In all cases measuring statistically the commensurability of extremely different motives and temptations, shown by money bribes and compensations. The final interest of the inquiry and its *truest* centre lies in the fact that the old religious motives of deterrence and of reward are ceasing to be efficacious and we have to consider what can take their place. I shall have a curious variety of facts to look up—some bearing on incidents in barbarous life—some in very various civilisations—some in our own. Also the power of illusion forms one very large branch. Then again the economic laws of value and their mathematics, and very much else.

I hardly know whether this random account will convey a provisional notion of what I mean, but it is the best I can at this moment give in a small space. Sometimes I lean towards making the *illusion* the dominant idea. But it is very inadequately thought out at present.

Very faithfully yours, FRANCIS GALTON.

We shall be at 5, Bertie Terrace, Leamington, on Monday, and are going in the meantime to Lady Welby near Grantham.

42, RUTLAND GATE, S.W. *December* 15, 1892.

MY DEAR BESSY, Very many thanks indeed for your capital account of Sir M. C.'s teeth. It is particularly appropriate for quotation (of course without names), as it concerns imagination in three different senses, Feeling, Touch and Sight. Merely as a story not vouched for, it would do as an illustration; but we scientific men always desire to be as careful as possible about our alleged instances, and I should be uncommonly glad to get some verification of the story. Is Dr Henry Gisborne alive and resident in Derby? If he is, do you think he is a sort of person to whom I could write, with the chance of getting an accurate reply? Is there anybody you can think of who could help me with further information in respect to this? I might for example write to the Editors of the Derby newspapers (do you happen to know the names of the newspapers?), and could at least get a sight of the original article at the British Museum. Sir M. C. would be the right person to ask!!! If you can give me any useful hints I should be truly obliged. I am so glad to hear that you are out of doors at length. All fairly well here.

Affectionately yours, FRANCIS GALTON.

The story runs thus: Sir M. C. of Derby made one of a week-end party. Retiring to bed, he suddenly missed his denture, and, coughing violently, he became conscious of having swallowed it and could *feel* it in his windpipe. The nearest doctor and surgeon were summoned. The doctor looking down his throat *saw* the end of the denture, and the surgeon *touched* it. Hurried preparations were made for an operation, and the patient was laid on a table in an adjacent room. Just as the surgeon and doctor were prepared to start, Sir M. C.'s valet came into the room and presented his master's denture on a silver salver; he had picked it up by the bedside.

Letter to Professor James Ward.

HÔTEL VALESCURE, ST RAPHAËL, VAR, FRANCE. *March* 24, 1893.

DEAR MR WARD, Thank you heartily for your careful and valuable criticism, which I have read and re-read and shall I hope profit by. The object I had in speaking about what was called the measure of the Imagination, at a Royal Institution lecture, was to invite criticism and to hear objections before taking much further pains in experiment, and to get opinions as to *what it is* that such experiments measure. (I had a good look at James' book first.) Probably we shall meet before long at Trinity, and I should be very glad of the opportunity then of talking a little further about it, and of submitting myself to your questions. It would be a great help towards clearing my own mind.

As regards a *measure* of familiarity, it does not seem to me an absurd notion. The maximum of familiarity with objects gives the complete sense of being *at home*. It is very interesting to analyse this feeling when returning to familiar haunts. Complete strangeness can also be imagined pretty easily, for every one has now and then dropped into very strange surroundings and the feeling is easily recalled. Between the two limits there must be intermediate conditions which it is possible, very rudely, to appreciate. About Weber or Fechner: I know only too well the inadequacy of the statement, but having first looked at James, Sully and a few others, I thought that the very brief statement, reserved as it was, might pass. I wanted chiefly to show that a spiral balance might represent with sufficient approximation and in a very conspicuous manner, the narrowness of the limits of the scale of sensation. I know well that many quite disagree with the view that increased sensation is produced by accumulation of increments, but for my own part I habitually use the imagined sense of waxing fatigue (for example), from zero up to extreme fatigue, as a standard whereby to judge how tired I really am on any particular occasion.

But I must not tax your patience further and can only repeat how very grateful I am for your criticisms.

It is lovely weather here on the Riviera. I, and my wife too, had both suffered in England from influenza and came out ten days ago, with the happiest results.

Very sincerely yours, FRANCIS GALTON.

What a blank Croom Robertson's* death has left!

The Oxford Honorary Degree.

To Francis Galton. From the Rev. Bartholomew Price, D.D.

PEMBROKE COLLEGE, OXFORD. *May* 22, 1894.

DEAR GALTON, I had yesterday the great pleasure of proposing to our University Council that in recognition of your long and excellent service to science and especially of your anthropological work, the Honorary Degree of D.C.L. should be conferred upon you at the ensuing Commemoration; and I have the greater pleasure of informing you that the proposal was received and carried most enthusiastically, and with such observations as would be most gratifying to you, were I at liberty to repeat them. Will you kindly inform me, whether you will accept the proffered Honour, which is, as you are perhaps aware, the highest of its kind that the University bestows on distinguished persons, whether its own children or *extranei*, so that I may report to the Council on Monday next. The Commemoration takes place on Wednesday, June 20, at 12 o'clock. It is usual for the Proposer to entertain his candidate at the time of the Encaenia, but owing to the death of one of our daughters, Mrs Price and I shall be absent from Oxford

* Professor Croom Robertson was an old friend of Galton, and a portrait of Croom Robertson and his wife was highly valued by him. It is now in the possession of the present writer who greatly appreciated the kindness and friendly aid of one of his early colleagues at University College.

at the time and we are in consequence unable to offer hospitality. You have however many friends in Oxford, and there will be, I know, no difficulty in this matter. I can say nothing more until in accordance with our standing orders your acceptance has been ratified in council.

Believe me, Yours very truly, BARTHOLOMEW PRICE.

The sole statements I have found about the conferment of the Oxford degree are the following:

Sophy Bree* reported on June 21, 1894:

The candidates for D.C.L. all followed up in their turn. Uncle Frank came last. He was very much clapped and then the R.P.C.L. (Professor Goudy) stated that Mr Galton was a cousin of the celebrated Charles Darwin (shouts and claps), while still a young man he made a long and dangerous journey of exploration up the White Nile and afterwards undertook a similar expedition in Southern Africa, obtaining the medal of the Royal Geographical Society in respect of the latter journey. He was also described as a distinguished meteorologist and anthropologist. In recent years he had devoted his attention to the study of natural selection and the descent of man—having propounded a theory of heredity which is now becoming recognised as of the first importance. All that of course was in Latin. The newly made D.C.L.'s were in their Doctor of Law gowns. After that the Public Orator made his Oration in Latin.

And again:

"At the dinner," Frank said, "Lord Rosebery made an effective speech. Lord Justice Fry, having to return thanks for the new Doctors, was funny about my composite portraits." L. G.

The degree was conferred at the same time on the Earl of Kimberley, Bishop Mandell Creighton, Sir Horace Davey, Sir Edward Fry (Galton's old foe: see above, p. 122), Captain Mahan of the U.S. Navy, Émile Boutmy, Prof. Mendeleef, Prof. W. M. Ramsay the archaeologist, John Henry Middleton and the Latin scholar Arthur Palmer, an all-round noteworthy batch.

The *Oxford Gazette* contains no report of the speeches.

The Cambridge Honorary Degree.

(*From* TRINITY LODGE, CAMBRIDGE.)
42, RUTLAND GATE, S.W. *May* 16, 1895.

DEAREST EMMA, The ceremony went off this afternoon and Grace, who was there, will I daresay write about it. The Public Orator made some amusing hits in his speech. He will send me printed copies (2 or 3) in Latin, which I daresay Archdeacon Bree, or some other friend, will translate to you. It was all very nice—quite a quiet ceremony and several old friends present.

We are on the point of returning to London after a very pleasant stay at this most hospitable house. Excuse more now. Bessy will be on her way to Alnwick†, so I send no message to her. Ever affectionately yours, FRANCIS GALTON.

Frank bore his part bravely and looked very well, the red gown very becoming. It is so cold to-day, I hope Bessy will not suffer on her journey.

Much love, Ever affectionately, L. GALTON.

42, RUTLAND GATE, S.W. *May* 21, 1895.

DEAREST EMMA, Here is a copy of the speech in the Senate House, of which the Public Orator has sent me a dozen. One is being forwarded to Archdeacon Bree—the ingenious *Latinity* of it will amuse him—and I said that when he next happened to be with you in Leamington you would probably be glad if he translated it to you. I will send a copy to Bessy also, very likely she will soon come across a translator.

Ever affectionately yours, FRANCIS GALTON.

* Sophy Adèle, a daughter of Lucy Amelia Moilliet and therefore Galton's great-niece; she is sister of Lucy Evelyne Biggs, Galton's companion in later years, and married the Ven. William Bree, Archdeacon of Coventry and Rector of Allesley.

† At that time the home of her son, Edward Galton Wheler (later Wheler-Galton), now of Claverdon Leys.

42 Rutland Gate SW
May 4/93

Dearest Emma

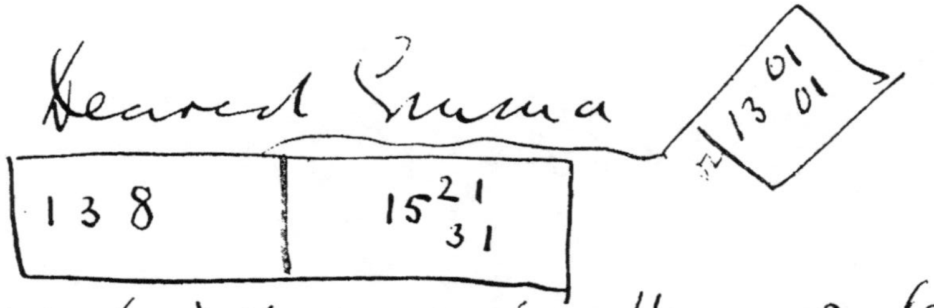

Which of these is the Home-ruled and which is the Home-ruler? see over leaf. This and other things of the same kind got from Sir John Evans at a Royal Society Club dinner — very trifling indeed! By the way, that Club reaches its 150th anniversary next Nov^r, and we are to have a dinner consisting of just the same dishes as at its first meeting, the account of which has been kept. Its price was 1^s/6^d & the pièce de resistance was boiled pork.

I am going to meet Louisa in ½ hour at the Academy, so write this in the mean time. Emily Butler has been with us for a week & left this morning. We go to Trinity Lodge next Thursday to Monday. There is such a row going on at the Geographical, & there may be an explosion on Monday next at its Anniversary Meeting. Every thing seems in confusion. The President & Secretary, both such excellen

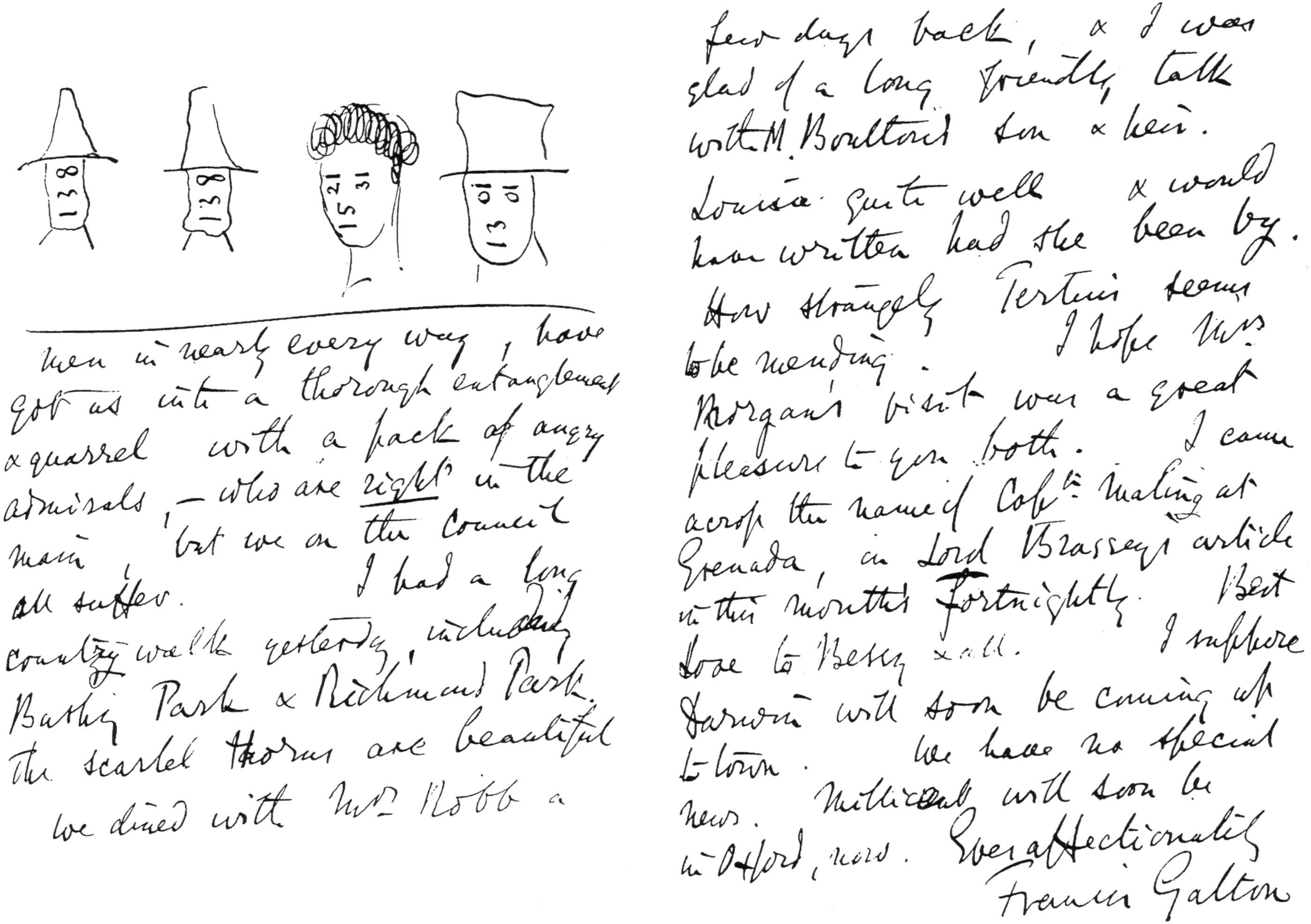

men in nearly every way, have got us into a thorough entanglement & quarrel with a pack of angry admirals, — who are right in the main, but we on the Council all suffer. I had a long country walk yesterday, including Bushy Park & Richmond Park. The scarlet thorns are beautiful. We dined with Mrs Robb a few days back, & I was glad of a long friendly talk with M. Boulton's son & heir. Louisa quite well & would have written had she been by. How strangely Tertius seems to be mending. I hope Mrs Morgan's visit was a great pleasure to you both. I came across the name of Capt. Maling at Grenada, in Lord Brassey's article in this month's Fortnightly. Best love to Bessy & all. I suppose Darwin will soon be coming up to town. We have no special news. Millicent will soon be in Oxford, now. Ever affectionately
Francis Galton

Facsimile of one of Francis Galton's weekly letters to his Sister Emma (aged 82) in 1893.

The weather is improving here and the lilacs, laburnums, etc. will be glorious when the sun shines. There was a chilly geographical river party yesterday to see the Franklin remains at Greenwich, which I did not care to join. What is Arthur Galton's address in Sydney? I owe him a letter of thanks for a published lecture which he recently sent me (unaccompanied by an address).

The following is the Speech delivered by the Public Orator on presenting Mr Francis Galton, M.A., F.R.S., of Trinity College, for the honorary degree of Doctor in Science*.

Sedes olim sibi notas hodie revisit alumnus noster, qui flumine Nilo quondam explorato, et Africa Australi postea perlustrata, velut alter Mercurius omnium qui inter loca deserta et inhospita peregrinantur adiutor et patronus egregius exstitit. Idem, velut alter Aeolus, etiam ipsos ventos caelique tempestates suae provinciae audacter adiunxit. Hodie vero Academiae nemora nuper procellis nimium vexata non sine misericordia contemplatus, e frondibus nostris caducis capiti tam venerabili coronam diu debitam imponi patitur. Tempestatum certe in scientia iamdudum versatus, ventorum cursus tabulis fidelibus olim mandavit, gentesque varium caeli morem praediscere docuit, laudem philosopho cuidam antiquo a Nubium choro Aristophanico quondam tributam uno saltem verbo mutato meritus: *οὐ γὰρ ἂν ἄλλῳ γ' ὑπακούσαιμεν τῶν νῦν μετεωρολογούντων.* Longum est avorum et proavorum ingenia magna in ipsorum progenie continuata ab hoc viro, Caroli Darwinii cognato, virorum insignium exemplis illustrata percensere. Longum est tot honores titulosque ab ipso per tot annos cumulatos commemorare. Hoc autem in loco, eloquentiae eius undecim abhinc annos conscio, instituti anthropologici praesidem non corporis tantum sed etiam mentis humanae mensorem appellaverim. Inter antiquos quidem celebratum erat illud Protagorae, omnium rerum mensuram esse hominem. Inter recentiores autem notum est hunc praesertim virum hominum omnium, imprimis pessimorum, mensuram ad amussim velle exigere. Ceterum plura hodie dicere supervacaneum est; constat enim ne optimorum quidem virorum a laudibus abesse debere mensuram. Duco ad vos virum de scientia anthropologica et meteorologica praeclare meritum, caeli et terrae indagatorem indefessum, studiorum denique geographicorum etiam inter nosmet ipsos fautorem insignem, Franciscum Galton.

Translation of Dr Sandys' speech by Archdeacon Bree for the benefit of Miss Emma Galton, and possibly of some of my Readers.

The Public Orator, Dr Sandys, in presenting for the honorary degree of Doctor in Science Mr Francis Galton, F.R.S., M.A., Trinity, referred to Mr Galton's early travels on the White Nile and in the Damara- and Ovampo-lands in South Africa, adding that the author of the "Art of Travel," "*velut alter Mercurius omnium qui inter loca deserta et inhospita peregrinantur adiutor et patronus egregius exstitit.* Like another Aeolus, he had also taken the winds and tempests for his province, and on his return to his former haunts at Cambridge he had doubtless looked with pity on the "groves of Academe," lately laid desolate by a disastrous storm. As meteorologist he had been the first to map out the course of the winds on an extensive scale, and had thus facilitated the forecasting of the weather; the high regard in which he was held in this department of science might perhaps be expressed in language partly borrowed from the *Clouds* of Aristophanes: *οὐ γὰρ ἂν ἄλλῳ γ' ὑπακούσαιμεν τῶν νῦν μετεωρολογούντων.* Descended from the same grandfather as Charles Darwin, he had himself written largely on subjects connected with heredity. His Rede lecture, on "The Measurement of Human Faculty," delivered in the Senate-house 11 years ago, entitled the former president of the Anthropological Institute to be called a measurer, not only of the human body, but also of the human mind. Protagoras had in ancient days taught the doctrine that "Man was the measure of all things." Mr Galton had in modern times taken a leading part in insisting on taking the measure of men in general and of criminals in particular. It was, however, superfluous to expatiate any longer on his merits; even the praises of eminent men had their limiting law and their proper measure.

* *Cambridge Reporter*, May 21, 1895.

Letters of Galton copied from the Originals by Maud Gardiner Odell.

"The first letter is an answer to a letter from my husband asking about possible observations and measurements, yielding desirable data, to be made upon infants. My daughter arrived Nov. 5 at Naples where we were spending the winter, my husband being on leave of absence from his work at the University of Colorado, U.S.A. Dept. of Biology."

42, RUTLAND GATE, LONDON. *October* 6 (?), 1894.

DEAR SIR, I cannot help you much. Preyer's books are probably within your reach at Naples. I have received from time to time from the United States pamphlets on the subject. There is now one on my table by a Mrs Shand (?)—California—Part III, this being on a child aet. 3, the preceding parts being presumably about an earlier age. (It is in a cover like the Smithsonian publications.) I have had so little to do with children in my life, that I have not interested myself in the inquiries about them and am therefore too ignorant to be an adviser. I should think that the observation of the increasing power of muscular co-ordination, and that of muscle with will, would be as good a clue as any to direct you. I suppose the colour-sense is developed quite early?

I heartily wish you would take finger-prints of the child at the earliest possible age, with a view of determining whether there is any alteration in the papillary ridges *during babyhood.* I have next to no data for investigating this. They are by no means easy to take, partly on account of the restiveness of the infant, chiefly on account of the very slight relief of the papillary ridges. In effect it is very delicate printing. You ought to use the THINNEST possible layer of rather fluid printer's ink, spread on a *polished* plate, and dabbing the child's fingers on it, dab them immediately after on smooth paper. Don't attempt to get any more than a lightish brown impression—Black is an impossibility. It is *clearness* that one wants. Unquestionably the most delicate impression of all, is in varnish thickly spread, that has been exposed sufficiently long to the air to have a *slight* pellicle over it. Dabbing the finger on this leaves a beautiful but transient impress, not so transient however as to prevent a cast being taken from it, if the plaster is at hand and in readiness to use. In regard to instruments for measuring the growth of the soft dimensions of a baby, I cannot tell you and doubt much if such measurements are ever to be trusted. You would have to exercise a strictly constant pressure.

Faithfully yours, FRANCIS GALTON.

P.S. I am not "Sir Francis."

To PROF. JOHN GARDINER.

Address to: 42, RUTLAND GATE, LONDON. *July* 21, 1895.

DEAR MRS GARDINER, Your letter and the most interesting series of prints of your baby's R II reached me on the Continent yesterday morning and I have already gone over them carefully twice, with the aid of such lenses as I have by me.

The general result is that about eight points of reference admit of being compared at different periods of growth and show in a very instructive way how the ridges become more and more sharply differentiated. I must wait till I get back, to study them as thoroughly as they deserve. They ought to be photographically enlarged—that is to say, the best prints and the most important of them. This I will do when I return home. On more than half of the days after birth, from the ninth to the thirty-fifth, on which prints were taken, namely, in the first 18 sets of prints, one and sometimes two prints are clear enough to study. *These are always the darkest* of their respective sets. From the ninth week to the thirty-first inclusive, six sets were taken, but unfortunately not one of these is distinct enough to be of use. I wonder what the cause of failure here can be? Perhaps the materials were not so good. That taken in the thirty-sixth week is quite serviceable.

What remains to be done, to thoroughly deal with this finger, is to get some really good impressions *now*, such as will show the *delta* as well as the core. For this purpose, if you cannot *roll* the finger a little, you might take some of the prints slightly from the *side* of the finger, the thumb side in this case. Also to very kindly let me have some prints, *one* set in *each* future year. The result will be to create a truly valuable series, and at present a unique one. I presume that R II means the right forefinger, so that R I would be the thumb, and R III the right mid-finger? Kindly tell me the baby's full name for future reference. If I publish anything about these, should you object to my mentioning names? If you dislike this, I would idenitfy them by initials. As regards the other digits of which you have taken prints, which you kindly offer to send me, I should be very grateful for them. Whatever is said about the R II which I have, would apply to these also. Your zeal is deserving of the warmest recognition. I can assure you that I fully appreciate and am grateful for what you have done. I would be greatly obliged if you would describe the method you adopt of getting the prints —how do you pacify the baby? How do you hold it? What printing materials do you use? I am very ignorant of baby-ways, but my assistant, who tried hard with his baby-granddaughter, found he succeeded best when it was *sleepy*. I could fancy drilling the child to a game of *patting*, and, at the judicious moment, direct two of its pats, the first upon the inked slab, the second upon the paper.

I hope to be back in England in time to receive any letter that may be written by you a week after receiving this. With kind regards to yourself and your husband and with every wish for the baby's health in whom I naturally shall always take interest,

Believe me, Faithfully yours, FRANCIS GALTON.

Note by Mrs Gardiner. When my daughter, Dorothy Gardiner, was six days old I sat up in bed long enough to take the prints of her fingers. After that prints were taken every day of all digits, for some time. Then every week—subsequently every month—and later on yearly until she was about seven years old. The prints were sent to Mr Galton—a few of them proving good enough for reference, but the majority of the early ones were not very good. I should judge from the letter above that I sent a sample of one finger only, before burdening Mr Galton with the great number taken. We returned to Boulder, Colorado in August 1895.

[Post-card.]

42, RUTLAND GATE, LONDON. *May* 7, 1896.

Just a line to acknowledge safe receipt of the *very good* finger-prints. You have quite acquired the art of taking them. In a few days I shall be free to photographically enlarge them and to send them with a duplicate of those that were accidentally destroyed, and will write then more at length. In the meantime let me say how grateful I should be for the prints of such other babies as you may hope to obtain repeated prints from, at an interval of not less than about a year, the object being to accumulate evidence for or against persistency during early childhood.

Red Indian dabbed prints of the three first *fingers* (fore, mid, and ring) of *right hand* only would be very acceptable. In that form, they would be comparable with all my other race collections. Those from school-children would be every whit as good as those from adults.

Very faithfully, FRANCIS GALTON.

[I had offered to obtain prints from American Indians, if desired, through the services of some University Students whose homes were close to an Indian School at Grand Junction, Colorado. M. G. O.]

Copied from the Original by Dorothy Gardiner.

42, RUTLAND GATE, LONDON. *June* 2, 1896.

DEAR MRS GARDINER, At last I have the pleasure of sending the photos of your baby's fingers. My photographer was busy about preparations for the eclipse, hence the delay.

Those I enclose are direct from the enlarged negatives; those I send separately by book post, are paper enlargements in the camera, from those negatives.

I have not regularly studied them yet. It will take time to go into all their details in the way I want, and I must defer it. You probably will like to examine them, and I think in doing so, you may find help from my book *Decipherment of Blurred Finger-prints* which my publisher will send for your acceptance.

A good way to mark an enlarged print is by a *fine* needle (not pin) prick and making a pencil circle ○ round the hole at the back, with a number attached to identify it hereafter. (You may prick through a blank paper underneath at the same time, so as to have a duplicate.) The prick holes do not in the least damage an enlarged print.

Believe me, Faithfully yours, FRANCIS GALTON.

42, RUTLAND GATE, LONDON. *November* 7, 1896.

DEAR MRS GARDINER, The prints of the baby, with Winifred Palmer, arrived safely, and I have carefully gone through, picked out, and mounted the most effective ones, for future photography. It is indeed difficult to take legible prints of such young creatures. The following seems the most hopeful direction for improvement. In your sets the only fairly clear ones were those taken on the 7th, 8th and 37th days, and all these happened also to be dark. I suspect that on those days the ink was in a more suitable condition than at other times; therefore that great care as to the right fluidity of the ink is an important condition of success. As you are doubtless aware the mixture of a very little "drying oil" makes a great difference in its consistency. Another point in these very delicate printings is to grasp the baby's finger *firmly* and to print from it *rapidly*. There are signs of its having moved in the great majority of the prints. The marks left by the ridges below the joint are often very sharp and clear, while those on the bulb are illegible. The last thing I would mention is the use of the same sort of smooth paper as the enclosed, which is employed in all those high class illustrated journals in which the delicate photo-process printing is used—such as *Harper's Journal*. I wonder whether the more or less dampness of the baby's hand has much to do with the success in printing? It is needless to say how much I should prize any more baby prints you may send me. A few, *well printed*, taken at an early age, and not necessarily of many fingers, would be the most welcomed. Of course with the hope of getting prints from the same fingers a few years later.

Very faithfully yours, FRANCIS GALTON.

[Winifred Palmer is the daughter of Charles Skeele Palmer, at the time of this letter and for some years Professor of Chemistry in the University of Colorado. I obtained prints of Winifred when she was (I believe) less than a day old, certainly less than two days, and continued from time to time as subsequent letters will indicate. M. G. O.]

42, RUTLAND GATE, LONDON. *July* 7, 1897.

DEAR MRS GARDINER, Thank you very much for the two sets of finger-prints. I am glad to infer from the firm, plump marks of your child that she thrives well. Both hers and Winifred Palmer's are very good records and I will put them carefully with the rest. But I cannot now have them enlarged as I am packing up for a summer on the Continent. Thank you also about the hope you hold out, of sending me next September some prints of American Indians. They will be very acceptable. I wonder if by any chance you happen to be acquainted with any authority (of a scientific bent) on American trotting horses and their pedigrees. I ask, because I have just been able to verify a law of heredity (which I proposed tentatively a few years ago) on a certain pedigree stock of hounds. If I could get the racing speeds of the pedigree stock of the trotting horses, with some completeness for three generations back, *i.e.* at least the grandparents and better the great-grandparents of the "subjects," I could make good use of them. The first notice of my paper (which is not yet published but will be published in the *Proceedings of the Royal Society*, in three or four weeks) appeared yesterday in *Nature*, p. 235. (That is the name of our principal scientific weekly paper.) Excuse my troubling you on the distant chance of your being able to help me. I know and have written to Mr Weston, Pittsfield, Mass., and I also know of the existence of the *American Trotter* by H. T. Helm, 1878. Also of yearly *Year Books* which up to 1896 contained the speed of nearly 13,000 trotters, but I do not know particulars about them. Mr David Bonner of New York (I do not know his further address) seems to be one of the principal authorities.

Believe me, faithfully yours, FRANCIS GALTON.

42, RUTLAND GATE, LONDON. *May* 30, 1903.

DEAR MRS GARDINER, Your letter of Feb. 23 reached my house while I was away for the winter. Now that I have returned, let me thank you very much for the enclosure of Winifred Palmer's, aet. 6 years, prints. They will do very well, but perhaps as the years go by you will kindly let me have another set when she is older and becomes more submissive to the printer. I should also be very glad of future prints of the other children so far as you can easily get them. There are sure to be *some* useful points of comparison, which can be utilised. Now to show my "gratitude" (in the cynical sense given to that word, of a "lively hope of future favours"), let me tell you my present needs. I am taking up finger-prints again, from a new and hopeful point of view and send printed papers by this post to explain. You will see that I want two things, of which the second includes the first:

(1) Prints of the two forefingers of many adults in quadruplicate and *rolled.*

(2) Prints as above of batches of relatives of all ages.

The circular speaks of a small outfit that would be willingly sent to those desirous of helping, but I hardly know how to send this to America at a reasonable cost. However I will make a trial, believing them to fall under the head of "Printed paper" (etc. or, ?) "Samples."

The tin box is this size. [Here follows a sketch of the box in plan and section, showing red india-rubber inker and tube of ink. Enclosed with the letter was a circular of instructions and explanations concerning a "Proposed Collection of Finger-Prints. By Francis Galton."]

It can be passed in England as a *letter*, in a "safe-transit" envelope, together with forms, and a printed envelope to return it, all for *two-pence*, in fact they only *just* exceed the $1\frac{1}{2}d.$ stamp. So I can send them readily in this country, if not abroad also. But you *have* the printing outfit, so would not want one for your own use, even if what I now send miscarries. I should indeed be greatly obliged for help in making the necessary collection. The problems, which it ought to assist in solving, are of high importance and the attempts I have recently made, with such limited material as I possess, give much hope.

Very faithfully yours, FRANCIS GALTON.

The envelope enclosed in the packet—for return—is not stamped because I possess no American postage stamps.

[Pursuant to request contained in this letter I obtained prints of a considerable number of relatives—parents, uncles, aunts, cousins, double-cousins, etc. in both branches of my own family, and sent them to Mr Galton in the books with which he provided his contributors. M. G. O.]

42, RUTLAND GATE, S.W. *July* 16, 1907.

MY DEAR MRS GARDINER, It is grievous to me that I shall miss seeing you and your daughter. I leave London on Aug. 1 for the country, my precise address there being *Yaffles*, Hindhead, Haslemere, Surrey, a house which I have rented for six weeks (the extraordinary word "Yaffle" means in the Surrey *patois* a "green woodpecker").

I have laid finger-printing aside now, as it thrives and flourishes in Scotland Yard, our centre of prison administration, but for all that I *think* something more might be done in classification.

The series of finger-prints of your daughter will remain a classic in the history of the science. It stands quite alone in its completeness from the first week of life—even from the day* of birth—to girlhood.

I *may* have occasion to run up to town for a day and *if so* will certainly endeavour to see you if you will send me your London address. Very sincerely yours, FRANCIS GALTON.

Don't call me "Dr" please—I hate the epithet, except on formal occasions.

[In my daughter's 13th year we visited for a few weeks her father's relatives in Scotland and England, spending a short time in London before we sailed for home. I hoped to be able to call upon Dr Galton, whom I had never seen, and wrote to him shortly before we went up to London, but because of his absence from town failed to meet him. M. G. O.]

* This is not quite accurate—I sat up and took the first prints of my daughter on the *sixth* day after her birth. M. G. O.

42, Rutland Gate, S.W. *November* 18, 1896.

Dear Mrs Hertz, You always send me valuable information, and this about Mozart is perhaps the most extraordinary of all*. There are plenty of instances in a faint degree of the mind working independently of the executive function of the hand, in carrying out an already determined plan, but none that I know of which is comparable in degree with that of Mozart. Thus in writing a letter, the forthcoming paragraph is being planned while penmanship is going on. Certainly some people feel the effort of penmanship very much less than others, so that they think ahead while writing as freely as a person who is copying a picture may be thinking of some design of his own. The detachment from noises and interruptions is not uncommon either. We most of us feel that when full of any subject and suddenly obliged to leave it, that we can easily pick up the dropped thread when free to return to it. I wish these wonderful people would submit themselves to tests and not leave the description of their performances to biographers. You do not mention Mrs Macdonell, I trust her health is better. Hoping to call soon and to hear more, believe me, Very truly yours, Francis Galton.

42, Rutland Gate, S.W. *December* 31, 1896.

Dear Mrs Hertz, Your welcome Christmas card, in the form of von Lippmann's curious and interesting pamphlet, has reached me in Cornwall where I am staying a few more days before returning to town. I will keep it for *regular* reading in the homeward train. Pray accept my best New Year wishes for you and yours. It is amusing that the young lady is already busy on books of travel and adventure. What savages we all are, in our primary instincts! Pray remember me most kindly to Mrs Macdonell and believe me,

Very sincerely yours, Francis Galton.

42, Rutland Gate, S.W. *December* 7, 1896.

Dear Meldola, Thanks many for the proposals on your post-cards which suggest further the word "phylometry†."

I see in Liddell and Scott's dictionary that φῦλον is just what we want—viz. primarily a "set of any living beings naturally distinct from others." It has been used for a swarm of gnats—for the races of birds, beasts, fishes, also in the sense of "the whole tribe of them," as applied to the Sophists. Further, as a race, a nation, or "a clan or tribe of men according to blood and descent."

* Extract from a letter of Mrs Hertz from 40, Lansdowne Crescent, W., dated November 17, 1896: "I enclose a leaf torn out (rather ruggedly) from the Programme *résumé* of the popular concert last Saturday. It struck me that the facts therein related concerning Mozart's celebration, if you have not already come across them, might be of interest to you. To me it seems little short of miraculous that he could write out that sublime composition, the *Ouverture* to 'Don Giovanni,' while his wife read aloud to him. Indeed the statement that he did so during the night before its performance excites much doubt in my mind. For he would have had to write the part of each instrument separately, and when could the members of the Orchestra have studied and rehearsed their parts? Nevertheless it seems probable that Otto Jahn, his biographer, a writer of repute and standing, took trouble to verify the main point, which is that he had the faculty of thinking out a composition in its full detail and completeness before he set pen to paper, and that he could then write it down correctly while devising a fresh composition, or while concentrating his mind on some quite different subject. Have you met with any other such surprising manifestation of the twofold simultaneous action of the brain?......Are you on the track of fresh discoveries about the animal whose behaviour and whose motives grow more and more bewildering and perplexing, more and more difficult to control, guide and regulate?"

† Meldola's suggestion of this word as a name for what we now term "biometry" deserves to be recorded. It is about coeval with the use of the latter term. When I adopted the word "biometry" for the science which applies the modern theory of statistics to the study of variation and correlation in living forms, I was unaware that Christoph Bernoulli in his *Handbuch der Populationistik* of 1841 had termed the study of life tables "Populationistische Biometrie."

What do you think? The question suddenly presses, as I find that the Committees of the Royal Society are formed at its Council next Thursday.

Failing a better name I would propose to the Council of the Royal Society that our present title be changed as above to "Phylometric Committee." What do you say? I wrote to Weldon and to F. Darwin in the same sense. Also I should propose to ask the Council to add the words "with power to add ordinary and accessory members."

"Accessory" is a phrase in use at the Royal Society to signify, as I understand from Harrison, either non-fellows or paid fellows or both. Do you agree?

Very faithfully yours, FRANCIS GALTON.

P.S. I have looked up the Minute of Council under which we were originally appointed. It was on Jan. 18, 1894—Minutes, p. 71: "...for conducting statistical inquiries into the measurable characteristics of plants and animals." The notice of our Friday meeting is gone back for *revise* so you will not get a copy before Wednesday. F. G.

42, RUTLAND GATE, S.W. *February* 15, 1897.

MY DEAR PROFESSOR K. PEARSON, You will not, I am sure, doubt that I fully share the view that the future of biology lies mainly in exact treatment of homogeneous statistical material. The first thing is to *get it*. Now the Sub-Committee seems to me better adapted than perhaps any other collection of men that could be named, to do this. They represent between them the departments of mammals, birds, fishes and insects. They know the conditions of rearing and the existing workers, and they have the confidence of the latter. I have already a considerable list of suggested experiments such as would be statistically serviceable. The details of each would be of course a serious problem, so to be arranged that neither sterility nor disease shall interfere with it; and, again, such that will lead to no ambiguous results. After Tuesday's meeting of the Committee it will be more easy than it is now to anticipate, but at present I am in high hopes that we shall ultimately succeed in the really important task of controlling, in a useful sense, a vast amount of existing work that is wasted for want of scientific sympathy, criticism and encouragement. It must always be borne in mind that we are dealing with human workers, who have their own ideas which must be respected and humoured, if we are to gain their cordial co-operation. We have, to speak rather grandly, *statesmanship* problems to deal with. I trust we shall often have occasion to consult with you as to the best of alternative plans. Just now, we must busy ourselves in finding out lines of least resistance in pushing forward our nascent work. Very faithfully yours, FRANCIS GALTON.

I had warned Francis Galton that his Committee, extended into an "omnium gatherum" of various schools of biological thought, would achieve nothing further in the way of "conducting statistical inquiries into the measurable characters of plants and animals"—a prophecy which unfortunately was only too soon and too fully realised.

42, RUTLAND GATE, S.W. *March* 14, 1897.

DEAR PROFESSOR K. PEARSON, Pray try and forgive my troubling you with a question. It is whether the enclosed problem is a recognised one and, if so, where I can read about it? I have a big batch of very promising statistics in which it would be very serviceable, and at present I do not see my way clearly in respect to it. Any guidance from you would be most acceptable. Very faithfully yours, FRANCIS GALTON.

If, when your proofs of spurious correlation reach you, you could spare me one copy I should be much obliged.

A Problem relating to Fallible Judgments.

Suppose two kinds of balls, A and B, which differ so little as to be often mistaken for one another when viewed somewhat carelessly, though they are surely distinguishable on minute scrutiny. A rather careless examiner, No. 1, is given a batch of 1000 of these mixed balls, in which there are known to be a balls of the A kind. He has to select out of them the a balls which he considers to be A. Then each of the selected group is marked with an A_1, of so minute a size as to be readily overlooked. A second examiner, No. 2, who overlooks the A_1 marks, proceeds in the same way, and each of his selected set is scratched with an A_2. Subsequent investigation shows that:

I. a_1 of the balls marked A_1 are truly A.
II. a_2 of the balls marked A_2 are truly A.
III. a_3 of the balls marked A_1 and A_2 are truly A.

Question. What is the trustworthiness, *when measured on a scale of equal parts*, of the three estimates defined by I, II and III?

Mem. The scale of trustworthiness is bounded, below at a zero point, of no trustworthiness at all, when $a = \frac{a^2}{1000}$, and, above, where precision is absolute, when $a = a$

F. GALTON.

42, RUTLAND GATE, S.W. *May* 20, 1897.

MY DEAR PROFESSOR KARL PEARSON, You were not, as I heard, at the Royal Society Soirée last night, where I had hoped to have thanked you sincerely for the book and for the exceedingly kind writing on the fly-leaf. It is one of the great pleasures left me, to know, now that I grow older and stupider, that anything I may have done has proved serviceable to others who, to misquote Tennyson, can "step from *my* dead self to higher things."

I was absent from London all the day-part of yesterday and have only very cursorily as yet looked through the book, but have seen enough to astonish me at its wide range and serious reasonings and at its substantial unity among apparent diversity. You must indeed have had difficulty in assigning a title to it. What an awful time to live in the 14th century must have been to most persons, with its plagues and endemic manias of flagellations, tarantellas and the like, and savage wars. No wonder that dances of death were popular. I look forward greatly to reading the two volumes properly. Very sincerely yours, FRANCIS GALTON.

GRAND HÔTEL, ROYAT. *August* 13, 1897. 5 a.m.

DEAREST EMMA, It is ill news that I have to send. You heard that Louisa* had been ill since last Sunday, when she packed up in good spirits and with much interest for a tour among the Dauphiné mountains, beginning with the Grande Chartreuse. But it was not to be. She was seized with a severe attack of diarrhoea and vomiting during the night, a repetition of what she, I and Mme de Falbe had all had in a lesser degree. Still Dr Petit thought little of it on Monday morning, even on Tuesday morning he was not anxious, but she grew steadily worse. The bile thrown out was exceedingly disordered and I think its presence throughout the body poisoned her. She had of course discomfort at times, but was on the whole drowsy. Yesterday she was evidently sinking. I had a nurse to sit up through the night, who awoke me at 2½ a.m. when dear Louie was dying. She passed away so imperceptibly that I could not tell when, within several minutes. Dying is often easy! I believe French formalities require very early burial, probably to-morrow, but I know nothing now. When the people are up and moving I shall hear all about necessary legal formalities, which may take time. This is written to catch the morning post to England. You shall of course hear again very soon. I cannot yet realise my loss. The sense of it will come only too distressfully soon, when I reach my desolate home. Please tell the brothers and sisters. I am too tired to write much, having had long nursing hours. Mme de Falbe is our one friend here, but she was in bed yesterday and to-day with a slight attack of the same malady. Her maid has been very helpful. The landlady is all kindness. The nurse (a *religieuse*) did her best, and so did the chambermaid, they and another woman got up in the night to do the sad and necessary offices. Dear Louisa, she lies looking peaceful but worn, in the next room to where I am writing, with a door between. I have much to be thankful for in having had her society and love for so long. I know how you loved her and will sympathise with me. God bless you. Ever affectionately, FRANCIS GALTON.

* Mrs Francis Galton, née Butler.

PLATE LII

Francis Galton, aged 75.
Royat, July, 1897.

Address, 42, RUTLAND GATE, S.W. GRAND HÔTEL, ROYAT. *August* 15, 1897.

DEAREST EMMA, I hardly know how much time has really passed since I wrote, for each day has been divided into two or three by intermediate dozes or sleeps and the last week has been terribly long. Dear Louisa was buried with simple decorum yesterday in the cemetery of Clermont-Ferrand. The day was lovely, the mountains looked singularly imposing, the English Chaplain, Mr Wilcox (of Battersea Park Road) officiated, and a most kindly and tactful clergyman, Mr Jennings, the clergyman of St Stephen's in Cheltenham, who is now copying documents at my side, came with me to the grave. The landlord of the Hôtel came also, and acted as a perfect courier in managing all the numerous details and formalities. A feeling allusion was made in the sermon of to-day, and appropriate hymns were sung. I shall, I trust, see you for a day before long and can tell more and answer questions. Mme de Falbe has written a full and independent letter to Spencer Butler, describing all she knew, and filling in some needed details. She could not help, or come *herself*, during the latter part of the illness, being then, and still is, confined to her bedroom by doctor's orders, but she sent a useful maid. You will easily understand how desolate I have felt, but thanks largely to Mr Jennings' tact, consideration and manly sympathy, I have already, perhaps, gone through the bitterest period, though I look forward with dread to the most painful task of distributing her familiar personalia, etc. Dearest Louisa,—I have very much to be grateful for, but our long-continued wedded life must anyhow have come to an end before long. We have had our day, but I did not expect to be the survivor. I got for the first time in touch with England yesterday, through receiving a telegram from Spencer Butler, who is still in London. I thought he had gone to the Engadine. I have had also to-day a telegram from Gifi. These telegrams are a boon to me. People generally do not (and I did not) realise that you can telegraph in English if you please. In any case the cost is only 2*d.* a word. I hope to get to-morrow, or at all events by Tuesday, any letter you may have sent to Grenoble. On Tuesday evening I propose to start home, arriving there on Wednesday evening. I am anxious to hear about yourself; it seems to me that I have not heard for a fortnight, but, as already said, I am astray as to time, and my papers are huddled up in disorder.

Of course Bessy will understand that in writing to you I write also to her and, through her, both to Edward and Lucy*. I had not written either to Darwin, Erasmus or Milly, but have done so to-day, and enclose the two latter letters for you kindly to address and forward. Excuse more for I must husband strength. Ever very affectionately, FRANCIS GALTON.

Examination showed the cause of Louisa's long ill-health and final death to be an extremely small stomach and an extremely constricted outlet due to her illness 19 years ago. The stomach was barely one third the natural size, and the outlet leading out of it no larger than would just contain a common lead pencil.

42, RUTLAND GATE, S.W. *October* 13, 1897.

DEAR MRS HERTZ, An absurd piece of ill luck has prevented my yet reading von Lippmann's pamphlet. I am just now sorting accumulations of pamphlets, letters, etc., and the pamphlet in question seems to have got into one of the heaps of unsorted materials, whence in due time it will emerge, but at present I cannot find it. I should be much gratified if he does not lose but reads mine, of which I enclose a copy.

There is to me no difficulty in fraternal variation. The wonder would be if brothers did *not* vary considering the multitude of unseen disturbing influences on the general tendency of like to produce like. In my theory, the prophecy is that so many *per cent.* of individuals having like progenitors, will be this or that, and it is the nearly exact fulfilment of the prophecy that the memoir is intended to show. The Basset hounds of the same family are by no means all of the same colour, but the *per cent.* law holds good notwithstanding. Imagine a pair, whose ancestors are all known, to produce 100 puppies; then, what I prophesy is that from knowledge of the ancestry I can tell how many of them would be T and how many N†. The "coefficient" expresses that number. It varies according to the case from 96 to 52 in my Table VI.

Very truly yours, FRANCIS GALTON.

* Mr and Mrs Edward Wheler-Galton. The latter appears as "M. L." in later letters, probably to distinguish her from other members of the family with the same Christian names.

† T = Tricolour, N = Non-tricolour.

42, RUTLAND GATE, S.W. *January* 4, 1898.

MY DEAR PROFESSOR K. PEARSON, You have indeed sent me a most cherished New Year greeting. It delights me beyond measure to find that you are harmonising what seemed disjointed, and cutting out and replacing the rotten planks of my propositions. We shall make something out of heredity at last; all the more, when new and more abundant data arrive for testing the soundness of each advance. I wish many more mathematicians would attack the subject. A mere statement of your results—what *can* be done—would perhaps help to make people understand that there *is* a science of heredity, approximately understood at present, but sure to be developed. Let me please keep the MS. for two or three days. I have gone through it superficially twice, but want further time to do it more thoroughly. I will write again. You are very flattering to me. Very sincerely yours, FRANCIS GALTON.

42, RUTLAND GATE, S.W. *January* 10, 1898.

MY DEAR PROFESSOR K. PEARSON, You overwhelm me with sentiments of gratitude. I cannot help feeling them partly in respect to your most flattering references to myself, but really and honestly chiefly in respect to your furtherance of the just understanding of the effects of heredity. The subject is so enormously important that my own personal interests in it are quite secondary. It is indeed a big work that you are carrying on and which you have advanced to a point at which the results cannot but impress the scientific public, being large and palpable, simple and consistent. I hope on those grounds that you may see your way to publish it soon. It will be a vast encouragement to those who collect and to those who furnish data, to be assured that there is a clear and very important object in view, in collecting them, and that their efforts will not be wasted. Such remarks as I might venture to make, would be of little importance to you. I wish however that you would not mind the appearance of prolixity, in expressing the first paragraph on p. 3 at greater length, so that its meaning should be unmistakably clear. Also, if you use the word "mid-parent of the 5th order" instead of "mid-ancestry of the 5th order," the word "parent" in that context should be defined. It is true that we say grand-parent, etc., but "parent" strictly means either a father or mother. I am stupid in *realising* the meaning of the new values of r', r'', etc. (p. 8). I wish you could somehow make the *rationale* of it clearer, as distinguished from the mathematical proof. In the comments, p. 22, on the contents of p. 21, would it not be well to be somewhat more diffuse, in order to do away with the mistaken idea that first presents itself to the mind, that they contradict the well-observed fact that pedigree stock are very "even," i.e. they vary little among themselves. It is the $\frac{\text{fraternal}}{\text{general}}$ regression that you speak of, and therefore the ratio may well be maintained, all the same. Your coefficient of stability will be of great use and importance. But *all* is so valuable. Very sincerely yours, FRANCIS GALTON.

The cross-heredity is a charming piece of work. It has just reached me.

42, RUTLAND GATE, S.W. *January* 25, 1898.

DEAR PROFESSOR K. PEARSON, The memoir you send, and which I return, is full of interest to me. The cephalic index seems an admirable subject for hereditary inquiry, making observations on school-children available, and it is excellent too for Christmas family gatherings. The index of conjugal fidelity in a race is delicious! I am very glad to see how closely theory and observation run together in all Indian kinships except the paterno-filial. Not the least of your many achievements is that of "enthusing" (as Americans say) such competent workers as you do. I hope Miss Cicely Fawcett will continue her investigations.

Very faithfully yours, FRANCIS GALTON.

Of course I shall be at the Royal Society on Monday. I am grieved that influenza still grips you.

42, Rutland Gate, S.W. *July* 1, 1898.

Dear George Darwin, Your small son has, I hear, a faculty about which I have been particularly interested in another child, namely the aptitude of identifying the perforated discs used for musical boxes. I wish you would talk it over with your wife, and perhaps make a few experiments and tell me the result on Wednesday. The experiments I mean, are by taking the pile of discs and pulling out one of them very gradually from among the pile until he recognises it.

Does he know them with equal ease, face upwards or face downwards? How many does he distinguish? At what age did he begin to do so? Am I right (do you think) in supposing that it is a similar act of memory to that of recollecting a hieroglyph or a scroll pattern, or the like, or is there any possibility of suggesting the tune, in the distribution of the holes? I should be very glad of some verbal information about this, as the case I have heard of in Northumberland seems to be a very curious one, hard to explain except on the hypothesis of a portentous memory of patterns. Do you think that *you* yourself could easily recollect and distinguish the discs? Can the other children? Ever yours, Francis Galton.

42, Rutland Gate, S.W. *November* 16, 1898.

Dear Professor Karl Pearson, Possibly you may intend going to the Royal Society "at home" on Thursday (to-morrow). If so, or otherwise, will you dine as my guest at the Philosophical Club? It is not necessary or even usual to dress. It would, I thought, have been possibly a breach of etiquette, had I written, as soon as I knew it was settled, to congratulate you heartily on the forthcoming award of the Darwin Medal of the Royal Society. It seems in every way most appropriate. I am delighted at the wisdom of the choice.

Very sincerely yours, Francis Galton.

The enclosed card will give needful particulars as to the Phil. Club.

7, Well Road, Hampstead, N.W. *November* 30, 1898.

My dear Mr Galton, I quite realise the difficulty about the term Reproductive *Selection*, but I sought in vain three years ago for a better, and failing to find one have used it ever since in my papers. I think also that it has something, not very much perhaps, in its favour. Evolution takes place by taking out of the community A, B, C, D, E, F,......X, Y, Z, certain members L, M, N, and putting them into a position of advantage for propagating their kind. Anything which contributes to this advantage is *selection*, a differential death-rate is Darwin's *natural selection*, it should be *Selection of the Fitter* as all selection in wild life is "natural." Selection by a differential birth-rate is my reproductive selection; it is selection of the most fertile. There is a third kind—selection by a differential pairing rate, individuals L, M and N pair, or on the whole pair, more frequently than A, B, C,.... This is also a possible progressive source of change. It can be demonstrated to exist in civilised man, I am uncertain whether it is actual as well as potential in wild life. All these three kinds of *selection* are factors in potentia of evolution, but the last two involve *no destruction*. A uniform, *non*-differential death-rate will still cause progressive change. Thus a selection of Celtic over Teutonic elements in a population might arise without any survival of the fitter, if (i) the Celts married equally frequently with the Teutons, but were more prolific, or (ii) if the Celts and Teutons were equally prolific, but the Teutons married less frequently than the Celts. In both cases we might speak of *selection*. In the former case we have selection by differential fertility, in the latter case by frequency of pairing. In both, to be effective, the fertility must be inherited or the relative tendency to pair, inherited. The former is what I term *Reproductive Selection*, the latter is—what? Please send me a name for it, before I find it absolutely needful to coin one.

Yours always sincerely, Karl Pearson.

42, Rutland Gate, S.W. *November* 30, 1898.

My dear Professor K. Pearson, It is not so much the word "Selection" that seems to be a stumbling block, as *Re*product*ive*. I did my best to think it out, owing to the fact that the Royal Society paper was sent to me as one of the Referees, and it was a duty to do so. What I then wrote was somewhat to this effect: (1) The termination of the adjective should accord with natur*al*, artifici*al*, sexu*al*, and therefore be "*-al*," or its equivalent "-ic," for

"-ical." Reproductive Selection conveys the idea of a Selection that reproduces itself, which of course is absurd. The termination "-ive" seems quite misleading.

The *Re* in "Reproductive" also, as it seems to me, misleads. It implies the substitution of a unit by a *similar* one. There is no advance, but it leaves things, statistically speaking, as they were. "Production" is much less objectionable. What is wanted, is to express production to more than the average amount, such as a prefix like "pre-" or "super-" might imply. But this might be dispensed with for the sake of brevity.

Two words occur to me as worth consideration:—"Progenic Selection" (I won't say Pre-progenic), "Genetic Selection" (I won't say Hyper-genetic). I am afraid that "superior fertility" cannot be expressed by any tolerable word that has "fertile" for its base and ends in "-ic" or "-al." "Proles" is a good word to *force* into an adjective, "Prolic Selection." I am afraid "Prolifical" would be too cumbrous, and "prolific" is too differentiated a word to use. Your idea, as now expressed by the words "Reproductive Selection," will hereafter become so important an element in all questions of Evolution, that before they are too firmly established I really think it most advisable to change them. It will be otherwise a continual stumbling block to new students. Pray forgive me for all this. Very sincerely yours, FRANCIS GALTON.

42, RUTLAND GATE, S.W. *November* 30, 1898. (2)

DEAR PROFESSOR K. PEARSON, I was stupid about "fertile," for, looking into a Latin dictionary, I am reminded of the word "fertus" for fruitful (ferta arva—fertile fields), so perhaps "Pre-fertal Selection" would be the best of all. I will think over your other question.

Very sincerely yours, FRANCIS GALTON.

42, RUTLAND GATE, S.W. *December* 4, 1898.

DEAR PROFESSOR K. PEARSON, The phrase "Prefertile Selection" does not sound right to me. Why not drop Selection and use a phrase, which *in full* would run "effects of differential fertility on race"? It would *currently* be shortened to "effects of differential fertility" (30 letters) or even to "Differential fertility" (21 letters). I left London on Friday for three nights in the country and left behind your last memoir by accident, which prevents me from judging how far some such phrases as the above could be substituted for "Reproductive Selection" without bungling the sentences. I cannot think of a couple of technical words that should express "the effects of early marriage on fertility," but are they needful? Yet that may not be what you want, rather "the effects of differences in age of marriage on race." What you say about a Committee to discuss and pass new words is *prima facie* very attractive. Some few men have a great gift in striking out good words. Huxley had it in an eminent degree. Sir John Lubbock used to take immense pains in giving names to species that would suit the genius of the French and German languages as well as our own. I return to London, to-morrow, Monday. Very faithfully yours, FRANCIS GALTON.

42, RUTLAND GATE, S.W. *March* 9, 1899.

MY DEAR EMMA AND BESSY, Thanks, Bessy, for your letter. It gives a very painful account of the Lloyds*. There is much to tell:—On Monday I called at Chester St and saw Evelyn Cunliffe. The fact appears to be that Douglas† is by no means in so exhausted a state as I had understood, but is able to sit up in bed and take food and even to get partly out of bed himself. Also that the ear and the lung have both got well. On the other hand, phlebitis has set in, which is of course very serious and the swelling of the wrist continues very painful. It may prove to be an abscess. This was on Monday. I have not heard since.

The Horse photos and measures ended very successfully. Thanks wholly to Sir Jacob Wilson a number of big and little difficulties were smoothed away, any one of which would have been fatal. On Monday evening I met him and the photographer at the yard in the Agricultural Hall. A level platform of three rows of flag-stones was laid down on sand shovelled in for the purpose alongside a temporary structure that served as a background. The camera was 30 feet off and a dark room was improvised for changing plates. Mr Reid, the photographer, brought 36 slides and during the process refilled 24 of them. Tuesday morning

* On the relationship of the Galtons to the Lloyds: see our Vol. I, Plate C in the pocket at the end.

† Sir Douglas Galton, the engineer; he died from blood poisoning in the same year. Evelyn Cunliffe was his elder daughter.

PLATE LIII

Francis Galton's Great-Niece Evelyne Biggs—Mrs Guy Ellis. Compare Plate XIV, Vol. I, and Plate XLVIII of the present volume.

was fair in Rutland Gate but as I approached the Agricultural Hall the fog began and worsened until at 9 a.m., the time for beginning, nothing could be seen at a short distance! However, in time, the day cleared, with the result that all the horses selected for making the final judgment were taken. The only mistake was in not securing a lighter background. My staff consisted of the photographer, his son who did the main part of the work, and a most intelligent stud-groom (whom he borrowed from Lord Arthur Cecil to help him) and two collegiates sent by the Veterinary College to make the measurements. There were others in the yard, besides the groom that led in each horse in turn. I was surprised at the facility with which they placed them. Of course some of the beautiful brutes stood on their hind legs and pawed in the air, and others kicked fore and aft, but on the whole they were hustled into place, and in every case stood on the middle row of flags which was only 25 inches wide. So all the photos are in *standard* position. I wanted to mark the position of the hip bone and did so with paper wafers, each the size of a shilling, with a dab of very thick paste in the middle, which was laid on with a little spud, that I cut from a pencil. It was held by its edge, clapped on the right place, and adhered firmly. They told me that the grooms were puzzled as to the object, but on the whole thought it was a mark of distinction, so they left them on and in the afternoon parade there were the spotted horses! It must have puzzled the spectators. I was standing about helping, on a coldish day from 9 to 2; then there was lunch, and afterwards the final judging, but by 4 I began to feel cold, and left before being formally introduced to the Duke of Portland, etc. It certainly *was* cold (to me). A friend of Edward's, Sir John Gilmour, to whom I was introduced, asked me for some particulars, but at that moment my teeth were chattering so that I could hardly reply intelligibly. The upshot is that I have got material for a useful little paper, but time will be needed to work it up. I shall have the photos* sent to me abroad, to work at when otherwise idle. Yesterday Frank Butler came for final instructions. He will act altogether for me, in emergencies, and will answer my letters, which will be forwarded to him. His address is A. Francis Butler, Esq., Haileybury Cottage, Hertford. I will take every care of Eva Biggs†. She comes to me on Monday. I want to tell Edward Wheler about the photography but have little time just now to write more. Will you send this, therefore, to him? Ever affectionately, with much love, FRANCIS GALTON.

AT SEA, PAST LISBON. *March* 21, 1899.

DEAREST EMMA, I begin now, as there will probably be hurry and sight-seeing to-morrow morning at Gibraltar. The sea has been unexpectedly favourable, but weather is so cold that I have used all my wraps the whole day and over the bedclothes at night. Eva and a very few other ladies have been squeamish and sick and she is not yet quite right, though sitting on deck. It is a wonderfully well-arranged steamer. We each of us have had the good luck of having a cabin all to ourselves, which, as a cabin is 6 ft. 3 in. long, the same or a little more in height and 6 ft. wide, is luxurious. With two in a cabin it would be rather hugger-mugger, at the best. The ship rolls so slowly, it takes 17 seconds to roll to one side and back again. There is no jar or smell of steam engines whatever; the ship seems propelled by attraction or some other smoothly acting force. This is the section as I understand it. Nobody but the ship's officers are allowed on the upper deck, but we walk and sit mostly at A, which is under the cover of the captain's deck and very pleasant to be in. We can walk along B but it is much narrower than A. The first class passengers are separate from the rest and walking all round their part at A is just $\frac{1}{10}$th of a mile, as I find; so 10 "laps" are 1 mile. They feed us over abundantly. Eschbach (the courier) makes a capital lady's-maid for Eva, and evidently knows all a courier's duties very perfectly. There are about 100 first class passengers, some pleasant to talk to. I think much of you all, also of the great sorrows left temporarily behind. So much for the present.

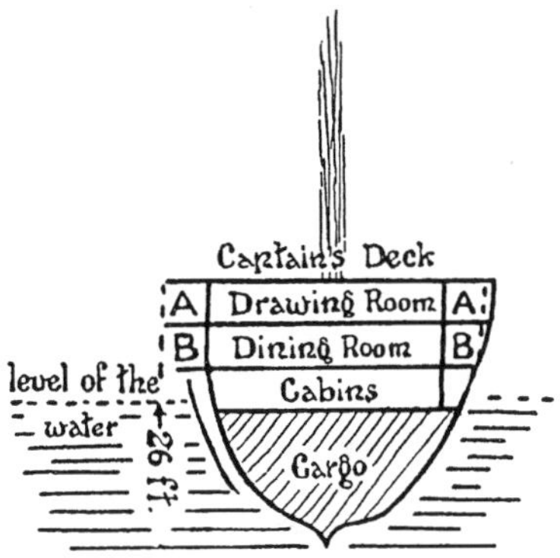

* These photographs have disappeared entirely, and Mr Reid informs me that after taking two sets of prints he destroyed the negatives. Alas!

† Galton's great-niece was about to travel with her uncle for the first time, and Spain and Tangiers were to be visited.

Wednesday midday. *March* 22.

Yesterday was calm, warm and enjoyable. We reached the neighbourhood of Gibraltar during the dark and got up quite early to see the grand outlines of the hills and a brilliant planet. Some time after the day broke and ultimately we landed at 7 a.m. Then we took a walk and afterwards a pleasant hour's drive under the big rock, then to breakfast at 9½ and a sleep after. It is quite warm, flowers in masses and green all about. To-morrow (Thursday) we reach Ronda late, sleep two nights there and reach *Hôtel Madrid, Seville*, on Saturday to stay there nine days, that is over Easter Monday. Eva is very bright and has been practically free from sea-squeamishness since yesterday afternoon. We both left the ship with some regret, having begun to enjoy sea life and having made various acquaintances. I quite see how pleasant it might be to take summer cruises on these big ships with a party of friends. Good-bye now, with best love to Bessy and to all. Ever affectionately yours, FRANCIS GALTON.

Note the 1*d*. Gibraltar stamp on the envelope.

Address up to April 9, *midday* post, Hôtel Washington Irving, Granada, Spain.

SEVILLE, morning of *April* 3, 1899. (Your letter of Tuesday, March 28, arrived last night, Easter Sunday, 5 or 6 days on the road.)

DEAREST EMMA AND BESSY, It seems so odd to have only just received a reply to my Gibraltar letter, for though we have been only 12 days in Spain, it seems to have been months, and we have been eight days at Seville, doing something fresh every day and getting a more complete change of ideas in a short time than I had thought possible. Eva is a capital companion and Eschbach is quite a first-class courier. Though old and half-blind, he always knows everything or finds out everything we want. He is always at hand and ingratiates himself everywhere. I would back him and Gifi, each in their way, against any men in their profession whom I have seen. The religious processions and church services were almost constant in the late afternoons of Thursday and Good Friday, and in Saturday morning's service the "veil was rent," pistols fired, bells rung everywhere and Lent was over. We drove out to see the bulls, which had just arrived and had been driven along with belled oxen to keep them quiet; they were in a paddock beyond the suburbs. All fashionable Seville was there, and the bull-fighters too. Nay more, Eschbach made friends with one of them and suggested that we should take him inside our little open carriage, to explain everything, which we did, to our mutual satisfactions. He chattered away and was most amusing and gave us a good lesson in elementary Spanish at the same time. He was quite a natural gentleman. Of course I went to the bull fight, which did not horrify me as I had expected; I found it full of interest. I won't go into details, though they differed in importance, as it seemed to me, from what others have said. The six bulls between them tossed and killed at least a dozen horses and the riders got ugly falls, but none were hurt. The bull heaves them in the air, rider and all, with his great force. It is not a rapid dash that he makes at them, but a murderous business-like push, working his horn deeply in. I don't mean that all four legs of the horses were lifted off the ground at once, but three of them were sometimes, and always two. Every one of the six fights had its peculiar features, and it is this variety of incident that makes it so attractive to Spaniards. Moreover there is no *cry* of pain, no visible sign of pain to curdle one's blood. The badly wounded horses still obey the bridle, showing that they are not in any *agony*. One must not read one's fancies into facts. The squeal of a scared rabbit affects my own nerves more than anything I saw in the ring, and the feats of cool daring and agility were marvellous. I am glad though, that Eva did not care to go. She had her experience, by lighting her mosquito curtains by accident, while dressing for dinner. The blaze was furious, but there is so little material in them to burn, that the body of heat was really small and insufficient to set a house on fire. It is like those futile attempts to light a coal fire with a newspaper only. She was neither hurt nor frightened, but was wet through by pouring the contents of two big cans of hot water and two jugs of cold water upon the blaze and partly on herself. She sketches much and makes many studies of heads, and goes about to churches in Bessy's beautiful black lace shawl as a mantilla, having been well instructed in the art of wearing it by an Anglo-Spanish lady, who vastly admired the lace. We have both been quite well, except that I was slightly out of sorts with a usual traveller's ailment for two or three days. As to my old cough, it has gone away, though the throat does not seem *yet* to be *quite* strong again. It betters every week. We leave here to-day (Monday) for Cadiz, cross to Tangiers by (seven hours) steamer on Wednesday, and then plans are uncertain for a few days, but we ultimately get from Morocco somehow to Malaga and thence on to Granada, which, as our time-table now stands, we should leave on April 14; but I dare say we may find it wiser to give more time to this forthcoming and most interesting bit of travel. It is perfect English summer here. I began writing this letter at 6 a.m. this morning, with the windows wide open. The sky has been cloudless for many days and we read with wonder about snow, not only with you, but at Nice also. It is a grievous affair about the Earl of Warwick s property. I will give your messages to Eva, but must close now for the post. With best loves to all.

Ever affectionately yours, FRANCIS GALTON.

Tell me about your own health when you write; please do.

Address next to: Hôtel de Rome, Madrid, Spain, but I shall not get there till about the 18th, and propose staying till the 25th, at least.

DEAREST EMMA AND BESSY, Your joint letters of April 1 reached me to-day at Tangiers, Thursday, April 6th. We were called out of bed yesterday when in Cadiz at 4½ a.m. and finally, in such a bustle and clamour, landed here about 2 p.m.; since when we have been busy sight-seeing. It is such a very Oriental-looking town with crowded streets of costumed natives; a most complete change after Spain. We passed Cape Trafalgar, and Eva made sketches as we did so, and has copied and will send herewith one for you, Bessy. What a historical part of the world we are in! Cadiz is a flat Portland Island, connected with the mainland by a long narrow strip of land, corresponding to the Chesil Bank. We had a breezy passage, calm sea at first and then abundance of "white horses." Among other things here, we saw a snake charmer who put out his tongue for the snake to bite, which it did very thoroughly, opening its mouth very wide and fixing on to it. Then he put out his tongue for us to see and sure enough there were the two bleeding punctures made by the two teeth. Then he chewed straw for a while, and putting out his tongue again—hey presto—it was healed.

I called to-day on the wife of the British Consul, Lady Nicholson, to whom Mrs Robb gave me a letter. He is the son of an old acquaintance of mine, Admiral Sir Frederick N. (no relation of Marianne's*)—such a beautiful situation and gardens. Sir William Dalby, the aurist, turned up to-day, and gave me a full medical account of Douglas's last illness. The details were much as I had heard from Marianne, but he did not think his sufferings had been so terribly great as she seemed to think, when speaking to me about them. He, his son, Eva and I have been to a Moorish coffee-house with singing, and in the middle of our cups were rushed out to see a Moorish bridal procession. The paving of the narrow streets is atrocious, but I have not yet had a tumble. My cough came on a very little in consequence of a draughty railway carriage from Seville, and it was fortunate for me that it did, for I was hesitating about accepting a very flattering invitation to the jubilee in July of a university in America. They wanted me to give three lectures or conferences, said their usual fee was £100 but begged me if I did not think that enough to ask for more, and assured me of various honours. The writer is a man I highly esteem, he is the President, but I am not strong enough; my voice might fail and I should disappoint. But I am sorry to refuse, having some new things to say that appear suitable for the occasion. Anyhow I *have* refused. I must close the letter now for to-morrow morning's post, and send Evelyne's sketch with her best love. The yellow in the sketch seemed to be pure sand. There is of course much more to tell that has interested us greatly, but it is hard to explain briefly. We are both in excellent health. Good-bye, best loves to all.

Ever affectionately, FRANCIS GALTON.

I am glad, dear Emma, of the fairly good account you give of yourself and hope you are now regularly in for spring at last. It is too hot here in the middle of the day for out of doors and we always have taken a long siesta then. It has been a very healthy life. You must not risk measles, though the risk may be very small.

* Marianne Nicholson, wife of Sir Douglas Galton.

HÔTEL WASHINGTON IRVING, GRANADA, SPAIN. Friday, *April* 14, 1899.

(But please address up to midday post of April 20 to Hôtel de Rome, Madrid, Spain. If a little later, I *think* the letters will be forwarded, I will certainly *tell* them to do so.)

DEAREST EMMA, Your letter of April 8th welcomed us on arriving here last night. We have continued to have great variety of interest and pleasure in the journey and are both quite well and happy. At Tangiers we stayed five days and made several acquaintances, among them an English lady whom I had long greatly desired to see, the widow of the late Sheriff of Wazan. My friend, Dr Spence Watson of Newcastle, wrote a book about her, long ago. She was a handsome girl (? a governess) some 18 years ago and the Sheriff of Wazan, who is a sort of rival Emperor of Morocco and of most holy Mussulman origin, but who affected European ways, met her. They mutually fell in love and married, she going to Wazan, continuing Christian and wearing European dress, but of course much shut up, and he remaining the religious and temporal head of a large and fanatical community there. She did her part with great tact and got on excellently. At length her husband died at Tangiers, and left her with two sons and an adequate property. She is now a plain, sensible, rather brusque but very kind, middle-aged and fattish woman. We quickly became great friends and she told us any amount of her experiences. The people kiss her hand and her *shoulder* which is the correct homage from an inferior, and she showed us the house and room where the Sheriff died and which her eldest son, for whom she has just found a correct Mussulwoman to marry, is to occupy. The trousseau box was gorgeous to look at. All this was quite a feature in our stay. On Monday we sailed to Spain again, opposite to Gibraltar, and went in seven hours to Malaga, where there are wonderfully beautiful gardens to be visited, all sorts of tropical trees and clumps of bamboos, but I thought them as muggy as they are beautiful. Yesterday, ten hours railway brought us here, to stay for three or four days. Then we go to Cordova where (at the Hôtel Suisse) I may possibly find some letters. After that to Madrid till the 25th, then to the South of France, then to Hôtel de la Poste, Clermont-Ferrand, Puy de Dôme, France, which I hope to reach about May 3, and home by about May 7, or a few days later. Eva is a capital companion, always cheerful and punctual and interested; moreover she always sees the good side of things and of persons. Eschbach continues to be perfect. We are idling this morning, as I have many letters to write, and the weather is a little dull and unsuitable to give an excellent first impression of the Alhambra, to which I am now close by. You will have missed Bessy during this week; give her my best love, of course. I am so glad the bicycle tour was a success. What a scandal it is about the Warwick and the Beaufort properties, and to think too of the Stoneleigh pictures! I gave your letter to Eva to read, so she knows of your messages and will write. I am very glad that Guy has a free passage to England and another chance for his career as a soldier. Amy Johnson's is a sad case. I trust she will be guided by her lawyer's opinion (Wm. Freshfield) before going to law. She told me the whole matter. What fun about Lady Harberton! I hope *Punch* will make something, good-naturedly, of it. About Lady Stanley, she was a kind friend. Louisa and I stayed some days with her, near Holyhead. Your "ups" will I hope increase and the "downs" diminish as the weather gets warmer. It was like midsummer in Malaga, but this place is 2000 feet high and cooler. Best loves, ever affectionately, FRANCIS GALTON.

[Enclosed with previous letter.] HÔTEL WASHINGTON IRVING, GRANADA.

MY DEAR AUNT EMMA, Your letters are such a pleasure, Uncle Frank gave me the one of last night to read. I don't suppose he ever mentions his cough, so I will tell you about it. It has never actually gone yet, but is much better and he looks very well and is tremendously energetic, the Spaniards all ask me his age, and won't believe it when I tell them; you should see his complexion when he is on the sea, it is splendid, just like Sophy's when she is very well. He is really a perfect person to travel with, because he never fusses or gets impatient *or* grumbles if we are kept waiting ever so long for food or luggage! I hear from Sophy as usual with accounts of you; a letter from her last night tells me you were "in your tea gown and very delightful" when you last met. Poor Mr Lloyd, I am so sorry he is still so ill. I wrote to Gwen from Cadiz, I felt so very much for her and was truly fond of Mrs Lloyd. I am so happy out here, I love the Spaniards, they are so kind and polite to us, but all the same they are poor creatures and not a bit strong-minded or intellectual, but so picturesque. We hardly ever see any English, if we do they are men and rather second-rate. Plenty of fat overfed Frenchmen! Very much love to you and Aunt Wheler, Your ever affectionate niece, EVELYNE.

Uncle Frank was a saint over my fire, wasn't it dreadful! I am ashamed of myself. I did not burn Aunt Wheler's mantilla, but my evening bodice and night-dress and dressing-gown.

Extracts from Letters of his Sisters to Francis Galton.

5, BERTIE TERRACE, LEAMINGTON SPA. Friday, *April 14th*, 1899.

Erasmus goes to-day to Ryde, and will stay there till June—He says—Vessels without sails to *steady* them, must roll—In his day, a three-decker scudding in a gale of wind, having so much top hamper, caused by three tiers of guns, took he believes nearly 30 seconds in recovering herself—I wrote him word, what you said, about the Packet rolling*....... E. S. GALTON.

Tell Eva that Gussie went with Sophy Bree† to see the Twins. It appears they can't speak English, only Hindustani. Margot sat on the rug listening to the conversation and hearing Gussie say "Sophy Bree and the Archdeacon..." Margot with a face full of fun mimicked her, saying "Soapy Bree and Archchicken" to their great amusement....... E. A. WHELER.

Address up to April 29 inclusive, to Hôtel de la Poste, Clermont-Ferrand, Puy de Dôme, France.

MADRID. *April* 21, 1899.

DEAREST EMMA AND BESSY, Your letters of Friday 14th were here on our arrival last night, which had been delayed two days by our taking Toledo by the way, instead of doing it, as was intended, during our stay at Madrid. Nothing can have been more successful than our tour thus far; perfectly healthful, full of interest, while Eva is a model companion with abundant artistic pursuits of her own; so, on the few dull days, I take to my arithmetical figures, and she to drawing human figures, and we are both happy. In an hour or two, we go to the grand picture gallery which is the last *great* sight left to us in Spain. It is grievous to come north already. The glorious vegetation of S. Spain is now left quite behind, and Madrid has a northern and Parisian look. But all good things must finish, and so must this long-looked-for journey. I gave both your letters to Eva to read; it was she (not I) who drew for you the sketch of Cape Trafalgar. What a budget of news you send.

* * * * * * * * * * * *

I chanced to see Lady Frere's death‡ in one of the few English papers that I have lately come across. It was very suitable that she should be buried in St Paul's. I am very glad that Darwin§ seems distinctly better. The coming summer will bring pleasure to you all. Eva asks me to say how interested she is to learn that the "bat" pattern, which Lucy is working, comes out well. If a *bat* is a symbol of sleep, a *mosquito* should be one of wakefulness. We have not however been much teased by them. There are none here, not even mosquito curtains, nor at Toledo, which has the repute of being the centre of Spain. We shall stay some four days here in Madrid (Hôtel de Rome), then a hateful railway journey of some 18 hours to Barcelona, after which all will be straightforward. There are two ancient feudal towns in France that I have always longed to see and which are on our way to Clermont Ferrand, viz. Carcassonne and Aigues-Mortes. Very amusing about the "twins and Archchicken."

With best loves, ever affectionately, FRANCIS GALTON.

So glad that Bessy enjoyed the outing.

Address up to May 1 inclusive to Hôtel de la Poste, Clermont-Ferrand, Puy de Dôme, France. I propose leaving Clermont-Ferrand early on May 5th and to be home on or about May 7th (Sunday).

MADRID, on the *point* of starting. *April* 26, 1899.

DEAREST EMMA, We have *done* Madrid and leave in three hours for Barcelona, thence by Carcassonne, etc., to Nîmes and to Clermont-Ferrand. Nothing could have been more successful

* See our p. 507 above.

† See the first footnote p. 494 above. Gussie, Augusta B. Stewart, Herman Galton's second wife. ‡ Wife of Sir Bartle Frere. § Galton's brother.

than our journey. Eva's many good points as a companion have made it very pleasant throughout, and Eschbach takes off all trouble. He trots us about and arranges everything and I believe would be prepared to wheel us like two babies in a double perambulator. We have been at least four times to the picture gallery, mainly to see those by Velasquez. Moreover we have picked up pleasant acquaintances, and a half-English lady, who is married to a high Spanish official, to whom I had a letter of introduction, has been most friendly. We gave her a drive in a smart carriage yesterday into the Queen's private park and she comes here to tea at 4 to-day; we start for the railroad at 5. I have a little really good English tea, which we learnt that she prized. She knows numbers of English persons whom I also know, and we much enjoy each other's company and talks. Moreover she tells Eva about art matters and parasols, boots, etc. The finances of the Spanish upper classes are seemingly greatly reduced through the war. An English sovereign changes for 30 Spanish francs (pesetas) instead of the customary 25, so the purchasing power of their incomes is reduced one-sixth by that cause alone. We failed to see either the Queen or the King, but have *done* the palaces and all very thoroughly. The Premier had had a long call at our friend's (Señora G. de Riano), just before we arrived to take her for the drive; she was however reticent on Spanish politics. We have a long railway journey before us; the train starts at 6 p.m. and does not reach Barcelona until 11 a.m., where we shall have dipped down from the highlands of Madrid, above 2000 feet, to the level of the sea and to mosquitoes, of which we have not had one specimen since leaving S. Spain. Of course I have engaged sleeping berths. Barcelona is said to be a beautiful place. In the bull fight here, that I saw, one of the six bulls leaped over the barrier twice, among the people behind it. Also two of the bull fighters were knocked over and one of them hauled himself clear of danger by laying hold of the animal's tail and coming out between its hind legs. It was a terrible looking business, but neither were really hurt and both did some very plucky feats after a little rest. Two of the horses were lifted wholly in the air with their riders, *all* four legs being in each case off the ground at the same time. A bull when he has been tired is not so quick as the *quickest* of the men, who will let him rush at them without any red cloak or other thing to distract his attention, but he seemed to me quicker than most of the men. Many bulls jump and bound in the air like buck rabbits. It is a very strange scene altogether, and certainly a fascinating one. I have ever so much to tell, but it would be tedious to you to hear details about places you do not know. How I wish you had health and strength still to enjoy travel. Eva begs me to send her very best love with mine to you and Bessy. I trust that Darwin's betterment continues. What a pleasant outing Grace* seems to have had.

Ever very affectionately, FRANCIS GALTON.

42, RUTLAND GATE, S.W. *May* 13, 1899.

DEAREST EMMA AND BESSY, It was so very pleasant getting your letters when I arrived here on Monday evening, but I wish, Emma, that your account of yourself was better. I have waited writing, wanting to propose a day to come to Leamington, but things do so press and I cannot even yet get through arrears. Besides, Eva goes home with her brother, Walter†, on Saturday. I want her father‡ to lend her to me a good deal, and she wants to come. I wrote, and so did she, a week or more ago to Sophy about it. Of course they are short handed at Ettington, and it will be difficult for her father to spare her, but if she could make my house a good deal her home and be with me again when abroad, it would help me a great deal. The people I have talked to, insist that I ought to spend future winters in sunny lands; that my throat and cough are well-known ailments of advanced life, and that there is no option but to go.

Of course, I shall inquire further, but this prospect has to be faced, so I have arranged gradually to drop my only two scientific ties to London, and to keep myself free to go next winter. Then again, of the brief six months between now and then, I may be ordered to Royat to give the throat more strength, for though all regular cough has long since gone, I feel the

* Tertius Galton's wife.

† The Rev. Walter Bree Hesketh Biggs, brother of Eva Biggs and of Sophy Bree, and Vicar of Ettington. See our Vol. I, Pedigree Plate A in pocket.

‡ The Rev. George Hesketh Biggs.

tendency is still there, and I might have a bad attack of it if I got a cold. One must submit. Forgive this long story. I am quite well *now* and full of engagements. We go to-day to lunch with Lionel Tollemache at the Crystal Palace Hotel. Next Sunday I go to Mrs Simpson (née Senior) who now chiefly lives near Guildford (I think you know her). Then there is to be a grand affair of three days, beginning on May 31, at Cambridge—"the Stokes' Jubilee." Then comes a big dinner in the City, at which the Duke of Northumberland presides. It is the Centenary of the Royal Institution and is given, I fancy, by the Secretary, Sir F. Bramwell.

Eva is a capital companion and I shall miss her much. She is exceptionally good-tempered, prompt, and inclined to see the best side of men and things, and she takes her part well in entertaining. Mr Henry, the Chief Inspector of Police in India, dined with me on Tuesday. He uses finger-prints in India (all India and Burmah), exclusively as a means of finding out whether prisoners have been convicted before, and he has got a law passed in India to allow the evidence of experts on finger-prints to be accepted in Courts of Law. He will read a paper at the British Association (which meets at Dover on Sept. 16) upon it. Hubert Galton's brother-in-law, H. Clifford of the Malay Peninsula, is in town. They two, and their wives, come to dinner on Tuesday. My news is much scattered; many small things difficult to bring into one. Lady Galton has gone to Himbleton for a fortnight or so, but will return in a week. Mrs Robb is as gay as ever. Eva and I went to pour out heart-fulls of gratitude for her useful introductions. I was sorry to miss Grace. I look forward to a Monday morning letter. I trust Darwin continues as well as he was when you wrote.

Best loves, ever affectionately, FRANCIS GALTON.

Address to: 42, RUTLAND GATE, S.W. *June* 3, 1899.

DEAR EVA, The grand doings are just over here at Cambridge. I talked to Miss Pertz (the artist) who asked much after you. Previously, in London, I happened to meet your friend Miss Julia Young, who did the same. I asked her to come and see you when you are with me. You must arrange to see both. The amusing thing was discovering a man whose face I knew and who kept looking at me. He was a chance acquaintance at Castellammare when travelling last year with Frank Butler, and he turned out to be Herkomer. I talked to him about you. We sat near together at dinner last night, and I asked him if there was any truth in the tale current about him and the posthumous portrait of the Hungarian baron*. He exploded with negatives, and I asked him to tell the story, which he did with admirable emphasis. His explosive denials had attracted attention some way up and down the table, and his tale excited roars of laughter. You recollect the story? He said it was told of many painters, especially, as I understood, of Bowles (am I right in the name?). Lowes Dickinson, the old portrait painter, was next to me and three good pictures of his (portraits) were hanging against the wall of the great Hall of Trinity just in front. The ceremonies and the swards and the trees and the red and all sorts of bright-coloured robes and the niceness of the people have been charming.

I shall be so glad when you come. L. Tollemache comes to town for an afternoon and holds a party on June 12, to which you must go with me.

Affectionately yours, FRANCIS GALTON.

M. de Falbe goes to Royat at about the same time as ourselves. Leonard and Horace Darwin go this month. We shall hardly overlap them.

42, RUTLAND GATE, S.W. *September* 9, 1899.

DEAR PROFESSOR K. PEARSON, I have been back three weeks, and on my road northwards saw Weldon at Oxford, and heard of a hitch in the way of granting Miss Lee the doctor's degree. A few days later a batch of papers reached me from the Registrar. The Joint Report of

* A Hungarian baron asked an artist to paint a portrait of his deceased father, sending him photographs and verbal descriptions. When the picture was completed, the son came to the studio to see it; and, looking for a time very sad and silent, said: "My poor dear Father, how you have changed." The story, perhaps, is of small humour in print, but it was otherwise, when Galton told it with the proper emphasis.

Sir W. Turner and myself had been referred to the original examiners and the whole matter on receiving their report was discussed by the Senate, who sent all the material to myself and I presume to Sir W. Turner also, asking certain questions. (They don't want to hear in reply before October.) I thereupon drafted what I had to say and, on returning to town last night (on the way to Dover), posted it to Sir W. Turner. Half an hour after this was done, your letter arrived!!

About the reaction time idea*; I send the only account on which I can lay my hand of the pendulum apparatus that I used regularly at South Kensington, and which Groves of 79 (?), Bolsover St made for me. We called it, from its shape, the "A" machine. The jar of sudden stoppage is there prevented by nipping a thread kept parallel to the rod of the pendulum by an elastic band. For your case, I should propose a heavy *frame* for a compound pendulum. The working part being threads with attached weights, whose periods of oscillation are a little shorter than that of the framework, so that for all the useful part of the oscillating they should never leave the frame. The frame should *retard* them. The nipping would be either by parallel-rule fashion, one pair of them for each string pushed separately, one pusher to each person, or by a vertical arrangement on some simple double-lever, pianoforte-key, plan. I find it most difficult to draw what I mean intelligibly, but it appears to work out quite simply and to require no skilled workmanship. You will know the formula for graduations. I have forgotten all about it, except that I got hold of some useful tables of Elliptic Functions to calculate them by. Please let me have my printed paper back as I have hardly any copies left.

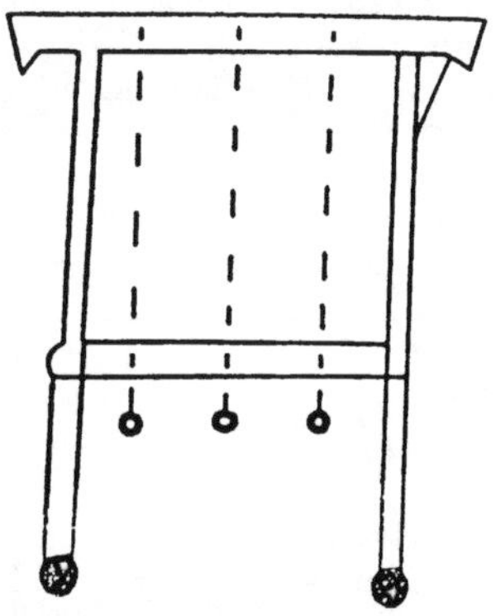

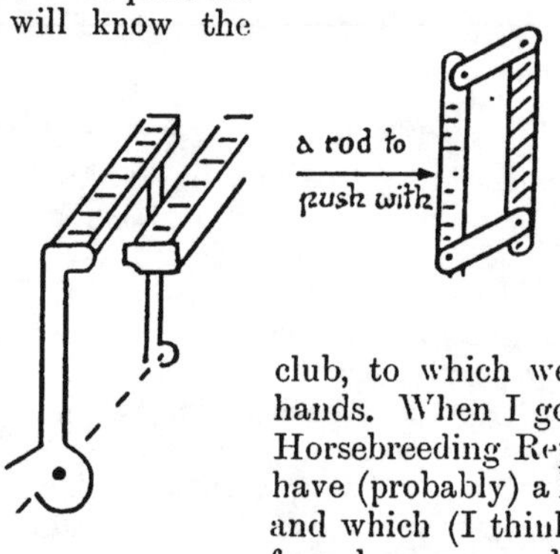

I am writing at a strange club, to which we are handed over while the Athenaeum is in the builders' hands. When I go home I will send you a copy of the Royal Commission on Horsebreeding Report, on which I have written a Report for Dover. Also I have (probably) a little probability paper there, very simple but rather pretty and which (I think) may be practically useful. It was only sent in three or four days ago, and may be crowded out.

I heartily hope you are strong and well again. I have been first for three weeks at Royat in France and then for two weeks in Switzerland, which were marvellously health giving.

Yours very sincerely, FRANCIS GALTON.

42, RUTLAND GATE, S.W. *October* 22, 1899.

DEAR PROFESSOR KARL PEARSON, I cannot suggest anything useful in respect to your paper (which I return), though if you were about to write afresh I should have been inclined to wish the "*Logic* of Chance" could be more developed. What a difficult subject it is to treat otherwise than technically! As to the forthcoming lecture at Leeds, let me suggest a diagram† such as the enclosed (see *both* sides of the paper). It would take people from abstractions down to realities. Also the topic of "nearness of relationship" would interest everybody. To show it off, the string might end in a little longish bag or bucket, into which the tip of the pointer could be slipped. You could then work the string high above your head and all the audience would see it. You will have a very intelligent audience at Leeds, judging from what I saw there some time ago. Very sincerely, FRANCIS GALTON.

* See our Vol. II, pp. 219–220. We have one of Galton's old Reaction Time pendula in the Galton Laboratory, which I purchased since his death from an instrument maker.

† A very rough model of a genometer: see Vol. IIIA, p. 30 and Plate I.

42, Rutland Gate, S.W. *October* 27, 1899.

Dear Professor K. Pearson, Sir H. Roscoe told me last night that Miss Alice Lee had got her degree. The mathematicians were however troublesome.

In your paper the wording of the most interesting experiment with the poppy capsules seems to me obscure. I am not sure that I even now understand it correctly. I have pencilled "obscure" on p. 33. Would it not be well to use some totally different word for the phrase *net fertility*? The word "zygote," though a direct derivative from the strict sense of con*jug*ation, seems to me unhappy. I have been seeking for occasion to protest against its use by Sedgwick in his British Association address. To speak like St Athanasius might have done, a *yoke* divides the persons and does *not* confound their substance; it applies to the stage when the spermatozoon approaches the ovum that is pouting to receive it, but not to the stage in which the nuclei of the two have become fused together, and which is that which it is desired to express. I have a parental weakness for my old word "stirp."

Enclosed I send a copy of my little British Association paper, just received, which may amuse you. Very faithfully yours, Francis Galton.

Hôtel Karnak, Luxor, Egypt. *December* 15, 1899.

Dearest Emma and Bessy, Your letters have come like the wind and have just reached me. I sent a provisional post-card yesterday and now send a proper letter. Particulars of Lord Methuen's serious repulse and heavy losses have just come here, not names of officers, only the numbers of them. He *seems* to have been out-generalled, and in other battles also the Boers *seem* to have shown more generalship than we have done. The Army is doing its best and we can't expect more. It is very very sad; inadequate intelligence of what the power of the Boers really was, and much else. May this terrible experience lead to good. I am glad that Lord Kelvin wrote his letter to the *Times*. It exactly allots due share to all concerned and emphasises what had already been expressed elsewhere. I am so sorry that Leonard Darwin failed in getting into the London County Council. Lucy* must be very pleased at her prize and commendations. How unlucky both she and you too, Bessy, have been with colds. As to the prize-winning cat, on this the third occasion congratulations are effete, so I send a reminder of the serious aspect of cat life. Cats must die. When they died in Egypt, at all events at Denderah, they were mummied with reverence; so were dogs. I was at Denderah (D on map) three days ago, and there picked up the mummied leg of a dog, but it might have been that of a cat, and cut off a scrap of the mummy cloth, which I enclose. It might be put between two bits of glass gummed round the edges. The map and pictures, which I enclose, will explain. Our vessel, the "Mayflower," is very like the "Puritan," there represented. It is comfortable having a big vessel with plenty of attendance all to oneself.

Some of the people, indeed most of them, are nice or fairly nice. To-day we had an excursion of seven hours including about 14 miles of donkey ride. I was lucky in beast and in saddle, and enjoyed it as much as any horse ride that I can recollect. The wonders are just unspeakable. All I can venture to say in addition to guide books is that the clearings of the very few last years have added immensely to what was to be seen before, especially to the many bright-coloured wall paintings and hieroglyphics. The unearthed bases of many columns have made them much more stately.

The only drawback here is that we are aloof from the natives. In a dahabieh one lived among them. On the other hand the convenience of river steaming is great. We start for Assouan to-morrow and get there in two days. Then wait four days there for a little steamer that plies between the 1st and 2nd cataracts, and ultimately return here to Luxor on Jan. 1. Then we go to Petrie for a week or so, and then return to Luxor for a stay of at least a week, probably more. I have made friends with a geographical Pasha, who promises to introduce me to people when we return to Cairo. You shall of course hear from time to time. It is rapidly growing dark, so I must stop. With all loves, ever affectionately, Francis Galton.

* Galton's niece, his sister Bessy's daughter, Mrs Studdy.

HÔTEL KARNAK, LUXOR, EGYPT. *January* 14, 1900.

Address now to Hôtel Angleterre, Cairo, Egypt.

DEAREST EMMA AND BESSY, We returned late last night from our most interesting stay of a week with Mr and Mrs Flinders Petrie. We had each a room with mud walls, nine feet long, seven feet wide and eight feet high, and a bedstead and empty packing cases for furniture. There was no regular door, but a mat, hung in front of the doorway, kept out the prowling dogs. It was on the desert sand, 150 yards from the palm trees, etc., and the floor of the hut was made by that sand. Every one had to throw away their own slops. A well was dug close by, to supply water. Besides our hosts there were three Oxford men who had grants for making researches, and a Miss Johnson (a lady doctor), the image of Miss Cobb in her early days, I think you saw her then, stalwart, merry and capable in every way. We dined on a table made of three rough deal boards and we ate tinned meats and jams, with bread made in the native way. No milk, butter, wine or spirits, nor potatoes nor onions. But every one seemed in the pink of health, and was at full work from day-break to at least 9 p.m. The quantity and variety of work were quite remarkable; the diggers had to be superintended, there were 130 of them in three parties; everything found was assessed and paid for to the workmen; it was drawn, catalogued and often photographed; bits had to be pieced together and every day some interesting "finds" took place. We had a very pleasant and instructive time of it, but life was very rough. No one wore stockings in the day time, on account of the sand. We were 2½ to 2¾ hours from the Railway Station (7–8 miles). All our luggage went on one donkey, who carried the donkey boy as well. Finally we started yesterday from the station in bright moonlight at 7.26, and reached Luxor (104 miles off) at 10.40, dusted through skin deep. But a *good* washing last night, repeated copiously this morning, has made us normally clean. Far more occurred than I can put down here. It has been to both of us one of the most interesting experiences of our lives. I am more than ever taken with admiration of Petrie, and his wife is as nice as possible. The costumes were astonishing at first, but soon the eye became accustomed to them. The Marquess of Northampton, who is cruising on the Nile with his hopelessly sick wife (Lady Ashburton's daughter), rode over on a donkey to see Petrie for a couple of hours, and there was much good talk. I had met him when staying Saturday to Monday with Sir John Lubbock (who I am glad to see is to be a peer), and have arranged to call when his boat reaches Cairo. He knows Egypt well.

Mat before door.

As regards future plans, we have the choice of two steamers to return to Cairo, they leave here Jan. 26 and Feb. 9. You will receive this letter about the end of January and I shall get your reply somewhere about Feb. 15 at Cairo, at the Hôtel Angleterre, which will (I think) be about the time of arrival of the Feb. 9 steamer from here.

The Nile is so low and shrinking so fast, that it will possibly stop the running of the steamers soon. It has shrunken in width, since we left a week ago, to about that of the Thames at Westminster, if I judge rightly; during the inundation it must be quite seven miles broad. Such a difference! There are very few English tourists on account of this terrible war, very few Americans and hardly any of other nations. The church to-day was not ¼ full. Doubtless more will come later. There were only four persons at lunch at this hotel, which has table-room for 60. We are sitting out of doors in its very pleasant garden, half orderly, half disorderly. Eva is painting studies of the changes in colour of the only remaining chameleon*. It was the biggest, the tamest and the most interesting. The other two escaped at different times when with the Petries, and were lost. I told you in my last letter that we had met the famous African hunter, Mr Barber. We talked then a good deal about Seton-Karr, who is his equal in that way, but whose adventures are in other lands. Oddly enough, I met Seton-Karr to-day, who had returned two days since from Omdurman and is on his way back by Cook's steamer. I have just waved a parting adieu to him. So much for ourselves.

Your nice letters dated Dec. 29 reached me when at the Petries (I think, but am confused). Mr Forsyth's death is another break with old days, and so in another way is that of Sir J. Lennard†. How well I recollect him at old Mr Hallam's. So the Cameron Galtons have left

* This chameleon was brought home, but died soon after. Its skin and some eggs it laid are in the Galton Laboratory.

† He married Miss Julia Hallam referred to in our Vol. I, pp. 179–80.

Paris. I am not surprised. I wish my house could be of some use to them, but until Chumley* has been operated on and is cured, it would be impossible to offer it. I will certainly call on the Miss Horners when I return. I trust Milly and her household have got over their influenzas and that Guy may be finding some opening. I will write to her. In the scanty newspapers I have seen there is no ill news of Bob. It is very good and plucky of M.L.'s brother to go out with the yeomanry. I feel very painfully the contrast between my enjoying myself lazily in this glorious climate and the sufferings of our countrymen at the Cape, but cannot think of anything I can *now* do usefully, except get thoroughly well. I am very glad that Darwin's cough is not worsened by the horrid weather you are having in England. Ours is sunshine from sunrise to sundown, but it can be *bitterly cold* on a still and cloudless night. It was so on three occasions at the Petries. I heaped everything on my bed with Eva's assistance, and next morning made a list of what I used. It was necessary to sleep between the blankets because the sheets struck cold; so a sheet was placed on the outside, tucked in at its top round the blankets to keep the fluff off. This was the section of myself lying in bed taken at my feet:

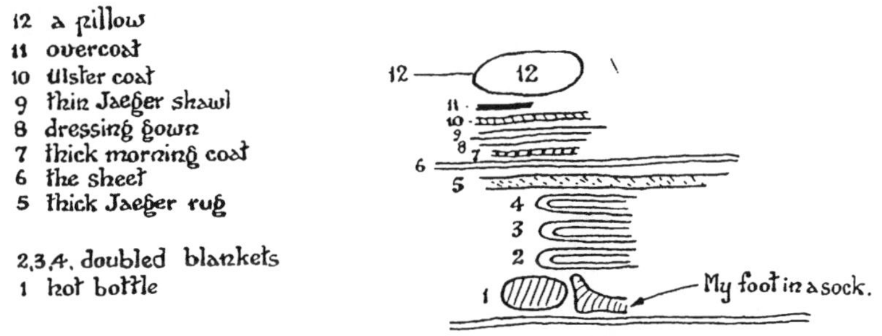

Besides this I slept in thick socks, in a jersey, drawers and in complete pyjama suit. Thus I felt warm, but by no means stuffy. The air is so nimble that it gets through everything woollen. Here, as in South Africa, skins and furs ought to be the best. I love your letters.

Ever affectionately, FRANCIS GALTON.

†We have had a very nice queer time in the desert, very healthy! Uncle Frank just a little pleased to give up teetotal ways and have a glass of wine! My best love to you both. E. B.

KARNAK HOTEL, LUXOR, EGYPT. *January* 22, 1900.

But address to Hôtel Angleterre, Cairo, Egypt, please.

DEAREST EMMA AND BESSY, Your letters of the 13th arrived, as I had hoped they would, to-day. We are all right, and have taken a bit of a walk this morning; only four miles, but the roads are *very* dusty and tiring. Donkey riding is the correct thing, but we wanted exercise. You doubtless got my letter (followed by a post-card) a week ago. Nothing particular has happened since. Of the few people here are Professor Macalister the Cambridge anatomist—he has gone to Petrie—Professor Sayce in his large house-boat (he comes every winter to the Nile on account of his chest), and Lord Northampton. Lady N. was at church; carried there and back in a chair by magnificent sailors in gorgeous dresses and sat in it by the door all the time. It is most piteous, having had fortune, beauty, rank, and high spirits and nice children, and now to be hardly alive except in the brain. She is powerless to move and her head is continually agitated by a shaking palsy. Her state is said to be hopeless. Sayce is the great orientalist and has been a thorn in Max Müller's side (who has been long very ill, but is now better). Macalister started on Saturday as we did last week, early in the morning, for Baliâna and Petrie. His wife and daughter were left behind to join him at Baliâna this morning on their return steamer, but a telegram came yesterday to say that the steamer has broken down, so they are at sixes and sevens. There is nothing like an hotel at Baliâna and Petrie's camp is a good seven miles off. How they will meet, I can't guess. A beautifully ornamental tomb, as fresh as if newly painted, has lately been got at here. It is not yet open to the public, the air inside

* A maid of many years' service at Rutland Gate.

† Postscript added by Galton's great-niece, Eva Biggs.

being very bad. Macalister, however, saw it and says it is more gorgeous than any other he has seen. But it is of late date, only as far back as Rehoboam; Abraham's time is thought here to be rather late. The interest now is in the people who lived here *before* the pyramids were built, ending with about 4000 B.C. There are beautiful flint knives of far earlier date, the most beautiful I have ever seen in workmanship and in art. The Nile is very low and is running out fast. One of the people connected with the irrigation told Lord Northampton that he expected it might become a mere chain of pools before the next freshet. Maud Butler returns to-morrow from Assouan, and will stay a fortnight at our hotel, which will be pleasant. I did to-day a somewhat silly thing. They imitate ancient Egyptian things, sometimes very well, at Luxor (mostly to sell as originals), so wanting a small seal I gave them my hieroglyphic to cut on an imitation "scarab" for 4/-. The man proved to be a poor hand and has made for me the enclosed, which is legible but very badly cut. However it will serve its purpose. Both Petrie and an Egyptologist (Dr Lieblein) approved it. I have no right to a cartouche,

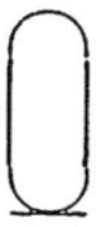

not being a king, but Maud Butler, whose pet name was "Queenie," might use one. We have had no war news to-day. How glad Bob* must be that he was not fatter, else the bullets that went through his clothes might have gone through his body. Nelson's cocked hat was once shot through; had he been a taller man, he would have died long before Trafalgar. I am glad that Gascoigne Trench is going out. He knows the work that is needed, and is still young enough. Guy's* recurrence of Indian fever will make it unlikely that he should be passed as fit for service now. I am glad that you all keep fairly well notwithstanding the wretched weather you have had. Give my love to Darwin and to Erasmus when you see them next. I am very glad that George Darwin receives those family mementoes. He is the best representative of the Darwin family, and had great affection for the Admiral, of whom he saw much at Malta (I think) when flag-lieutenant to the Admiral's ship. Edward Wheler must feel the war fever in his veins, from his brother-in-law's going out so pluckily and from his many neighbours doing the same. I see in the newspapers a quoted chorus of disapprovals of Arthur Balfour's speech, which I myself like very much.

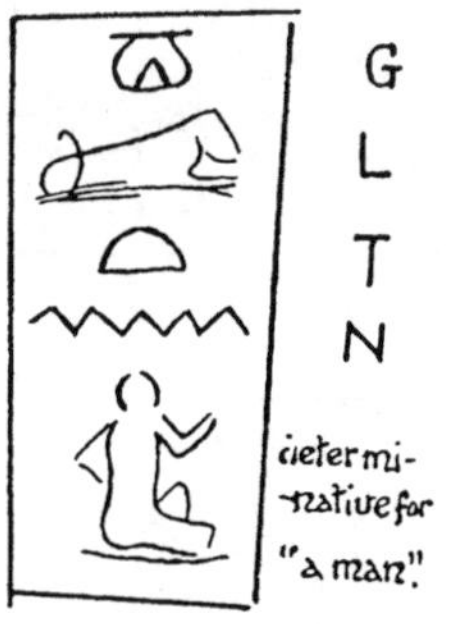

Professor Sayce has just called and taken us off to tea in his boat. It is the largest and broadest on the river, its yard-arm is *134* feet long, so three of them end to end would reach far higher than any English Cathedral; I think Strasburg is only 400 feet high.

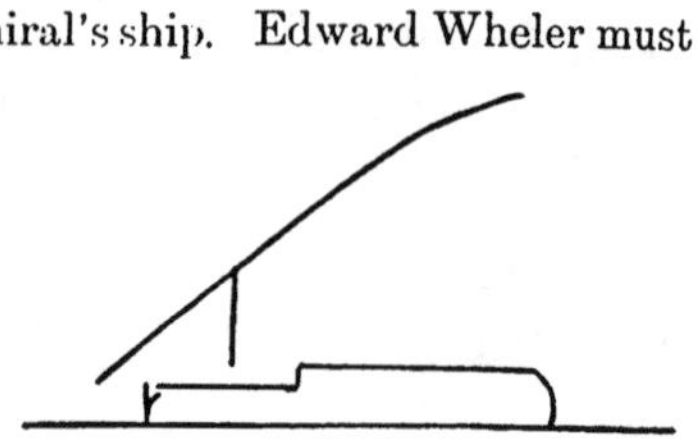
(Professor Sayce's boat.)

Eva sends you a drawing of her only surviving pet, with her best love. It is about $\frac{2}{3}$ scale.

Ever very affectionately, FRANCIS GALTON.

Address still to HÔTEL ANGLETERRE, CAIRO, EGYPT. Sunday, *March* 4, 1900. Posted *March* 5.

DEAREST EMMA AND BESSY, We are still at Helouan (Tewfik Palace Hotel) but the above is our address. The last letter I had from you was dated Feb. 16; it was received Feb. 22, and was answered the same day. We are quite well, but are bothered by the difficulties in the way of simply camping out in the Desert, which I thought had been overcome, but are still going on. According to what an excellent dragoman now assures us, there is always a risk with the Bedouins unless elaborate and costly arrangements are made. We shall hear more from him after his inquiries. There has been something of interest nearly every day since I wrote. On Friday I drove with Professor Schweinfurth in one carriage, and Admiral and Mrs Blomfield in another, across the desert and along valleys for two or three hours. Then we picnicked, botanised and geologised for four hours and then returned, after seeing (1) an ancient barrage, built of stones, in the time of the Early Pharaohs, to dam the water when it ran down the creek, (2) some *true* Jericho roses, of which I send a few (see further on). If you dip them in water they begin to expand, almost instantly, into a true flower. The *false* Jericho rose is the one usually so called, but it is merely a seed-vessel with dry fibres grasping it, and which expands imperfectly and slowly.

* Sons of Galton's niece, Mrs Lethbridge.

On Saturday, the 24th, my short geographical speech* came off, quite successfully. I will use the letter in French received to-day from the Secretary (in evidence) to wrap up the Jericho roses in. Sunday and Monday were days of *heavy* rain. Cairo was flooded and the desert was quite wet. We had tea with a Syrian, by name Makarius, who is a literary man and a printer, both in Arabic and in English, and whose acquaintance I made last autumn at the British Association. He showed me an Arabic periodical that forms a fat annual 8vo volume, and which describes what goes on in the scientific world everywhere. There was a chapter in last year's volume about my latest work (the "Ancestral law," as people call it). We go with him to-night to hear some Arabic music. Tuesday we walked to see some big quarries of white stone, whence files of camels take the stones all day long to the Nile. On Wednesday I had a lunch and a tea party; Maud Butler and her companions came, also Eva's cousins with three children, and Mrs Procter. On Thursday we (Eva and I and a friend) went on donkeys about six miles, to see the wonderful quarries from which the stones were cut, which formed the Pyramids. The stones must have been rafted across the Nile, when flooded. From my window I can see at least seven large Pyramids (including those at Gizeh). I am told that it is possible to count seventeen of them. On Friday Eva and I made a desert expedition by carriage, and then onwards on foot. Yesterday we went for the day to Cairo, to do things, and to-day is Sunday. Schweinfurth and Professor Sayce (whose boat is 2½ miles off) come to lunch with me to-morrow. The weather has now turned hot, with a southerly (sirocco) wind, of which this month of March is sure to have plenty. They call it the *Khamsin wind*.

Those Jericho roses—they will make a letter unsafe, as the post office people may think they are something valuable. So I have enclosed them in a separate packet, which may or may not reach you, and I send the crumpled letter of the Secretary in this—tear it up.

The above was written yesterday. We went in the evening to an Arab concert. The singers were five Syrian Jewesses. The room had a gallery round it with muslin draperies, behind which the native ladies sat. The few European ladies and all the men sat below. Eva was taken up to see the native ladies and says they had very good and pleasant manners and some were very picturesque. They were all powdered on the faces, and the eyes and eyebrows were much painted; not much perfume. Yesterday Mr W. Bearcroft introduced himself. His father was the clergyman at Hadzor†. He is on the engineering staff of the railroad. He had heard that the Cunliffes (Evelyn‡) were on the point of going, perhaps had already started for Cairo.

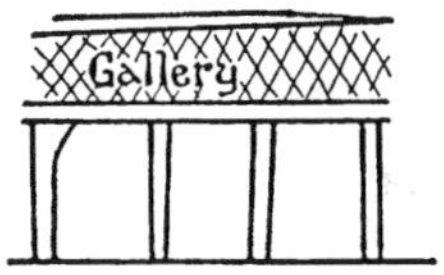

I am anxious for home news of all sorts, for Gifi also is a little later than usual with his letter; so also is Frank Butler. I only know that Chumley *has* been successfully operated on. I hope that Darwin is recovering steadily, and that you, Bessy, have lost your cough at last. Mine is practically gone for present purposes, but I know that bad English weather would soon bring back that peculiar abomination. As for you, dear Emma, you do not often tell me about yourself, so I imagine ups and downs. I hope Erasmus is now quite right. Bob Lethbridge has not apparently been in the late heavy fighting. I wonder how soon the regular fighting will be over, and armed occupation begin. This is only a sort of diary, you must please interpolate many affectionate thoughts in my bald matter-of-fact story. Ever affectionately, FRANCIS GALTON.

The following Postscript from a letter indicates that Galton had by midsummer exchanged the desert for Pall Mall.

THE ATHENAEUM, PALL MALL, S.W. *June* 29, 1900.

P.S. I am enjoying this afternoon at the Club, and my favourite (but unwholesome) afternoon provender is just set down at my elbow, viz. tea and muffins, with a muffineer and a *large* napkin to wipe buttered fingers on.

* See our Vol. III^{A}, pp. 158 and 159.
† See our Vol. I, p. 53 and Plate XXIX.
‡ See p. 506, second footnote, above.

42, RUTLAND GATE, S.W. *October* 29, 1900.

MY DEAR "CHATTELL" EVA, I am delighted that you are now to be altogether transferred to me and to take charge of my household henceforth. You weren't transferred *quite* as a "chattell" (I don't know how many *t*'s or *l*'s there are in the word) as I said in my letter to your father "if she acquiesces...." So you will now have "42, Rutland Gate" at the bottom of your visiting cards. I am very glad we shall meet so soon.

Violet left this morning. It was pleasant having her. She will get a sight of the C.I.V.'s to-day. I have not been sight-seeing. It rained heavily till near two, and the ground in the park must now be sloppy. I forgot to tell you that one of the first persons whom I met after you left me, was James Knowles (the Editor of the *Nineteenth Century* and originally an architect); it was he who built Tennyson's house. He told me much about it, which I will tell you. He was staying some weeks at Hindhead this summer and was curious to learn about the Townshends. Knowles was a great friend of Tennyson, and of many notabilities—rather *Boswell-y* in his disposition.

Guy Lethbridge made his appearance yesterday, looking very nice and gentlemanly. He has been horse-buying on commission, in Ireland. *Tommy and Grizel* is wonderful. I finished it last night, after eleven o'clock. The characters "grow up" quite naturally, so it is an exact sequel to the other book. I took Milly and Amy to see Julius Cæsar (last representation) on Saturday, and learnt immensely. 1. Julius Cæsar is made so egoistic and vain as to be odious to the assassins, or to most of them, and to be insufferably arrogant. So they *hated* him. 2. Cassius is *not* a pale thin student-like man, but vigorous and powerful (which his story of saving Cæsar from drowning justifies). He is a lean, bilious man, full of energy and hatred, and a very d—l as an enemy. Ever affectionately, FRANCIS GALTON.

Poor Walter Butler is at death's door, but his state is not hopeless, quite.

BRITISH MUSEUM, LONDON, W.C. *July* 19, 1901.

DEAR MR GALTON, Here is the result of our experiment. How do you like it? I do not doubt that with more careful preparation one could increase the area of sharpness a little, but probably not very much. The developed neck irresistibly suggests shoulders, and the best way of restoring it to intelligibility is to bend the print backwards into a cylinder. It is curious and pretty the way in which the square pedestal has come out.

Yours very sincerely, ARTHUR H. SMITH.

Photograph enclosed.

As by the aid of a panoramic camera the whole view round a hill top may be photographed on a single plate, so the idea in the above experiment was to take on a single plate a continuous picture all round a statue or bust. The result is shown on the accompanying plate.

HÔTEL DES ANGLAIS, VALESCURE, PRÈS ST RAPHAËL (VAR), FRANCE.
Nov. 28, 1902. (We stay on here for quite a week longer.)

DEAREST MILLY, If you can only let your Knole Lodge and get the pretty Prestbury! I am so glad you are strong again.

I am *quite well* too. The asthma left me more than a week ago and the bronchitis went a little later, so that—pity my sense of *loneliness*, at missing the habitual cough! Even a grumbling farmer could hardly beat that. My room was stuffily carpeted, so notwithstanding the pure outside air I had violent bouts of asthma every night. So I had the carpets taken up, and a large sackful of straw that had been spread beneath them for warmth went with them. I feel sure on reflection that all my worst coughs have been connected with well warmed and stuffily carpeted rooms. So I am about to take strenuous measures at Rutland Gate. The floors of the dining and drawing rooms and of my bedroom are to be parqueted. The very old paper of the drawing room is to be stripped off and the walls painted white, like the staircase, and carpets abolished in favour of rugs. So I hope to be able to spend more months out of the 12 in my own house than hitherto. "Hope springs eternal......"

PLATE LIV

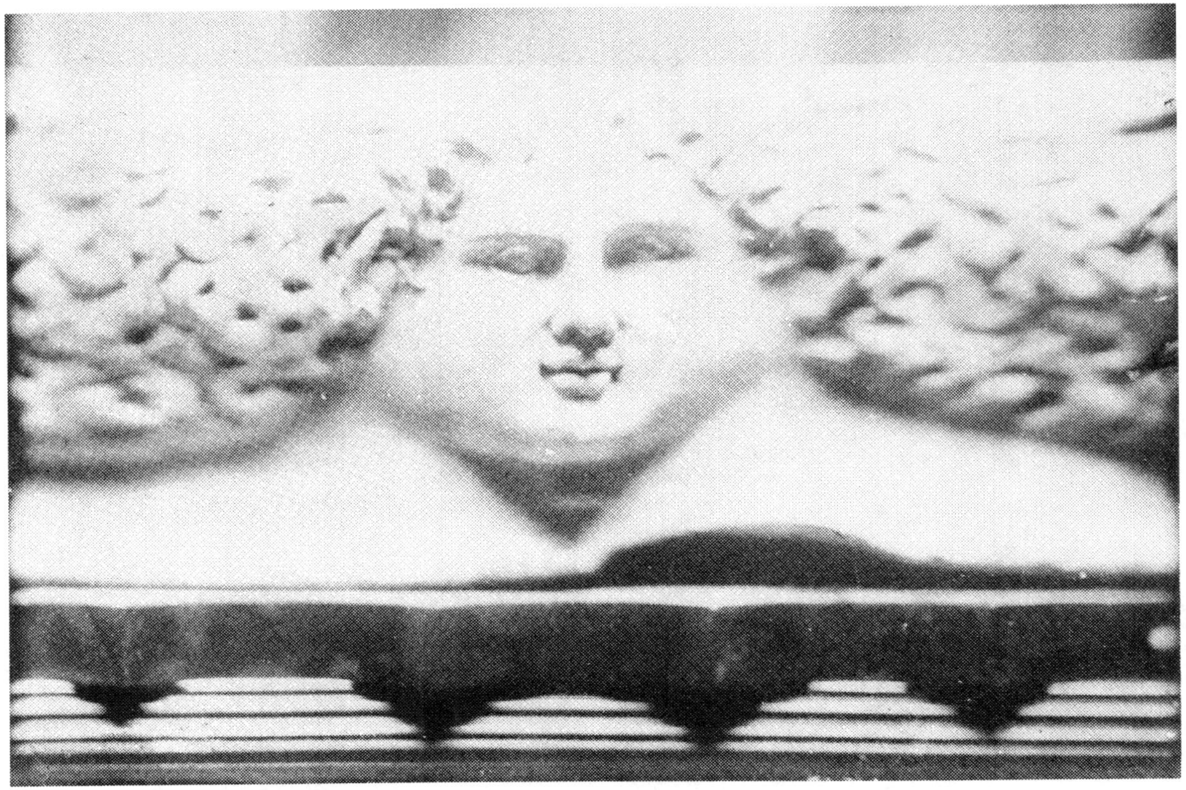

Experiment of Francis Galton and Arthur H. Smith on the exhibition of all aspects of a bust on a single negative. The "all round" photograph of a bust in the British Museum.

Tell Bob * I am sorry not to be in England to welcome him and his wife when they call. It is glorious weather here for the most part, and there are nice people in the villas about, but it is early for the visitors and we are the only two in this big hotel. There have never been more than two others, of which we are glad. Eva does much painting and seems as happy and as well as can be. I have as much work as I can do (which is very little), and am quite happy too, and can accomplish a good four miles walk without fatigue. (Alas, I *have* accomplished a measured *40* miles, but *with* fatigue, in old days.) Hearty thanks for your congratulations. I am particularly pleased with the Hon. Fellowship of Trinity College, Cambridge.

Ever very affectionately, FRANCIS GALTON.

With Eva's love, and both our loves to Amy †.

HÔTEL DE L'EUROPE, ROME. *January* 8, 1903.

DEAREST MILLY, Twelfth day is past, but it is not too late to send hearty New Year wishes. On Twelfth day just 50 years ago I first made the acquaintance of Louisa and of her family party. We were married in the autumn of that year.

The post has this instant brought me tidings from Emma of Darwin's death ‡. It is more of a shock to me than I could have expected, for many happy incidents of early days crowd the memory. His was a complex character, veins of clay and veins of iron and gold. He was loved by many and admired by many—not, as you know, by all. The most pathetic figure in the funeral cortège would be William Yeates, if paralysis enabled him to attend it. Darwin used to have a terror of death and was extremely moved if he heard unexpectedly of the death of any one he knew. Now he is initiated into the secret and has passed the veil. He is well out of suffering and the sense of incapacity with absence of hope for a better bodily condition. If his infant son had lived and grown up healthy in mind and body, how different his life would have been. I am sure that a candid retrospect would judge his to have been an exceptionally useful one. I can't write more on this sad event.

We are most pleasantly situated in Rome and most healthy. Two days ago we had a glorious afternoon on the Palatine among the recently exhumed foundations of the vast palaces of various Cæsars. The overwhelming might and magnitude of ancient Rome struck me more than it has ever done before. I hear that your desired lease is not yet signed, that Frank has gone to Durban, but no news about Guy's suffering nerve. When you write—after Darwin's funeral is over—please tell me what your own family news is, and what seems to be the consensus of opinion about Darwin. Emma will I am sure send me Leamington newspapers. I should think that Eddy would much regret his death. Love to Amy from both of us as well as to yourself.

Affectionately yours, FRANCIS GALTON.

I address to Blenheim, but believe you are with Bob.

GRAND HÔTEL ROYAL DE SIENNE SUR LA LIZZE, SIENA. *April* 8, 1903.

DEAREST MILLY, (En route) Alas, your letter of the 12th, or which Emma forwarded to me on that day, and which I am sure was a particularly nice one, is *lost*, utterly gone astray. Where it now reposes, I have not the slightest idea. We were in two hotels in Ischia and our letters had been addressed to a third, which was not then open, and I suspect that the letter came to grief between those jealous three. And I have been so anxious to hear of you, more especially of late, now that your African sons are probably back. But I am on my way homewards, hoping to be in London on the 20th, *via* Bologna, Milan, Cologne and Brussels. We spend Easter Sunday in Milan. Our tour has been most interesting, with interludes of ailments from sewer gas both in Rome and Naples. Our rooms were high up at both places, and the foul air came up the rain-pipes under our noses. But we are fit now, and look forward to the grand Easter music in Milan Cathedral. It was to have been Cologne, but Eva had to spend four days in bed. The glories of the South are great, when there is sun to show them. I have never seen greater beauty of rock, sea and sky than on this journey. Panoramas from the mountain tops both of Capri and of Ischia. Rows and sails round cliffs and a drive on a marvellously beautiful new road, round the peninsula from Amalfi to Sorrento. Also, we have had some simple life

* A son of Mrs Lethbridge.

† Amy Lethbridge, Galton's great-niece.

‡ Galton's eldest brother.

and pleasant friends; also Vesuvius erupting. It has been an Indian summer to my life. Now it is time to be home to greyer skies. I hear that you sent a sketch of the new house, Prestbury or Edymead. I shall hear much when I return. But I share your anxiety about the boys. I trust all has ended better than you seemed to have feared. The Italian papers are alarmist about strikes everywhere, at Rome and in Holland especially. I suppose we shall get through. The last and only time that I was in Siena before, there was a threat of a universal railway strike, during which, and while reinforcements of soldiers could not be sent, the mob were to sack Milan. But the scheme got wind. Siena was put under martial law while Frank Butler and I were there, and the riots at Milan were quelled, but not without blood. It is curious how soon an army of conscripts feel themselves detached from their countrymen and become ready to fire on them, if ordered by their officers. I have nothing of interest to tell you, but am burning to learn more of *Radium*. What with it, with air-telegraphy and with Röntgen rays we have suddenly become impressed with the magnitude and prevalence of unseen agencies. It will greatly change the view-point of ordinary materialists. As I understand it, if you constructed a suitable carriage for radium, radium could climb a hill. Fancy self-acting locomotives. Expensive though to make. Love to Amy and to you all at your home.

Ever affectionately, FRANCIS GALTON.

Correspondence with F. H. Perry Coste.

42, RUTLAND GATE, S.W. *July* 20, 1903.

DEAR SIR, I have spent $2\frac{1}{4}$ hours in *partly* deciphering your letter, and as I cannot easily spare more time now, have sent it to a typewriter, who will probably make out some phrases that still puzzle me. I was too boastful in saying that I had mastered the art of decipherment; when the typewritten copy reaches me I will answer it in full, but some things I can say now.

I feel thoroughly your difficulties and your zeal in overcoming them, as regards catching the subjects to print from whether fathers or children, and am heartily obliged for your pains. The new Olivers shall be underlined with red in their pedigree. I will study the latter carefully to-morrow. All your pedigree work has thus far seemed very good to me. I fear that I am not likely to find any one at Penzance, who would take trouble for me about the Olivers there. The suggestion as regards Scilly is valuable. I will bear it in mind.

Partly from difficulty of decipherment I fear that I have not rightly caught your question as regards the finger-prints of the "odd" parents. Every family pedigree must have an alien fringe, but the odd parent ceases to be alien *if* he has children, but only then. It would be indeed interesting (if easy) to compare the surnames (as from parish registers) at 50 years interval in some small but conservative place. In those I have seen they change much. You may recollect Doubleday's book, written three-quarters of a century ago, in which he declares that all the families of parishioners, who occupied any notable position, as a rule die out. But the laws of fertility puzzle everybody.

Pray tell me in good time whenever you want forms or schedules. I shall be from home on Wednesday, Thursday and Friday next. The typewritten copy will I hope be ready by my return, when I shall be able to answer further. Very faithfully yours, FRANCIS GALTON.

F. H. PERRY COSTE, ESQ.

42, RUTLAND GATE, S.W. *July* 22, 1903.

MY DEAR SIR, The typewriter (Dickens's grand-daughter) has performed miracles of decipherment and placed your valuable letter on clear record. Will you very kindly correct it where necessary and insert a few omissions. I return your letters for the purpose. Please let me have all back. I shall not have returned home till Friday evening. Let me tell an anecdote. The late Sir George Gabriel Stokes was a member for many years together with myself of the Meteorological Council. We protested against his handwriting which was perhaps half-way as cacographic as your own. One day he informed us with a mysterious air that his writing would henceforth become remarkably improved. And so it was! He bought a typewriter and used it ever afterwards.

In great haste for an early train. Very faithfully yours, FRANCIS GALTON.

F. H. PERRY COSTE, ESQ.

42, Rutland Gate, S.W. *July* 29, 1903.

Dear Sir, The packet of 50 Puckeys has just come, together with the very legible letter for which pray thank Mrs Perry Coste most sincerely on my behalf. I will bear your suggestions about scrutinising pedigrees well in mind. As yet I have not found time to begin a careful revision, etc., of them, but shall be free very soon to do so. I find your notes of relationship, thus far, perfectly clear. Thanks for the hint about the chamois leather dabber. I *dab* with the india-rubber, which I keep scrupulously clean. The small children make beautiful prints when the ink is spread *thinly* and evenly and when the children are submissive. It is a good plan just before pressing the child's finger on the paper to direct its attention to the window, then its curled-up finger relaxes at once, and a good print is taken.

The "odd" persons are acceptable. As I said in my circular (or in a revised re-print of it) I am just now glad of a large collection of unrelated persons in addition to the related ones. They are wanted for the first purpose to which I alluded, of getting a *natural* classification. You are indeed carrying through a big work*; it is most useful to myself. It is impossible not to see evidences of finger print relationships when outlining the patterns prior to a more exact study of them. I have now greater hopes than ever of extracting *much* good out of this inquiry.

Very faithfully yours, Francis Galton.

F. H. Perry Coste, Esq.

42, Rutland Gate, S.W. *August* 21, 1903.

Dear Sir, I had written to the Vicar of Lizard before your second letter arrived, and have had a very courteous reply from him, but he declines for want of leisure. I shall treat Polperro folk as a much intermarried group, and this fact comes out conspicuously in the much greater frequency of arches in their finger-prints, than occurs in the population at large. Their prints have not yet been strictly isolated in my preliminary statistics though they will be. At present, a large infusion of them is sufficient to raise the arch-frequency of a mixed lot.

Enclosed I send more schedules and forms. On Monday we move to a house I have taken for four weeks certain, viz. "Manor House, Peppard Common, Henley-on-Thames." It will be better to address letters there for the present. One reason for my going there is to be in the neighbourhood of Prof. Karl Pearson who also has taken a house there. Moreover Prof. Weldon comes down each week-end, Friday to Monday, so biometric affairs can be discussed and especially some problems connected with these finger-prints before I finally commit myself. The ins and outs of Statistics are as you well know singularly intricate and apt to mislead inquirers. Very faithfully yours, Francis Galton.

F. H. Perry Coste, Esq.

Manor House, Peppard Common, Henley-on-Thames. *August* 28, 1903.

Dear Sir, (1) The large contribution of Curtises, (2) the schedules, and now (3) your letter of yesterday, have all reached me. I am working as hard as I can at my material with a niece to help, and am gradually getting it into order. As yet I do not see my way to discuss more distant relations, *as such*, than first cousins, but propose to deal with batches of inter-related persons as wholes. It is most remarkable to note the frequency with which particular patterns affect such groups, and I shall before long get this into a numerical form. But there is much to be done before I can even attempt this. Thank you much for your offer of help in getting such members of the odd parents' fraternities as it may prove desirable to have. I will bear this in mind. You also say that you could get "200 or 300 more finger-prints" in order practically to exhaust the population of Polperro. Of course I should be *most grateful* for such a large contribution, but I am diffident in asking for so much. The utility would be many-sided. They would be welcome merely as prints, to establish the first of the objects of the inquiry. They would perhaps include *something* of the native Polperro types, and they would serve in some degree as a "control" series. Should you really brace yourself up to this great additional labour I for my part should be greatly obliged and should value the finger-prints highly. Please in that case take particular care *never to use too much ink*. Your fault as a "finger-printer" is blottiness. This

* Mr Perry Coste had taken the finger-prints of nearly the whole population of Polperro, a Cornish fishing village with much inbreeding.

cannot occur if the ink be spread over the tin box with thinness. Whenever too much ink has been accidentally squeezed out, the superfluity should be removed by dabbing the over-inked rubber on waste paper. The best of all your impressions are of children. One of a child of 2 years and 8 months is as good as can be. I will give your message to Prof. Karl Pearson. I feel inclined to let Dr Appleton senior stand over until I have got some results and know more exactly than I now do what I shall want ultimately. This place is exceedingly grateful, pure air, breezy commons, and geographical height.

Very faithfully, still with many thanks to your Wife, FRANCIS GALTON.

F. H. PERRY COSTE, ESQ.

MANOR HOUSE, PEPPARD COMMON, HENLEY-ON-THAMES. *September* 1, 1903.

DEAR SIR, First, accept my congratulations on the domestic event and best wishes. Enclosed is a copy of the Oliver-Toms' fraternity. If you want the part that I have omitted pray tell me. Do not think of putting yourself out now, by making and sending the big series of which you so kindly spoke. I propose to make a temporary halting place where I am now, and to work up the material thoroughly so far. I have quite enough for provisional results, viz. 865 sets. Let me reiterate how strongly I feel my obligation to you. It is a grand collection that you have made for me and, whether for individual lines or as a group of nearly related persons, it will give me abundance of work and I will do my best to do justice to the large material so laboriously obtained by you for me. I can assure you that I realise the difficulty of printing from the worn fingers of perhaps unwilling and often stupid fisherfolk, but what prompted my remarks was chiefly that in some sets of prints some are good, others blotted *in parts*, and others again densely blotted all over. One always tries to work up to a high ideal! Small has sent me three full books taken during his holidays in North Cornwall. They are perfectly beautiful, but they are taken from a non-labouring class. I have not yet noticed the occurrence of any of your Polperro names in his lists, but have yet to examine them thoroughly. I hope you will finger-print the baby as soon as its mamma permits. I have an enthusiastic correspondent in America who began to finger-print her own baby six days after it was born, and did so on every day of that week. From the many dabs I was able to select a complete and very fair set, which I enlarged and have compared at intervals with prints subsequently taken from the same child who is now 5–6 years old. There is no change anywhere. It will become a classical case. Very faithfully yours, FRANCIS GALTON.

F. H. PERRY COSTE, ESQ.

Letter of Miss Emma Galton to her brother, Francis Galton.

5, BERTIE TERRACE, LEAMINGTON SPA. Thursday Evening, *Dec.* 17, 1903.

DEAREST FRANCIS, We long to hear that you are in Sicily, for the weather is not genial, but I see by the Papers the floods have been very bad in Rome and in Venice, but I do hope you and Eva will have a good journey. Your account of Mr Herbert Spencer's cremation has interested us much, and Erasmus has just been here and read it with much interest, and I have shown your letter to some of my callers, and all were so interested about the ceremony. Cameron* called this morning—he and his wife Lucy came on a visit to Grace Moilliet†. Cameron was at the funeral at Hadzor yesterday (Wed.)—a very long affair, Incense and Bells and many Priests—Charles Galton officiated—Hubert, Howard and his wife Maud, were of the party who attended. May Barclay's‡ maid (Mrs Cowie) has written to Temple to say how *very kind* Hubert and Howard§ have been to them; May had left £100 to Mrs Cowie who had been 29 years with her, and £50 to Mrs Beal, the cook, who had been about 29 years with May, but Hubert and Howard have promised Cowie an annuity of £50, and the cook an

* Ewen Cameron Galton, son of Robert Cameron Galton and grandson of Francis Galton's uncle, John Howard Galton of Hadzor (see our Vol. I, Pedigree Plate A, and Plate XXIX).

† Wife to Tertius Galton Moilliet, Galton's nephew, son of his sister Lucy.

‡ Mary ("May") Barclay Galton, the only remaining child of Hubert John Barclay Galton. Her mother was a Barclay.

§ Sons of Theodore Howard Galton, Francis Galton's first cousin.

annuity of £26, and to Ann the housemaid an annuity of £26. The servants will stay on till January 15th. Grace had thought of all sorts of things for the Camerons, but this morning a letter came that Mr Serocold was *very ill indeed*, so Cameron left at 3 o'clock and his wife was off at 10—as they will have to go to the Riviera to see Mr Serocold, as they think he will die—having had two operations and being very feeble.

Edward Wheler writes how very cold it is, and so much snow at Alnwick. My House smells of Puddings and Cakes—and now the Mincepies will be begun to be made. Bessy will have a large party on Xmas Day—the Studdys* and a Nephew Studdy and other relatives, I believe Edward and M. L.† The Darwin family have had to pay some duty on Breadsall Lodge‡, left by Sir F. Darwin to his unmarried daughters. He died 40 years ago, and Aunt Darwin 34 years ago—and they know nothing about any receipts. They should write, as Annie Sykes did, to the Papers. With much love, yours most affectionately, EMMA GALTON.

Letters to W. F. R. Weldon.

42, RUTLAND GATE, S.W. *November* 12, 1903.

MY DEAR WELDON, C. Herbst's book fills me with shame at my ignorance—When will *you* wake up or another Darwin arise, to consolidate and co-ordinate the mass of scattered evolutionary material accumulated of late years. I have dipped deep into the book at several points, but feel myself too ignorant of the creatures spoken about to have any hope of mastering it. Besides, as you say, I read German with difficulty, and don't like it or take to it.

Your phrases—(1) blue shadows on a white road, (2) prevention of bronchitis better than cure, (3) purple irises in Sicily while Piccadilly is in mud slush—have been so "fetching" that I have already provided myself with a foreign Bradshaw and am thinking much more definitely than before of a late December start. If so, it would probably be, first, direct to Genoa—then a week on the Italian Riviera—next a day or two at Naples, and then to cross to Palermo the first fine night. I am trying to find out the relative merits of the Palermo Hotels, and whether the Igea is really good or only costly and out of the way. Also I heard of a new hotel at Pòrto Fino (near Rapallo), said to be suitable for a week's stay, to rest and to acclimatise. I *must* treat myself as somewhat of an invalid.

G. Brodrick's death makes me very sad. He was an old friend, back to the 'forties or early 'fifties, and a sincere one. I have few left even of his generation, far fewer of my own. Yet I don't want to die, except perhaps now and then, when in a gouty and pessimistic frame of mind. I send back the shells and the book separately.

Very sincerely yours, FRANCIS GALTON.

HÔTEL TRINACRIA, MESSINA. *March* 12, 1904.
En route to Lipari—*don't* address as above.

DEAR WELDON, Your interesting letter greeted me here on arrival, after a glorious week at Syracuse and previously at Girgenti. You are by far too kindly a critic. The first sentences of the intended circular are rather illogically arranged and will be altered. You should see, as a curiosity, the corrections made by F. Howard Collins who did so much for Herbert Spencer and has been very helpful often to me. He is ruthless, and has (I hear) scored the proof all over, as is his wont. Some of his revisions are however always valuable.

About the cancer cells—Have not any full series of experiments yet been made on *transplanting ova* in all stages of their development? It ought to be done both in warm and cold blooded animals. Immature spawn in frogs and fish. The separated contents of an ovary, each grafted into some different part of the body of a mouse, guinea-pig, fowl (not neglecting the vascular comb) and especially on the parts (? glands) that cancer most frequents. *Something* ought to be found out in this way.

* Lucy Studdy was sister to Edward Wheler and both children of Galton's "Sister Bessy."

† Mrs Wheler, wife of Edward Wheler, now Wheler-Galton.

‡ Breadsall Priory was the home of Erasmus Darwin after his marriage to Mrs Chandos Pole. It was afterwards Sir Francis S. Darwin's house. I do not know whether Emma Galton made a slip in calling it Breadsall *Lodge*, or whether this was a separate house on the estate left by Sir Francis Darwin to his daughters.

As regards your problem. I dare not now trust myself to analysis and to criticism of formulae; but have no doubt from elementary considerations that your results are sound. The upward and downward movements of P depend equally on those of A and of B. If those of A range less widely than those of B, the latter will on the whole have the predominant influence. If A does not move at all, the movements of P will be wholly due to B.

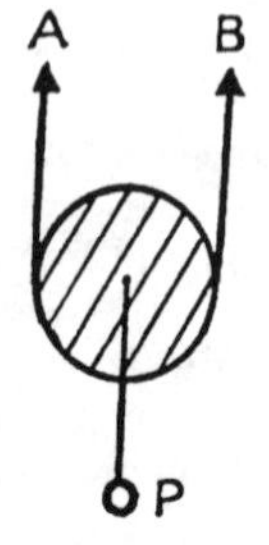

Again, the connection between small variability and small Arithmetic Mean value is clear (for symmetrical curves of frequency at least) on the supposition that negative values are impossible. A quasi-albino race (one with a small A. M. value) cannot produce individuals who are whiter than white, but if it sometimes produces such as are dark, its curve of frequency must necessarily be humped up against the axis of Y, and its positive tail can hardly be thicker at its root than in the figure. In this case MQ_3 is small (though larger than MQ_1) and $\frac{MQ_1 + MQ_3}{2}$, which is a fair measure of the S.D., must be smaller still than MQ_3.

On re-reading, I fear my explanation may be found less lucid than it should be, but you will probably understand what is meant. I have written it under epistolary difficulties of table and light. I have amended it, but not well.

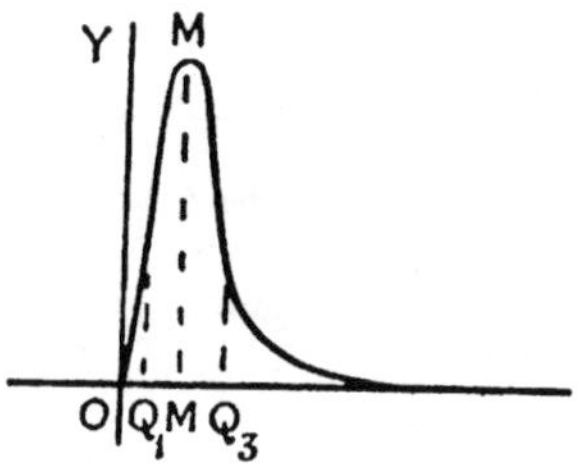

We are looking forward with keenest interest to a stay of some days in the Lipari Islands, among sulphur, pumice-stone, two active volcanoes (Vulcano and Stromboli) and deported Camorrists and Maffeists. They are allowed much freedom during the day but are confined and locked in at nights. There are no robbers among them, only murderers, one of another, and they are said to be very interesting and communicative. You, with your fluent Italian and Italian sympathies, would make out a good deal from them. We are fortunate in having introductions to the principal among the few honest people who live there, as to the officer now in command and to the agent of the chief landowner. It will be a funny experience. I expect to be housed like a pig, and *not* to be treated as a convict; but even they receive 50 centimes from the Government a day, which they supplement by working for wages. The weather is very variable, some sun *every* day; glorious sun most days. Now and then gales of wind.

The post that brought me your letter brought one also from K. Pearson, so I am posted with your Easter plans. I had hoped to be home just before Easter, but expect now to be delayed abroad a few days longer. Kindest remembrances from both of us to you both.

Very sincerely yours, FRANCIS GALTON.

42, RUTLAND GATE, S.W. *April* 10, 1904.

MY DEAR WELDON, I was so very sorry to miss you, and by only five minutes, yesterday. I did not *dare* to read your letter till this morning, being rather dangerously overworked and fearing disaster. With a lot of correspondence I *begin* with the least important, to ensure this not being overlooked, and end with the important. So your letter came the very last.

As regards albinos of all kinds, there is evidently an unusually close correlation between the soma and the germ (generative cell); total absence of colour in the one going with total absence in the other. When albinism is confined to the eye, the correlation is less close, but still close-ish. Perhaps the day will come when the mean correlation between soma and germ (generative cell), in respect to certain exceptional qualities, will be studied. What a puzzle it all is! The mice will be mines of facts. Those three beautiful volumes by Amari! I am ashamed to accept so valuable a present, but will do so, and read them through, and be more and more saturated with gratitude. I have a strong leaning towards Saracens.

Eva Biggs will have told you our news. We both go into Warwickshire to-morrow for three or four days, but to different places. I to my dear old sisters, 96 and 92 respectively* (5, Bertie Terrace, Leamington); she to her sister, Mrs Bree. The cold is as much as I can bear, but I am getting acclimatised again to my native country. It would amuse you to see F. Howard

* Bessy and Emma Galton.

Collins's revision of my paper. No boy's exercise at school could be more scrawled over. But some of his suggestions are good. As soon as I have finished this letter, I will take it finally in hand and post it to the printer.

If you care for a bit of pumice-stone you shall have some. All good pumice-stone comes from Lipari. There is a white mountain wholly composed of it, and convicts cut galleries into its sides to get at the choicest bits. Kindest remembrances to Mrs Weldon and to the Pearsons.

Ever sincerely yours, FRANCIS GALTON.

Alas! since writing this he has gone to bed with a temperature of 101. I think he won't go to Leamington to-morrow. It is, O, so cold—with snow in the wind. (E. B.)*

42, RUTLAND GATE, S.W. *June* 20, 1904.

DEAR MRS HERTZ, You ask me a difficult question about probable purchasers of Roger Bacon's *Magnum Opus*. The combination of scientific tastes, history of science tastes, purchasing power, and possession of library room, is rare. I have from time to time thought who might be suggested, but always in vain, so much so that I do not venture to send even the five *least un*likely names. The book seems to be more suitable to a public library than to any but a few very exceptional book collectors. I am ashamed of being so helpless. The physically scientific peers and baronets, who are Fellows of the Royal Society or of other societies, *might* be circularised, but from what I know of them I should doubt much success. It is too archaic a book for their wants, and they are hard pressed to keep their knowledge up to date. Very faithfully yours, FRANCIS GALTON.

42, RUTLAND GATE, S.W. *July* 5, 1904.

DEAREST EMMA, You may like to have an authentic copy of my "Eugenic" lecture. I have just received the usual few advance "Author's copies." The lecture and the long (wishy-washy) discussion upon it, will be published in due time by the Sociological Society. It is well printed, anyhow.... Ever affectionately, FRANCIS GALTON.

MALTHOUSE, BIBURY, FAIRFORD. Friday, *August* 25, 1904.

MY DEAR BESSY, Your letter was very welcome, I feared you might not have shaken off the illness. Milly and I have been corresponding about the inscription for dear Emma's grave†. I enclose two, marked (A) and (B). The (A) was the one to which her letter refers, I have just written out the shorter form (B) to see how it looks, but I prefer the (A) as being more interesting to the reader. How do you like the *words*? We have made many trials. You will see Milly's approval in her letter enclosed where I have marked the passage. Of course the proportions would have to be carefully attended to. Would you care to leave the matter *at first* quite in my hands, as Edward was disposed to do? If so, I will take much care to get a really good design that in respect to appearance shall be as *nice* and *simple* as possible, and I should be truly gratified if I might be allowed to defray the entire cost, in case my proposal should in the end be accepted. To get the nicest and simplest result one must consult persons of real taste. Very little changes of proportion make vast differences in effect. Then the material has to be considered; but I will not say more now, beyond that I wish to do everything with all the best advice I can get, and that I see my way to get it.

We went to the Cameron Galtons to tea. It is a curious place, very large in some respects, greatly cramped in others. The surroundings are mean, the gardens are very extensive, and the place is curiously rambling. Its history accounts for it, it was in part a wool-merchant's store, and that part has been pulled down by previous owners and its place otherwise utilised. My map (see p. 528) is I fear very incorrect. The house has excellent rooms, but the place gave me the idea that two persons could not pervade it; it has, however, great capabilities and I dare say they will settle happily. I hope Edward enjoyed Loxton‡. I go there to-morrow (and return here on Tuesday evening). Eva then goes to Adèle Bree and returns also on Tuesday. We made an expedition yesterday to join our two Professors at tea in a country town. I drove, they

* Postscript by Galton's great-niece, Eva Biggs.

† Galton's sister Emma died in 1904.

‡ Loxton, the quaint home of Erasmus Galton: see our Vol. I, Plate XXIX.

and their wives and Eva bicycled. Then we talked "shop" and other things to our hearts' content and separated after two pleasant hours. We did this every Saturday last year. How the autumn

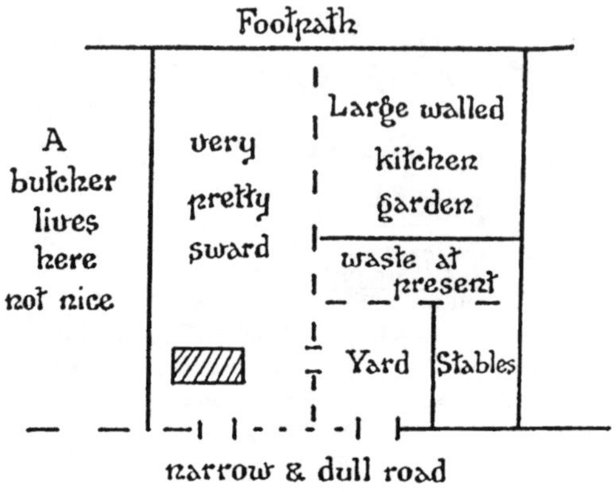

creeps on! I grieve at the departing summer. Give my best love to the Studdys—Eva would join if she were in the room. There is such a handsome old manor-house here. We went over it yesterday morning. A clear trout stream runs by its side.

Ever very affectionately, FRANCIS GALTON.

MALTHOUSE, BIBURY, FAIRFORD. *September* 6, 1904.

MY DEAR MILLY, I am anxious to hear about your eyes. How are they? Any news from the Cape? Now that dear Emma is gone, the family is like a wheel that has lost its tire. We must contrive means of keeping in closer touch. Bessy and I write every week. You and I must do the same. I went to Loxton a week ago. Erasmus most hospitable, but what an uncomfortable life it would be to most; but he takes real pleasure in it and it suits him to a "T." And the quantity of occupation that he gets out of it is surprising, for he does not a little foreman's work, besides agent's work and, loving to do things substantially, takes much time over each. Then he keeps minute accounts and reads books and does kind things, and so, although he sleeps little, the day is full. It is very pleasant having the two professors, Karl Pearson and Weldon, within reach. Weldon has astonishing energy. He cycled over last Sunday from Oxford, 28 miles, taking Pearson by the way. He walked here, some 5 miles, and talked till past 8 p.m. and then cycled back the 28 miles, and does his hard professorial and *other* work all the same. They two went to Cambridge and had a (verbal) fight with Bateson and his followers on Mendelism. There was a pretty long account of it in the *Times*, out of which some rather savage phrases of Bateson had fortunately been left. They both, with wives, etc., come here to-morrow to tea. I had one of the Master of Trinity's charming letters about the British Association. Balfour, the Duke of Devonshire, and Lord Avebury's party were his house guests. He describes Balfour as a miracle of *detachment*, full of interest in high subjects, fresh, delightful, showing no sign of the wearying work he had gone through nor of the serious foreign anxieties of the moment. And he was immensely pleased with the Aveburys, five of them; he, she and three daughters. I look forward much to coming to you on the 15th (we sleep in London on the 14th). Please tell me if this itinerary is right? Paddington dep. 12.25, Newton Abbot arr. 4.59, dep. 5.45, Bovey arr. 6.6, or should we take a fly from Newton Abbot? We can stay *over Monday* night, leaving you on Tuesday 20th, if that suits? Love to Amy and to all of the party who may be with you.

Ever affectionately, FRANCIS GALTON.

CLAVERDON LEYS, WARWICK. *September* 23, 1904.

DEAREST MILLY, Claverdon is so pleasantly changed. Edward and his wife bring in so many new interests, and they play their parts so well. I spent three hours in Leamington on the way here with Bessy, who looked singularly well; but she has rheumatic pains rather severely. She lives now at No. 5 and expects to migrate there altogether, being warmer and brighter and even fuller than No. 3 is of old associations on account of my Mother. There has been difficulty in dealing with the accumulation of dear Emma's things, but Edward tells me that all her books were most methodically kept, all bills paid up to almost the last, and everything was so neatly stored. I called on Temple and saw the other two maids there; she was very lachrymose, then cheered up, then, when I went away, became all tears again. As to you, so to me, she bitterly bemoaned her sudden isolation. Anyhow she has a charming house with a little grass plot and summer-house behind and a narrow plot in front. It is close to Bertie Terrace, on the same side of the main road, and nearly opposite to, but short of, the Post Office. I go again on Monday for the greater part of the day to Bessy, and hope to find Eva there. She is, or was,

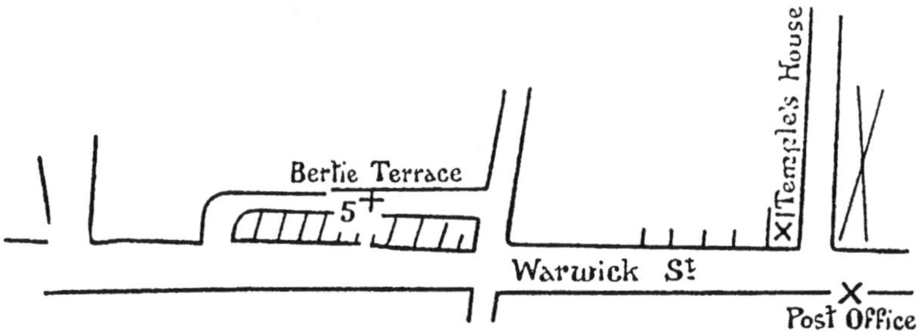

not sure whether she could leave London by Saturday morning. I had not time on Wednesday last to go to the cemetery. We lunched to-day at Wroxall Abbey, with Edward's brother-in-law. All so hospitable and family-like. I called on Grace Moilliet, but she was out. Gussy* is at Himbleton with Lady Galton†. Yesterday I went to Wootton Wawen Church where Darwin is buried and Mary‡ and Mrs Phillips and others. I liked all the memorials much, Darwin's included, which had been hardly criticised, but not deservedly, as I thought.

I trust, dear Milly, that your eyes are really mending. That horrid pamphlet tried them, I know. I wish I had never shown it you. It is so pleasant now to realise your surroundings and to think of you amidst them. It was a very delightful visit indeed to me and to Eva. Grimspound was grand and the moor most striking and beautiful in many ways, and the air felt so healthful. Give my best love to Amy. Also to the Captain and his Wife§. We expect to be back for good early in October. Ever affectionately, FRANCIS GALTON.

42, RUTLAND GATE, S.W. *October* 8, 1904.

DEAREST MILLY, I am to blame for letting the time slip by without writing. All is well here, but I have been full of my "Eugenic" plans, so full that I have not even written up my little pocket journal for a week. I stayed last Sunday with Lady Welby at Harrow, who is a most zealous friend, and have consulted with Sidney Lee, the Editor of the *Dictionary of National Biography*, with Barron, the Editor of the *Ancestor*, with Professor Weldon, with Branford, the Secretary of the Sociological and with Lionel Robinson, a general litterateur, and see no less than seven different ways of making the first effort, between which a wise choice has to be made. Every one of them seems hopeful at present. The iron is kept red hot on the anvil, and I am simply continuing to write concise family biographies, like that of the Stracheys which you saw (No, you didn't, it wasn't ready then). They too are giving me friendly help, and writing out ideas and getting them into shape. I cherish every day before winter, with its too faithful bronchitis, sets in. Lucy Wheler is staying a week with us, and is massaged every morning. She and Eva are quite happy, and shop, and see art things together. Bessy seems to be going on very well and will have all the changes of furniture made while she keeps on at No. 5. No. 3 will soon be, perhaps *is*, advertised to be sold. How quickly events move. I think I must have told you of my stay at Claverdon and of the tree thinning. It was all very pleasant there, where also events are moving quickly. I am curious about the newly discovered drive

* Augusta B. Stewart, second wife of Herman Ernest Galton: see our Vol. I, Pedigree Plate A.

† Marianne Nicholson, Sir Douglas Galton's wife.

‡ Mary Phillips, wife of Darwin Galton.

§ Captain Guy Lethbridge.

on the moor. I have often thought of those pleasant ones that you took me. We must fall into regular days of correspondence. I always used to write to dear Emma on Saturdays, and will to you. The lettering in the design for the bronze tablet has been improved and approved. It is now being engraved. I have ordered photos of it before being mounted on the stone, and will send you one. Give my love to Amy, also to Hugh* if he is still with you. I shall be glad of tidings when you next write (? on Friday) about Fred and Frank.

Ever affectionately, FRANCIS GALTON.

42, RUTLAND GATE, S.W. *October* 15, 1904.

DEAREST MILLY, You send good news of Frank. I trust the IF may go well. I have had an eventful week in fixing and carrying out the most hopeful of my numerous alternatives. You know that I have long been putting by a reserve of money for scientific purposes, either during lifetime or after death, which now amounts to a good round sum. Armed with the intention of bestowing £1500 of this in aid of "Eugenic" research, I determined on the University of London as the best of the 7 or 8, so on Monday I went to the Principal, my friend Sir Arthur Rücker, to talk the matter over. Now the University has the reputation of being a slow-moving body that requires everything to be done (1) through formal notice to their Academic Council, (2) through Committees appointed by the Council, (3) by adoption of the Report of the Committees by the Council, (4) by ratification by the supreme body, the Senate. The Meetings are fortnightly or monthly, so you may imagine the time any new piece of policy requires to go through, in the usual course. Now, as to what *has* happened in this matter. I went on *Monday* to Rücker, fired my proposal; then it turned out that the Academic Council met that very afternoon, and that as a "*matter of urgency*" my proposal could come on. So then and there I wrote it. It *was* proposed and accepted, and a good Committee of three important men, *plus* Rücker and the Registrar as officials, and myself, were appointed to meet on Friday (yesterday). On the day before, Rücker, the Registrar and I carefully drafted the details of the proposal to lay before the Committee; we met yesterday, improved and passed it, to go before the Senate on the 26th, when I have no doubt it will be confirmed. You shall have full details when it is. The result is that the £1500 is for 3 years (£500 a year) to appoint a "Research Fellow in National Eugenics" at £250 a year (the term is neatly defined, and so are the duties). Also an assistant at £100 to £120 a year, who may become titled "Research Scholar." All precautions are taken for superintending them and superseding them if they don't work well, and rooms are to be assigned to them. Also many academic advantages, too long to explain, are to be given them. Also a prospect of extending the Endowment beyond three years, if it is found to answer. So much helpful good-will exists, that I feel the seed is planted in good soil. Whether it will grow and flourish is another matter; very much depends on the holder of the Fellowship. But with inquiry and with advertisement, I have hopes of attracting a fairly high university man with lots of energy and sympathy and general intelligence, who sees in it an opening to future work of a more paying character. It has undoubtedly many attractions in that way, and the salary is as good as an ordinary college Fellowship.

Sibbie and Frank Butler† are with us for three nights. We had a particularly nice dinner party last night for them. John Murray, the publisher, told many anecdotes. Lady Pelly was there, and very helpful; so were the Rückers, and the Coleridges (*she* the novelist), etc. I have heard no more of the bronze tablet and do not expect news yet, but will write to the man in a week. Good-bye, Eva's love and mine to you all. Ever affectionately, FRANCIS GALTON.

42, RUTLAND GATE, S.W. *October* 22, 1904.

DEAREST MILLY, Yes, the weekly letter must become an institution. You must have glorious tints on the moors. Eva and I spent a day at Peppard Common where we had such a pleasant time last year, partly to see the woodland colours, and Professor Weldon joined us. One of our then neighbours was Sir Walter Phillimore, the Judge, whose daughter married the son of my old friend Mrs Hill. She, the daughter, had a bicycling accident a few days ago and was killed instantaneously by an omnibus. I have just been to the first part of the funeral service, held in a church in Sloane St. It was very affecting to see how many old retainers,

* Frank, Fred and Hugh, Galton's great-nephews, the three youngest sons of Mrs Lethbridge.
† Nephew of Francis Galton's wife, Louisa Butler.

PLATE LV

(i) "Sister Emma" (Miss Emma Galton).

(ii) Francis Galton, Secretary of the Royal Geographical Society, 1856–63.

(a)

(b)

(iii) Mrs Tertius Galton (née Violetta Darwin, Mother of Francis Galton and Aunt of Charles Darwin), in later years. (a) From a photograph taken at Leamington. (b) From a water-colour sketch in the Galton Laboratory.

I suppose, were there, in obviously deep sorrow. I walked out with Mrs Leonard Courtney. The last time we met was at the cremation of Herbert Spencer, when her husband delivered the beautiful, simple and forcible farewell to him. It will be nearly a fortnight before you get full tidings of Frank. You said that Guy was learning finger-print work; I suppose, how to read off and classify. They will of course have plenty of prints for his purpose available at the prison? I fear I could not help him with specimens as mine are all classified already. My fellowship affair comes before the Senate of the University of London next Wednesday, so nothing *could* appear about it in the papers before Thursday and then probably a mere notice under University intelligence. They will advertise for candidates, and that may attract notice; also inquiries will be made privately, for they do not bind themselves to select from those who answer the advertisement. At the best, it is "buying a pig in a poke," for so much depends on points of disposition and capacity that can only be guessed at, however elaborate the descriptions may be. I will tell you the results of course. I lunched last Tuesday with the Principal of University College, to see what rooms they could allot there for the "Fellow." It will shortly become an integral part of the London University, instead of being as hitherto a separate College. The professors are such a strenuous lot; I had coffee after lunch with them. Everything was simple. They, or the chemists among them, make the coffee; a big brew out of which each ladles his own cupful. I had a chat there with a charming professor, Sir William Ramsay, just back from a lecturing tour in America. He does not rate American science in *his* branch any higher than others have done in theirs. They have a few good men, mostly imported, as Professors, but not much that is indigenous. Edward Wheler and M. L. were here three or four days ago*. He fills his place *uncommonly* well and I am proud of him. He does real good work. Love to Amy. I do wish that your eyesight were better. Eva is off to-day to Constance Pearson. Ever affectionately, FRANCIS GALTON.

42, RUTLAND GATE, S.W. *October* 30, 1904.

DEAREST MILLY, Your autumn must be glorious. We two, Eva and I, had a glorious day out in Surrey. The trees were everywhere a uniform gold; no red whatever, but gold, gold, gold. I have never seen the like before. There were, as I heard, beeches on chalk soil, some three miles from where we were, that flamed in red, but I saw none of them. The portrait of me, by Charles Furse, which Eva insisted on having done *for herself*, is come. It was painted at his house last autumn, but not *quite* completed to his taste; so it was agreed that it should remain with him, to be retouched in the spring (he being full of work and obliged also to spend many weeks at Davos in the winter, for health's sake). He came back and was overwhelmed with orders for pictures and it was agreed that I should again stay with him this autumn. Well, as you know, he has suddenly died, leaving a large number of unfinished pictures. But mine is practically finished and is now here. It is an *excellent* piece of work and would hold its own in *any* gallery of pictures; besides, it is *very* like. It is Eva's property. She won't tell me any further particulars, but keeps it as a secret, which I respect.

The Fellowship arrangements are being rapidly pushed forward. While writing this, printed copies of the requirements have reached me, of which I enclose one for you as a memento. You see now (1) that everything is done in the name of the University and (2) that the word "Eugenics" is officially recognised. I am very glad of all this as it gives a status to the Inquiry, so that people cannot now say it is only a private fad.

Mrs Eustace Hills was *not* the lady you met. Was she not Mrs Hills, Judge Grove's daughter, and mother-in-law to Mrs Eustace Hills? Judge Grove was one of the very kindest friends I ever had. It was at his house, hired for the shooting season, that dear Louisa was suddenly taken so alarmingly ill with violent haemorrhage from the *stomach*. Mrs Hills, then Miss Grove, was so very kind and helpful. It laid the foundation of an affectionate friendship between them. That illness was many years before the end of dear Louisa's life. The cause of it was never properly explained. Lecky's remarks on Gladstone are in the preface to his second edition (the last one) of *Democracy and Liberty*, tell Amy. The photo of dear Emma sitting in her drawing room is excellent; perfectly life-like and domestic; perhaps her figure is a little

* Writing to his aunt, Emma Galton, in October 1898, Edward Galton-Wheler remarks: "I most thoroughly enjoyed being at Uncle Frank's. He is the best of hosts, always hospitable, and one feels it 'Liberty Hall' where one can do anything one likes."

stiff, but the whole is a valuable memorial. Ethel Marshall Smith* dined here the other day. She is quite an altered person, so radiant, healthy looking, and (how shall I phrase it?) *expanded*. You heard of Edward Wheler's retriever getting a second prize? Her breed is too gentle a one for your purposes. What a relief this morning the news is *re* Russia!

Ever affectionately, FRANCIS GALTON.

42, RUTLAND GATE, S.W. *November* 12, 1904.

DEAREST MILLY, *Si da* †, You must be anxious about Frank ‡ still. It was an awkward business. Guy ‡ will soon be with you. All is well here. Eva has been three nights in the New Forest, with her brother at Emery Down, and bicycled gloriously with him. Sensible girl—she made him take a short rope and tug her thereby, up hill and against the wind, like a trailer. I have been busy in relation to the new Fellowship. We four who form the Committee met yesterday to consider applications, and selected the three most promising to see next Friday, and probably then to elect. They are all good in somewhat different ways, and I am happy in the prospect of getting the best. A newspaper cutting came this morning, fuller than usual. You may like to see it, but do not trouble to return it. The photograph of the tablet for dear Emma, which has been engraved some days past, ought to arrive to-day. I trust the whole thing will be completed and set in place very soon, perhaps by the end of next week. I am grieved at the death of Emma Phillips§, for I saw so much of her between 45 and 55 years ago. There was something very nice and cheerful and sympathetic about her when at her best, and then a sudden wave of shyness, indifference, and dominant sense of self would come over her, and she was an altered person. It was very odd. I wonder what sort of a person the heir to all the strictly entailed property of Edstone is. Beyond knowing his name, which I have forgotten, I have heard practically nothing of him. He is Irish, and was hardly ever in Warwickshire. Somehow or other I missed seeing the graves of Aunt Sophia and Mr Brewin. There is much that is radically wrong in our British aesthetic sense, or peaceful burial grounds like that of the Friends in Birmingham would not be so rare||. I often marvel at the way in which an artistically minded person succeeds in turning a mere plot, with no particular natural advantages, into a beautiful garden. The Japs do this. This horrid, horrid war! Did you see some weeks ago of a Russian and a Jap locked in death. The Russian had gouged out the Jap's eyes and the Jap had bitten through the Russian's throat. However, dogs delight to bark and fight, and the same delight lies at the bottom of much human nature. Many loves.

Ever affectionately, FRANCIS GALTON.

P.S. I overlooked *your* P.S. T—— ¶ has turned rather silly, posing as a lady and calling her niece and Mary "the maids." She sits doing nothing in a grandly furnished drawing room, and in a house furnished far beyond her station, and I understand gets laughed at. Her head is turned. She told me that after what she had been used to, she could not have endured going to a smaller house.

Blessed be Higgins for his paste. [The P.S. was pasted to the sheet.]

42, RUTLAND GATE, S.W. *November* 28, 1904.

DEAREST MILLY, I had to omit my weekly letter, being in bed (mostly) all Saturday and Sunday with cough and cold, no asthma I am rejoiced to say, such as I always had when my bedroom was carpeted. Your rats sound almost alarming. There used to be a professional rat-catcher, who gave himself a high name, and who walked about London in a brown velveteen coat with silver rats sewn on to it as ornaments. He was a picturesque figure, and knew it, but he has long since disappeared—gone to the "rats," I suppose. I am so glad to be at home and not away in a comfortless place, this cold weather.

* Ethel, daughter of Cameron Galton, married Mr Marshall Smith.

† *Si da* is Galton's abbreviation for "sister's daughter." His niece Milly, Mrs Lethbridge, was his sister Adèle's daughter. See above, p. 446.

‡ Sons of Mrs Lethbridge, Galton's great-nephews.

§ Sister of Darwin Galton's wife, Mary Phillips, and coheiress of Edstone.

|| See our Vol. I, p. 52 and Plate XXXII.

¶ A pensioned servant of Galton's sister Emma.

You are doubtless an admirer of Wordsworth's "We are seven"; the following will serve as a pendant to it:

Dramatis Personae: Dirty boy, alone in an Edinburgh garret. Philanthropist Visitor. Ph.V. Where's your mither? Boy. Oot charing. Ph.V. Where are your brithers? Boy. Twa are oot begging. Ph.V. What ither brithers have ye? Boy. One in the Univarsity. Ph.V. Maybe he's studying for the meenistry? Boy. Na; he's in speerits in a bottle; he wa' born with twa heids.

Applicants for the Fellowship begin to be heard of. A very likely man is almost certain to apply. I had three hours' talk with him last Thursday. There are already five others, possible or actual candidates. Forgive this short letter. I have arrears to get through and am not yet wholly fit. Best loves to you all. Ever affectionately, FRANCIS GALTON.

42, RUTLAND GATE, S.W. *December* 3, 1904.

DEAREST MILLY, I trust you are satisfied with your "Lethbridge rat-research dog." My *Research Fellow* is still unfixed, but I hope daily to hear from the present favourite whether he will formally apply or not. He is now in France. What good news it is that Frank has got a permanent appointment of a kind that he likes, and apparently on his unassisted merits. I wish I were fit to go to S. Africa with the British Association next autumn, but of course that is out of the question. The people who *do* go will have a hard and busy time of it, and must I fear take nearly all of their Science with them, for there is not much of it there—at least only few signs of it. If George Darwin's health stands the work, it will be very congenial to him, for the most important feature will be the survey, as proposed, of an arc of the Meridian, to join the Cape surveys with the Russian, *via* Egypt. Geodesy is one of his special subjects. My past week has been one of coddle, until I am aweary of fires and blankets, which make me cough. We go to Branksome Hotel, Branksome Park, Bournemouth, on Monday for a few days, where "I may heal me of my horrid cough." It is a sort of "Island-valley of Avilion," with Poole Harbour on one side and the sea on the other. We are now looking forward to leaving for the South, somewhere early in February. I can't easily get away, and doubt if it be wise to get away, earlier. *Then* the almond trees are in blossom and spring is in the Southern air and the days are lengthening, and winter is past. Of course your rats are only the invading Hanoverians, not the more gentle and graceful black British ones. The latter are apparently almost extinct, under the action of blind Eugenics. Ever affectionately, FRANCIS GALTON.

42, RUTLAND GATE, S.W. Saturday, *December* 17, 1904.

DEAREST MILLY, What villains they must be in Pretoria! If you hear, please tell me whether, and how, the finger-print system acts out there. Without a bureau manned by a really capable man and a couple of clerks, it would probably get into a complete muddle, so far as classification is concerned. Is there an Identification Department? I have not lately seen anything of the Scotland Yard doings, but I believe all goes on swimmingly. Dear Emma's gravestone is not even yet put up. Edward Wheler has seen it in Leamington, at the yard of the man to whom it is entrusted and likes it much, but there are certain details which delay. I send you a photo of the inscription, which you will like to keep, all the more for having helped in drawing up the words. The Galtonias at either side are utter failures*. The artist has no excuse, for he was supplied with many drawings; but accuracy is not the strong point of artists. They think as much of shadows as of substances, and a bandbox casts as black a shadow as a block of granite. (That metaphor might be worked up!) Hugh will delight in Rome. I am very glad that Fred is now so strong and happy. The last rose of summer—the last rat of the year! You will have to keep and pet him or her. But the large probable families of rats are appalling. I heard that all the hives full of Ligurian bees in England, for

* Few things pleased Galton more than the naming in 1880 by J. Decaisne (Professeur au Muséum d'Histoire naturelle, Paris) of the *Hyacinthus candicans*, from South Africa, the "Galtonia." It is one of the most beautiful and hardy bulbs, shooting out a spike five feet and upwards in height. It differs much in habit though less in floral construction from our ordinary hyacinths. I well remember Galton's delight at finding two or three Galtonias growing in a bed of the garden of the house I was staying at, when the biometricians were at Peppard in 1903. See above pp. 523, 530. It was characteristic that he should place it on his sister's tombstone.

many years, were descended from a single queen bee, sent by post to England from the Riviera. Is it possible? I am not sorry to remain several weeks longer in England, being not strong enough now for the risks of an ordinary journey. We hope to be off in the second week of February. Things go on here in a humdrum regular way. No real advance just now. Loves to you all. Ever affectionately, FRANCIS GALTON.

*Galtonia**.

Flores hermaphroditi, regulares, penduli, bractea membranacea stipati, longe pedicellati, pedicellis summo apice articulatis.

Perigonium corollinum, candidum, campanulatum, limbo 6-fido, patente, laciniis planis vix apice papilloso-incrassatis, exterioribus oblongis, interioribus obovatis basi angustatis.

Stamina biseriata, subaequalia, tubo ad faucem inserta, inclusa, filamentis subulatis, glabris; antheris oblongis, dorso medio affixis, oleaginis; polline aureo.

Ovarium sessile, oblongum triloculare, loculis pluriovulatis septis glandulis nectariferis minimis; ovula biseriata, anatropa.

Stylus cum ovario continuus, erectus, obsolete trigonus stamina superans v. subaequans; stigmata tria, sessilia.

Capsula sessilis, oblonga, membranacea, reticulato venosa, loculicide† trivalvis, polysperma.

Semina ovata, mutua pressione angulata, testa membranacea, nigro-fusca; albumen carnosum; embryo cylindricus longitudine albuminis.

Herbae bulbosae, Africae australis incolae. Bulbus tunicatus. Folia pauca, magna, linearia, erecta v. patula, crassiuscula, glauca. Scapus metralis. Flores racemosi, inodori, albi, speciosi, bractea membranacea integra v. inferne lobulata stipati; pedicelli in floribus virgineis reflexis, fecundatione peracta, erecti, summo apice sub perianthio articulati.

* Galton (Francis), auteur du "Narrative of an Explorer in South Africa," London, 1853.

From "Note sur le Galtonia (*Hyacinthus candicans*), nouveau genre de Liliacées de l'Afrique australe," *Flores des Serres et des Jardins de l'Europe*, Tom. XXIII, p. 32, 1880.

42, RUTLAND GATE, S.W. *January* 1, 1905.

DEAREST MILLY, This is my first letter in 1905, written with a new pen and in a new suit of clothes. Also I feel a new man, the cough having apparently gone with 1904. A very happy New Year to you and all yours. I was so glad to hear what you told me about Frank. All your sons and your daughter are so much liked. It must be a great pleasure to you. I got out this morning for a long drive (for me) round Regent's Park, without being tired. I suppose it has turned cold with you as with us. The N. wind has driven the fog away, and we saw some sun at last. If life that has no history is happy, mine now must be supremely so, for I have no news whatever. I got to the Club yesterday; people seemed *older*; even Lord Avebury who was boyish for half a century looks at last rather old, the hair changing from colour to colourless. Dear Emma's gravestone is not even yet put up. Bessy tells me that the grave is prepared for it and that she has seen the tablet, which the stone mason brought to her, but there has been some delay about the stone itself, which is due from Portland. I wonder if it is quarried by convicts, or do they only quarry stone for Government works? This terrible Jap war! and the soldiers freezing with cold. How they do quarry mines! Fancy the explosion of two tons of dynamite. It was, I think, one ton that blew up in a barge some time ago in the Regent's Canal—or was it only gunpowder?—and shattered all the windows near and sent the tigress in the Zoo into hysterics. It must have been only gunpowder, or the canal would have been destroyed, and much besides. Love to you all and regards to the rats if more than one remains. Ever affectionately, FRANCIS GALTON.

42, RUTLAND GATE, S.W. *January* 8, 1905.

DEAREST MILLY, It must have been a great shock to you, that horrible accident close to your gate. Poor fellow, even if drunk, the punishment to him and his family exceeded apparently his sins, by far. I have often wondered and talked with people about what the results would be if our sympathies were vastly keener, or to put it in another way: What should

† Mr. V. Summerhayes, who has kindly looked for me at the specimens of *Galtonia* in the Herbarium at Kew, informs me that the capsules seem to be dehiscing for a short distance both loculicidally and septicidally, along six sutures in all. The capsule then seems to act as a censer or pepper-pot mechanism, since dehiscence apparently never goes beyond the upper third.

Galtonia (*Hyacinthus candicans*) from tropical South Africa.

be the colour of a clergyman's dress? It must be suitable for marriages, christenings, sick-beds and deaths. One suggestion was *violet.* My mother was fond of saying that she had had a much happier life than most, but that if she were given the choice of re-living it she would rather not. It is all very queer and no thinking about it gets one any "forrarder." Do you happen to recollect that skit of Voltaire, when describing the range of knowledge of his almost supernaturally informed Zadig. "...and as to metaphysics, he knew all that has been known since the creation, *c'est à dire très peu de chose.*" That bitter Monday last upset me in another direction, viz. gave me gastric catarrh, three or four days of sofa and slops. Eva was able to leave me on Wednesday for two nights at Allesley. She saw the cemetery at Leamington where the stone had just been placed, and she saw Bessy, whom she reports as looking exceptionally well and happy. I got out in a "growler" both yesterday and to-day. I was sorry to hear of your attack. Next Friday is the day of electing the "Eugenic" Fellow*; I shall be very glad when that is finished off. But though it will be practically settled on Friday, confirmation is formally needed by two bodies, (1) the Academic Council, (2) the Senate, before which the election cannot be final. I don't foresee the slightest difficulty in all this, only a week or two of further delay. Best loves. Is Guy with you? Is Hugh on his Swiss tour? I gather that Amy is with you.

Ever affectionately, FRANCIS GALTON.

42, RUTLAND GATE, S.W. *January* 15, 1905.

DEAREST MILLY, I am grieved for Fred's mishap. When you learn more, do send me a post-card to say if it is simple or compound, and an ordinary or a bad fracture. An ordinary simple fracture is not such a very bad thing and need not lose him his appointment. I hope it is not worse. My own small malady is better. Slops, hot bottles and bed are my prescriptions, but at this moment I am writing on my lap, well wrapped up in an easy chair. On Friday I got to my Committee for an hour and back straight to bed. We unanimously agreed to recommend a man who will be formally elected by the Senate on the 25th, and I am perfectly satisfied, and so we all are. He is not the man I had chiefly in view, but his merits came out stronger and the drawbacks to the favourite became more conspicuous, so there was no doubt in placing him first. It is better not to mention names till the election is final. All this is a very great relief to me. Much is going on independently now *re* Eugenics. You will be glad to get the other half of your "pair of scissors" back. The Arabs somewhere have a list of things which are in pairs and cannot work singly, and which they say must have been created so at the beginning. The only one I recollect is a pair of tongs. *With* them a blacksmith can make everything, but he cannot make *them* without another pair. Your garden, birds, and possibly rats, will all show signs now of the incoming spring. Snowdrops ought to show soon. What lies the Russians will tell. That in Stössel's memorandum about the number of Russians in Port Arthur, was not one half of the real number. What an ingenious idea that of painting the surrender of Port Arthur on *kites* and sending them over the Russian lines. I am assured that there is no fun extant equal to that of flying meteorological kites from a swift steamer equipped for the purpose. It is easy to explore the air in that way for much more than 1 mile high. It requires a great deal of skill and constant attention. Much has been done and is doing in that way. They are shaped quite differently to common kites, something like Venetian blinds, and carry no tails. They require a steam-engine to wind in the wire rope that holds them, and they are sent up in *tandems.*

Ever affectionately, FRANCIS GALTON.

42, RUTLAND GATE, S.W. *January* 16, 1905.

MY DEAR WELDON, I should dearly like to see your views about the nature of dominance and their effect on Mendelian theory. If you really do send them, be assured I will read them with all the care I can. Can you explain (in a way) each necessary step in the imaginary case, say, of only three sorts of interfering germs?

I have spent days, some wholly in bed and others mostly so, by strict doctor's orders for gastric catarrh now. Really I am rather liking it, and don't object to slops for food. Hot bottles are delightful companions—I regularly have two. Very sincerely yours, FRANCIS GALTON.

* The reader will have observed that Francis Galton here and in several earlier letters uses the adjective "Eugenic." Perhaps he already saw the fun of this; but several years later he solemnly warned me that I was not to allow any one to speak of the Eugenics Laboratory as the "Eugenic" Laboratory.

42, Rutland Gate, S.W. *February* 5, 1905.

Dearest Milly, I shall be glad to hear next Saturday how you have tided over your many small calamities—indeed rather big ones. It was a great pleasure finding Bessy so unusually well and bright. She will be now at Claverdon. Thanks to Eva's dragonship, I managed it all without fatigue, including a sight of Edward and his wife and of Erasmus. But after returning, and not I think in consequence of the trip, I got poorly and the Doctor kept me in bed all yesterday and to-day up to the afternoon. Just a slight feverish attack and need for a dose. He tells me I may keep an engagement of lunching quietly to-morrow with Major Leonard Darwin. I want to hear all the latest news about George Darwin's preparations for South Africa. He has a particularly strong staff of associates, as Presidents of the several Sections of the British Association. Schuster has been here frequently and is working away. He gets into his rooms at University College to-morrow, and spends half of each week there and half at his home in Oxford. Our Committee meets on the 10th to arrange particulars. I have already drafted an "unauthorised programme," which will be read with my other paper at the "So so" Society on the 14th, Schuster going on with it if I break down. I shall try some of Warren's (£10,000 a year) method. You know, he had to lecture at Leamington when at the height of his fame. He awoke with a stomach attack. His wife gave him some brandy. As he travelled down he felt no better and took more. He went to Jephson who said—take a couple of glasses of port. At length the lecture-hour came and he was got somehow into his seat on the platform, where he sat with eyes shut and arms folded. The chairman arrived late and at once began with a modest disclaimer of his own power of speaking, but "that does not matter as you will now hear the eloquence of our distinguished guest, Mr Warren." Warren sat still; his neighbour nudged him, saying "Warren, get up." With difficulty he did so. Then, looking round the eager audience with bloodshot eyes, he simply uttered the words "Bow, wow, wow" and collapsed back into his chair. About the Darwins, Mrs Litchfield has just sent me a charming two volume Life and Letters of her mother*. It is privately printed. The second volume is particularly interesting. I have taken salon-lits from Calais to Bordighera on the 20th. We leave London on the 16th and stay at Calais in the meantime. Love to all of you.

Ever affectionately, Francis Galton.

Insurance Data.

During the course of this year (1905) Francis Galton endeavoured to move the Institute of Actuaries to undertake, or prompt the Life Insurance Companies to undertake, an inquiry into the heredity of disease. To the outsider the proposal seems not only of great scientific interest, but of the highest commercial importance to the business of life insurance. The biometricians had shown definitely that length of life and general health were inherited characters. Galton's somewhat slender data indicated that certain diseases tend to run in families (see our Vol. III^{A}, pp. 70–76). My own more numerous family schedules are convincing in their evidence that most broad classes of disease, whether as cause of death or of ailment during life, have familial incidence. But when we remember the variety of familial relationships, and these for the two sexes, the range of age groups and the number of even broad classes of disease, it will be recognised that the full data for a thousand families, covering fifteen to twenty thousand individuals, are far from adequate to obtain a definite numerical answer to such a question as the following: A.B., of age *a*, has a certain number of relatives C, D, E, F, ... who died at ages *c*, *d*, *e*, *f*, ... of diseases belonging to certain broad classes, and a certain number of relatives C′, D′, E′, F′, ... of ages *c*′, *d*′, *e*′, *f*′, ... who are now suffering from

* Mrs Charles Darwin's *A Century of Family Letters* issued some years after to the public.

particular diseases. What is A.B.'s expectation of life? The inheritance of various types of disease is a subject on which there is very little medical literature and that not of a kind from which a numerical estimate of duration of life can be based. The present system by which Life Insurance Companies vaguely select the better lives by aid of their medical officers is wholly out of date, and even if it can be made profitable to the companies is not just to the insured. Every life has its individual expectation, and its corresponding premium, and from the standpoint of the insured it is unfair to reject a life because the insurer is too ignorant, or too inert, to obtain the knowledge requisite to insure it at a reasonably approximate rate. The fact is that insurance companies as now run are in the bulk commercial enterprises, having little regard for the needs of the population as a whole, unless those needs are such as with little scientific inquiry can be turned to easy profit. The time is ripe for the State to take over not only the insurance of the handworker, but of the whole community. It possesses in its records of births and deaths material from which, with labour and scientific oversight, an approximate picture could be made of how the entire population in its classes and families lives and dies. Such must be the basis of any insurance scheme fair to the individual, whatever be his health or his family history. And if there must be a profit made out of life insurance, as there certainly is at present, it is surely best that it be made by the State, rather than by commercial companies. The State would at least enforce the medical examination of annuitants as well as of the would-be insured..

Galton often referred to the importance of measuring the expectation of life with due regard to the susceptibility of the family to various types of disease which have high mortality rates at special ages. He considered it not only of value for scientific life insurance, but also fundamental for a right development of Eugenics. He consulted on the matter the well-known actuary Mr W. Palin Elderton, who at a meeting of the Sociological Society had stated that possibly the insurance offices had material for the measurement of the heredity of disease. Mr Elderton, after a very careful consideration of various proposals, suggested an appeal to the Institute of Actuaries.

42, Rutland Gate, S.W. *January* 22, 1905.

Dear Mr Elderton, If I could see my way a little further I should be glad to take steps to give effect to your suggestion about obtaining Eugenic data from Insurance Offices.

Can you help me with a little information? 1. Are the records kept for any considerable time after the death of the person insured? 2. What size number of them could be in likelihood obtained? 3. Could permission be easily got to have them copied? 4. If so, to whom should I apply? 5. What should you imagine would be the cost per 100 of obtaining copies? 6. Could I get 2 or 3 samples (without names)? Very faithfully yours, Francis Galton.

January 25, 1905.

Dear Mr Galton, I think I had better deal with each of your questions separately: 1. The records are kept for various periods depending on the practice of the particular office; in some cases for more than thirty years after death. 2. If you could get many offices to join, you would be able to take out thousands of cases, some records, however, giving little information.

3. I fear I can't say whether permission to copy would be easily obtained; I fancy most offices would insist on a member of their own staff being employed, as much of the information in the papers is confidential. 4. Application would really have to be made to each office. 5. The cost would probably depend on the time taken, which would vary with the accessibility of the material, some of the papers being stowed away in awkward places. 6. I would try to get samples if you like from my own office.

With regard to (4), (5) and (6) would it not be a good way to try to get the offices to combine to investigate the data at their disposal? If offices could be got to see that the data would be of practical use (which is the case) they would be more willing to agree, and would probably bear some or all of the copying expense. The difficulty is how to approach them. This might be done through the Institute of Actuaries, the Life Offices Association (a body which is a collection of Insurance Officials who meet for consideration in connection with practical routine) or the Life Offices Medical Officers' Association (a body formed from the medical examiners of assurance companies).

If the Institute of Actuaries could be induced to issue a circular to the offices asking if they would contribute, I think assurance companies would more willingly hand over their particulars than to a private individual, even if it were known that the collected statistics would be investigated by private individuals.

I enclose a draft card which with slight alterations might be adopted. It will show the particulars you can get.

I will if you like mention the matter officially in my own office (the "Guardian"), but I fear we could do little for some months as we have our quinquennial valuation on hand which means that the whole staff is stopping late over that, and additional work is quite impossible at present. I could mention the matter to one or two people in other offices if you think a preliminary sounding would be a good thing. Of course, you will recognise that I am merely expressing a personal opinion in my letter, but I shall be only too glad to help you in any way I can.

Very faithfully yours, W. Palin Elderton.

On the basis of Mr Elderton's suggestions Galton drew up an address to the Institute of Actuaries which ran as follows:

February 11, 1905.

To the President and Council of the Institute of Actuaries.

Gentlemen, Permit me to address you and call your attention to a serious actuarial need, namely of better data than are now available for computing the influence of family and personal antecedents on the longevity and health of individuals.

A vast quantity of appropriate and trustworthy material appears to be stored in Life Insurance Offices, out of which authenticated extracts might be furnished for the purpose of statistical discussion. (To avoid suspicion of breach of trust, names might be replaced in the Forms by register numbers, the keys to which would be confidentially used for the purpose only of determining relationships between persons assured.) A Form on which the extracts might be entered is enclosed in order to save lengthened explanation. It might doubtless be improved. I am assured that no person or Society would be more competent to arrange the details of such a scheme, or to bring it more weightily before the notice of the various Life Insurance Companies, than your own.

My justification for interfering in the matter is that the desired information would be especially serviceable for my own inquiries into what the University of London has now recognised under the title of "National Eugenics." On this account I am prepared to pay such moderate preliminary expenses as may be needed for an experimental trial, being not without hopes that the Insurance Companies may hereafter contribute to what will be of use to themselves. In the event of a *primâ facie* approval, I would ask the President and Council of the Institute of Actuaries to appoint a Committee to consider it in detail, with instructions to report on what it might be useful and feasible to obtain from Life Insurance Companies, on what would be the probable cost of the extracts at the rate of so much per thousand, and on the desirability of further action.

A rough draft of this letter had been made, when a passage in the recent Address by your President was brought to my notice, which gives hope that this proposal may meet with a still more favourable reception than I had ventured to anticipate*. FRANCIS GALTON.

FORM (suggested by Mr W. Palin Elderton).

"Register Number" of the life assured......

Date of Assurance.........
Age at Entry...............
Date of Death .... Age......
Cause of Death ....

FAMILY HISTORY

	If living	If dead		
	Age	Age at Death	Cause of Death	Reg. No. if any
Father Mother				
... Brothers				
... Sisters				

Remarks on personal antecedents prior to date of assurance.

* Galton's appeal made twenty-five years ago has led up to the present day to no investigation of this basal problem; it is doubtful if it will do so as long as the chief assurance work is done by commercial companies who can select enough first class lives to pay ample dividends on their invested capital.

CASA ——, BORDIGHERA, ITALY. *March* 2, 1905.

DEAREST MILLY, This stationery that we find drawers full of here is grand, is it not? It is all put at my service. I am grieved at your account of Guy and sympathise all the more from my somewhat similar afflictions, but the coming spring is in favour of him and this change of climate is fast curing me. But the weather here is far from paradisical. It has been so for $2\frac{1}{2}$ days in all, but rainy and often chilly all the rest. The past frost and drought have made cruel havoc with the gardens, so the spring show of flowers will be far poorer than it has been within recent memory. We are lodged luxuriously. It was very lucky for us that the lessor of this villa had to leave it for some weeks in order to seek a divorce from her husband in Edinburgh, and that she wanted cash for the purpose. So I made an offer of $\frac{2}{3}$ of what she asked and got the house and two excellent servants, all in perfect order, till the end of March. I shall try and get, first the Arthur Butlers (he and his daughter) who are at San Remo, and then Mrs Litchfield*, who is at Cannes, to come here successively, each for a week. The garden is such a nice, rambling luxury, with good shelters against wind. It is mostly sold already to be broken up and built over in the spring. The Italian railroad services are greatly hindered by the methodised obstruction on them, adopted in lieu of strikes. There seems so little public spirit in Italy, that strikers of all kinds are free to bully the public. I dare say they would retaliate interference with the knife. I constantly wonder how society can be carried on by people who are so abject as most people are. I have been reading Hodgkin's account of the slowly perishing Roman Empire, and the pictures of depravity in it are horrible; yet the Empire was long in dying. There are capital books about here; some in this house, others at friends' houses, and others again at a *good* subscription library. I have been hearing folk-lore tales lately. Thus, a shepherd was missing; his sheep returned, but not he; three or four days passed and the relations consulted the priest, who said, he will come back soon but you must ask him no questions. He did come back, silent and altered, but at length told his tale. He saw a cavern and went in and found a joyous company, dancing and feasting, who made much of him. This modern Tannhäuser remained as he thought many days and at length entreating to go, they conducted him out, but on taking leave said: "If it was not for what you have in your pocket, you never could have got away." It was a piece of salt, a bit of what he had taken with him to give to the sheep. Salt is supposed to have many occult virtues. Another story was about a man falling in on a particular saint's eve with a masked procession carrying lighted tapers; the last who passed him gave him his or her taper, which he took home and put in a drawer; in the morning he found it was a dead man's finger. So he consulted the priest, who said, wait a year and then go again, and give back the finger to the man who gave it you, but take a *tom-cat* with you. This he did, carrying the cat in an apron. When all was over, the cat was *dead*. The priests must be full of these legends, and ought to be very suggestive too, if they are always consulted and must give appropriate advice. I have been reading again White's *History of Selborne*. Besides all its natural history, merit and charm, what beautiful English it is. It was written about 1770, the time when my father was born†. Loves to you all.

Ever affectionately, FRANCIS GALTON.

CASA ——, BORDIGHERA, ITALY. *March* 10, 1905.

DEAR WELDON, Alas for optional Greek!—We are now in brilliant sunshine and warm weather, and I sit most of the day in a wooden shed in the garden, where I get through a fair amount of work. I wish you and Mrs Weldon were in this pretty villa to enjoy it also. You have a capital subject in working up latent characters in races that apparently breed true. *Shorthorns* ought to yield useful facts. It is interesting about the $\frac{13}{91}$ mice; $\frac{1}{7}$ instead of $\frac{1}{4}$. The enclosed cutting was sent to me; though the conclusion is rubbish, you might like to see the alleged facts if you do not already know about them.

What you say is quite a new idea to me, that the loss of power by the embryo to regenerate the whole from a part is unconnected with the loss of power in the adult to regenerate lost limbs. Certainly a remaining piece of begonia leaf does not renew the lost part. Your book when it comes out will be full of interest. I don't expect to go farther South than here. The place suits me perfectly and I want to get as well as I can, and not to fall back into invalidism

* Charles Darwin's eldest surviving daughter.

† White's *History* first appeared in 1789, Tertius Galton was born in 1783.

by rash acts of fatigue, etc. But I envy you Ferrara, about which I have been lately reading in Hodgkin's big work on the last days of the Roman Empire and of the Goths, etc. I have this villa only until the end of the month, but shall try for a prolongation of my sub-lease. We are quite at home here, having many friends about. I see that K. Pearson *has* delivered his three lectures, but detailed news does not reach me here. Oh! this blessed Riviera (when it *is in good humour*) for invalids. It is almost worth having been ill to enjoy the balm of its air. With both our kind regards to you both. Ever sincerely, FRANCIS GALTON.

Be sure to remember me to the Pearsons when you write. Schuster seems energetically at work. 43° 46′ = Bordighera, 44° 50′ = Ferrara.

CASA ——, BORDIGHERA, ITALY. *March* 11, 1905.

DEAREST MILLY, I am indeed grieved at your continued anxieties. The coming spring is however all in favour of your invalids. We have had three or four days of perfect weather here, and I have sat out most of each day in a wooden shelter in the garden and got through a goodish deal of work there. Carnival with its mild tomfoolery is happily over. It was got up by a *socialistic* town-council of all things. There is a superstition against it still, on account of the earthquake having come nearly 20 years ago, on (?) Ash Wednesday, owing to the sins of carnival during the preceding week. I am not sure of the exact logic, but it is something like the above. The owner, from whom we took this house, has lost her divorce suit; the Judge considered the action void of just foundation. *She* is much liked here. I know nothing, and care less, about the ins and outs of the case. *He* is "adored" (so an old Scotch lady told me) in Edinburgh, so presumably there are faults on both sides. I do not know whether she will let us prolong our lease until Easter, but I shall ask permission, not knowing any more suitable place to go to. I wish your invalids felt the blessing of returning health as I do, but I am not up to further travel now, and intend to risk nothing needlessly. One of the doctors here is a very interesting Italian, Agnetti by name. He was born in humble life at Parma, did well at College, became doctor, and settled here, much disparaged by his already settled competitors. There was then a government movement in favour of introducing suitable plants, and people having gardens were invited to help. Agnetti had a small plot and distinguished himself by what he did in planting and reporting, so much so that he was made "Commendatore,' which gave him considerable social position. He doctored me when I was here before and I thought him a particularly capable and pleasant man. Now he has become fired with political zeal and has been elected representative for Parma. So he is now "Onorevole," a much coveted distinction, and sees his way to combining parliament in Rome with physic here. It seems odd. I have not yet seen him, only messages have passed. He was full of the Italian quinine treatment and had good stories about it; one to the disadvantage of Koch, the Prussian, whom the Italian doctors hated for his arrogance, but the story is too long to tell properly. Briefly, Koch looked at a patient who seemed dying (in a ward placed at his disposal) and simply said: "Let his body be kept for me when I come to-morrow." The Italian physician thought, after Koch had gone, he might fairly intervene, so he injected quinine into the man's vein. When Koch called the next day the patient was sitting up in his bed devouring a hand-full of macaroni!! Of course the Italian doctors were delighted at Koch's stare of astonishment.

Ever affectionately, with loves to you all, FRANCIS GALTON.

CASA ——, BORDIGHERA, ITALY. *March* 12, 1905.

DEAR SCHUSTER, Enclosed are heavy but important letters, every one of them for you to read, and those to Miss Kirby, Miss P. Strachey, Sir J. Crichton-Browne for you also to forward. You might like to correspond with Dr Urquhart, and even with Miss Philippa Strachey and the rest. If so, write on official paper and enclose my letter with yours. I send stamps. You will see about the latter in the Strachey biography. She is very accomplished and *might* give useful help.

I am strongly inclined to think that, as Dr Mott has the insane in hand, you would do well to *concentrate on the feeble-minded*. My reasons are based not only on what I hear from you about Miss Kirby, but especially from what was told me a week ago by a most intelligent lady-doctor, who keeps a "home" in London (in Wimpole Street I think), Dr Lillias Hamilton, of Afghanistan celebrity. She was nurse and doctor to the late Ameer during five or six years.

Well, she was two years at a home for the feeble-minded, and explained to me the careful loving way in which the lady nurses inform themselves of the patients' family history—and their wide-awake scientific knowledge too. She promises to send me information, and I rely much on her. If you see your way to act on the lines mentioned in my letter to Miss Kirby, it seems quite possible that you might do a really big and useful thing, that would be your *cheval de bataille* on which to win the approval of the London University. You will have *zealous women* to work with, and the aid of women who are zealous (and wisely directed) is invaluable. Think well of this.

The refusal of the Life Medical Officers Association seems to finally extinguish our hopes in that direction. Dr Urquhart opens other fields. Don't *merge* your work in Dr Mott's. If he is working hard in his own province, be chary of trespassing.

Mr Eichholz is a first-rate man. I mentioned him to you as having given by far the best evidence before the Physical Deterioration Committee. By all means cultivate his acquaintance and seek his help. The Jews are a singularly well looked after body. I have seen a little of their organisation and know how thorough it is. Very faithfully yours, FRANCIS GALTON.

March 31, 1905. On and after April 10 to May 1, we shall be at Villa Stratta, Bordighera.

MY DEAR WELDON*, Your photos sent to Miss Biggs are wonderful. I wish you had brighter and warmer weather. We have lots of sunshine but of course nothing of historical grandeur. You have justly convicted me of gross geographical error. Another, one is liable to make, is to suppose Dover to lie nearly south of London. "Humanum est errare" and I feel at times very human in that respect.

Your mice give an unending problem. It is grand to have five generations. I don't believe anybody would have appreciated your work more than Mendel himself had he been alive. Dear old man; my heart always warms at the thought of him, so painstaking, so unappreciated, so scientifically solitary in his monastery. And his face is so nice.—I can't give you any useful hints. I wish I could. I am just a learner, and bad at that now. During the last week or fortnight I have been busy with my "Measurement of Resemblance," and am getting it into Royal Society paper state. It comes out all right. The only question with me is whether to wait, or to give it only in a theoretically complete form. In the first case, I should illustrate it photographically and provide apparatus to show; but I feel I have not power now to do such things properly, so I shall probably content myself with the theory for the present, and give minor illustrations.—Schuster seems eager and thorough. He has had a week or so of old work on skulls to revise arithmetically, but he has done that. He has useful relations too, whom he can get to give some help. There is quite a large, vacant, and promising field of work, anent the "feeble-minded." Very capable and enthusiastic ladies work up the family histories and are anxious to be of use. With a little intelligent direction they ought to be of much use. We shall see.

The sensation of the Riviera is the motor-boat competition. The boats will all arrive at Monaco to-morrow, and the show and races are to go on for more than a week. On April 10th we change our quarters, having rented Villa Stratta, Bordighera, till May 1st, and then home. Kindest remembrances to Mrs Weldon. Ever yours, FRANCIS GALTON.

VILLA STRATTA, BORDIGHERA. Easter Sunday, 1905. But post nothing here later than Thursday next, April 27.

DEAREST MILLY, All things come to an end, Riviera residence included. We leave next Monday morning, with many regrets, but still desirous of change. All visitors feel the air less good about this time, and begin to go. We propose to return leisurely; it is difficult to fix by which way, on account of Italian railway strikes. Your May 22nd ought to be a charming time for Brittany, if not still too cold. It is a land unknown to me, which I keep as a preserve to go to, some future day. I do not realise yet where Paramé is†. I happen to know a good deal

* I much regret the paucity of Galton's letters to Weldon. I have all Weldon's letters to Galton, but few of his letters to Weldon have survived, and those only by being mislaid, for Weldon systematically destroyed all the letters he received. I doubt the legitimacy, or at any rate the wisdom, of such destruction, especially in the case of men as noteworthy as Galton and Huxley.

† On the coast slightly east of St Malo.

PLATE LVII

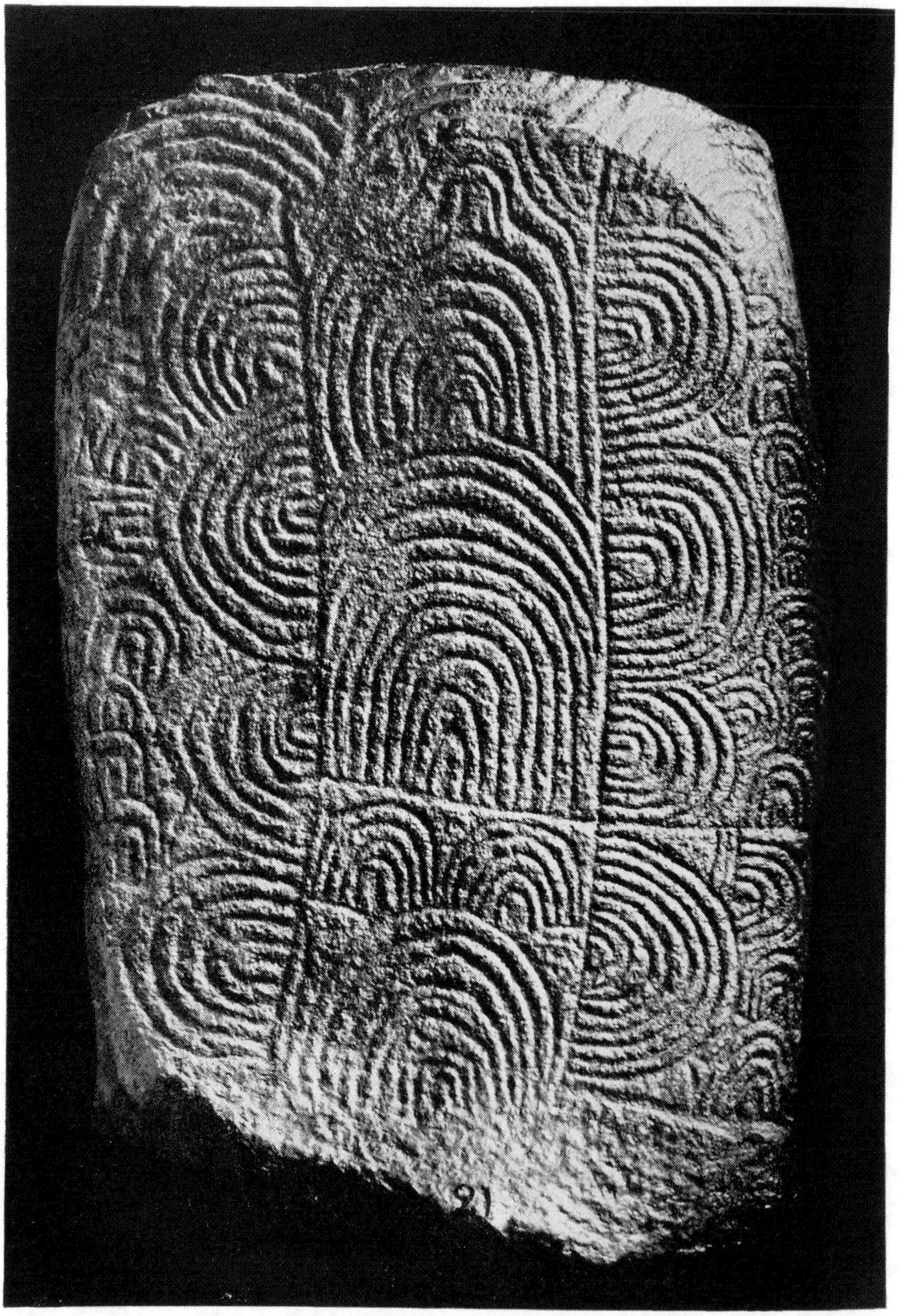

Sample of the conventionalised Finger-Print Ornamentation on the Stones at Gavr'inis, from the series of photographs in the *Galtoniana*.

about Gavr'inis and have photographs of the big stones—casts of them are in the Museum of St Germain. They are cut apparently as conventional renderings of the marks made by a bloody thumb or finger on a flat surface. They are certainly not exact copies of any real finger mark, being far too regular, but their patterns seem clearly to be based on the general appearance of one or more. The museum authorities allowed me to have the photos to examine. My resemblance problem hangs fire, for the makeshift apparatus I have been using proves inadequate, and I must get some (of which I possess the essential parts in London) properly fitted together. There are many alternative ways of carrying out the same principle and I am somewhat bewildered which finally to adopt. The subject too has many ramifications and I ought to show many illustrations. So the whole thing must wait awhile and mature. The greenery with you in England seems little short of what it is with us. There are however not many deciduous trees here to judge by. One horse-chestnut is in bloom, but the mass of the verdure is olive, palm and orange. What a sight a flourishing orange garden is! One understands their ancient name of golden apples. How pleased you will all be with your holiday trip. Best love to you all in which of course Eva would heartily join. Miss Cuénod asks after you. Do you recollect her at Vevey? Ever affectionately, FRANCIS GALTON.

Letter of Erasmus Galton to his brother Francis.

ROYAL VICTORIA YACHT CLUB, RYDE, I. OF W. *May* 3, 1905.

MY DEAR FRANK, I am so very glad to hear you are now quite well and on your way home. Yesterday was bitterly cold, but this morning we have sunshine and all appearance of summer coming on. Your idea about fruit trees is excellent in theory but not in practice. Fruit, to be first class, must have sunshine and room. Fruit trees planted as you saw them at Loxton have plenty of it, and have two wide avenues and two narrow ones, so that carts, bush harrows, and mowing machines may pass between the trees, in fact everything can be done by *horse* cultivation in place of manual labour. For instance, hay is cut, made, stacked and finished entirely by machines. Turnips are cultivated in rows of from 28 to 32 inches apart, cabbages still wider to allow horse hoes to work between, one horse and one horse hoe easily doing the work of twelve men. The Royal Agricultural Society's *Journal* of this quarter gives a long account of fruit farming, which I think you should read before sending in your paper, which paper I enclose in this letter for your re-consideration. I would advise sending it to the Royal Agricultural Society *Journal* or to the *Field*, to which papers I have sent a few articles which they accepted. Ever very affectionately yours, ERAS. GALTON.

P.S. Bessy has been so good as to tell me every fortnight about you.

42, RUTLAND GATE, S.W. *May* 13, 1905.

DEAREST MILLY, My letter is belated, for you have no Sunday delivery, but there is nothing to say. Eva and I go to Claverdon on Monday, for four days or so. We are nearly square again at home. There is now a mahogany rail put into my house, from the ground floor up to the second floor, up which I pull myself like an orang outang, and find it very handy in descending also. You will be very glad to be off and enjoy spring and change in Brittany. I feel now as though the past winter were a half-forgotten dream. The first letters almost that I opened on returning, were to say that the Council of the British Association had nominated me as President next year at York. They were very kind, assuring me that I need not attend Committees on account of my deafness, and might absent myself much, leaving the duties to a Vice-President, but I dared not risk it. The social duties are what chiefly knock me up. I think I could get through the Address, but even that, with my uncertain throat, would be a doubt. So I refused at once. Something of the same kind occurred to me before, and not only once, but I am conscious of many limitations to my strength, and then, as now, declined. It is a bore to renounce the opportunity of having so good a pulpit to set forth one's fads; it is in fact a unique opportunity for addressing all men of science and the public as well. George Darwin will have a very fatiguing time in S. Africa. He has to give two addresses, one at Cape Town and one at Johannesburg, and the travelling will be very long. It is a great way, and by slow trains, to the Victoria Falls. I fancy more than 48 hours each way, and there is ever so much more to be done. The *Diplodocus* (big beast 90 feet long, when measured along the undulations of his back) is at last on view. I shall call upon him to-morrow. You may have seen in the papers an account of the public presentation of him to the Natural History Museum yesterday. My Eugenics Research Fellow has been grinding on, but possibly he needs more *go*. Statisticians, like the children of Israel in Egypt, have not only to make bricks but to collect materials. Here it is that men differ so much in their success. The most hopeful line just now seems to be in the direction of the feeble-minded, about whom a Commission is now sitting. Several eager and capable ladies are engaged in the work, and they seem desirous of scientific guidance, so I hope something may be done there. They are to have a big meeting next month and are preparing their programme of work. I am so very glad that many of your family anxieties are over. Amy will, I trust, improve under the sky of Brittany. It is said to be a rainy part of the world, but it cannot always rain. At Marseilles and at Paris it *poured* while we were there, two nights at each place. I ate a *Bouillabaisse* at Marseilles which had been an epicurean dream for years. They say you ought never to eat it unless you have a spare day to get over the effects. It contains a vast variety of shell-fish, as well as other fish, which may be half poisonous. However, mine proved particularly digestible.

Affectionately, with many loves, FRANCIS GALTON.

42, RUTLAND GATE, S.W. *June* 13, 1905.

DEAREST MILLY, Your card of this morning gave great relief. The weather is all in your and Amy's favour now, but the "flu" is a nasty thing. My sister Bessy seems to have had a touch of it. Temperature only 100, but continuously, or almost so, for a week. She has been in bed at Claverdon. Since Saturday I have not heard. I think they *were* anxious about her. We, thus far, are all right. To-morrow I go to Cambridge where there is the function of degree-giving, lunch and dinner, which I hope to digest. A few days ago I was invited and went to a big *Statistical* dinner, at which when the visitors' healths were drunk, after talking about me, the proposer said I should leave my mark—he would not say on the *foot*-prints, but—on the *finger*-prints of time! Rather forced, but it did for an after-dinner speech. About a week ago, Eva and I went to the Farm Street Roman Catholic Chapel, to hear "Father" Galton preach. He is not the Bishop of Demerara, but Charlie Galton. Two of Theodore's sons became priests*. He preached uncommonly well, with singularly good articulation, as though he were fond of the sound of every word he uttered. He would be an excellent master of elocution. The chapel itself is one of the most beautiful and decorous I have ever seen. The congregation most reverent, and the music perfect. As you will have heard, and perhaps experienced in Brittany, we have had the rainiest week almost on record, to greet the King of Spain. I passed Windsor to-day and saw the King's flag flying. They are making ready a royal wedding for a king-to-be, but of only half the kingdom—Sweden—that he expected to have. Good-bye, love to you all and may you all pull happily through this hateful scourge of influenza.

Ever affectionately, FRANCIS GALTON.

42, RUTLAND GATE, S.W. *June* 25, 1905.

DEAREST MILLY, What a very gallant act of Guy! I wish it had been some millionaire whose life he saved. It was such an English act too, unselfish, single-handed and prompt, while others were "disposing themselves" to launch a boat. I cannot realise how with only one arm such a feat could be done, though I know he used to be an excellent swimmer. I suppose that the water was not deep and that the rough sea was not dangerous to a man accustomed to water, and able to keep his wits cool, and that Guy was able to touch ground and to push. It would have been most dangerous had the drowning man retained enough vitality to grapple. A sea bath is usually ruinous to clothes and watch. I hope he had nothing on or with him that suffered much? The excitement of this family event may have harmed, or may have helped, Amy in her convalescence. I sincerely hope the latter. Much has happened here during the past week of "Eugenic" interest, but it relates chiefly to administration, which was more than my "Fellow" could manage, together with research. So a readjustment of duties has had to be made and there will be a Lady Secretary. Also Murray, the publisher, will publish for us (on the half-profit plan) books of families on the same principle as that little pamphlet you saw, but on a substantial scale. There is material for one now, that Schuster has put into order, but to

* Theodore Howard Galton, Francis Galton's cousin.

which I shall have to write a preface. We have taken the Rectory at Ockham for six weeks, beginning with August 18th. Ockham is in Surrey, north of Guildford, and we have friends near, especially a very old and kind friend of Eva, who has suddenly become quite blind and whom she wishes to cheer and read to. I am afraid that my sister Bessy has been much pulled down by her influenza. The doctor has compelled her to stay in bed more than she likes. She is very calm and cheerful, but feeble. I hope to see Edward* to-day, who is in London about cattle shows, etc., and with whom it was fixed to go to the Zoo this afternoon. Penelope† dined here yesterday. I had two Syrians to take care of. One of them took great and hospitable care of us when we were in Egypt, so I had a little party for them. I wonder when and where we shall next meet. Amy's perfect recovery must be your primary occupation. I am for the present drifting aimlessly, but with a great deal of work to do ahead, for I must now "boss" these Eugenic matters a good deal, to make them "hum" as the Yankees are pleased to say. Arthur Butler and his daughter slept here the night before last and the Master of Trinity came yesterday morning. There had been a great function at Haileybury College. Lyttelton, the old Master, taking his leave, before going to Eton. Arthur B., you may recollect, was elected long ago to re-create Haileybury on its present footing and lost his health finally in doing so. I will keep this letter open till to-night in case there be anything to add about my sister Bessy. She is on the sofa: no anxiety.

Ever affectionately, with loves to all, and with much respect for Guy, FRANCIS GALTON.

CLAVERDON LEYS, WARWICK. *July* 31, 1905.

DEAREST MILLY, Your letter was full of information. I came here on Saturday. Eva writes from "The Log Hut, Teigncombe, Chagford" very happily. She has "a dear old curtseying woman to wait." Thank you much for your sympathetic telegram and writing. My precious Eugenics has now been advanced a notable stage in University recognition and ought to prosper. Murray is in full swing printing. I go to the Lakes to-morrow, touring about through once familiar scenes till Saturday, when I get to Highhead Castle, near Carlisle, for a stay of four days, or five, with my old friends the Hills‡. A letter there would rejoice me. I saw Bessy this morning, wonderfully well in face and talk, but rheumatic. Otherwise she would have been here in Claverdon, whither I have brought a calorif§ for her amusement. Edward and I have been constructing a mechanical "toss-penny." I want to illustrate what I have to say in my preface about Statistics that "chance" means merely the result of unrecorded, and by no means necessarily of unknowable, influences. The example I take is that of tossing a penny, which is typical of a "chance" result but which few would deny is the result of pure mechanism. I thought it would be well to see what sort of influence on the results would follow by using a machine, how far the chance could be reduced to a certainty. So we made a machine, and though it is a little shaky and uncertain, on one occasion it gave two sequences amounting between them to 48 "tails" out of 50 throws. A minute change in adjustment greatly alters its action, so a good "toss-penny" ought to be as well made as a gunlock. The new owner of E——, Mr Z——, is here to-day. He is quite a stranger, young and perky. Emma Phillips|| never saw him, which is rather unfortunate. But he will not enter into possession yet. Indeed he could not for two years, as the house is let. He comes from the Colliery district near Bristol. Love to Amy. I trust she is now settled in her new bearings. Ever affectionately, FRANCIS GALTON.

Archiv für Rassen- und Gesellschafts-Biologie, BERLIN-SCHLACHTEN-SEE, VICTORIASTR. 41.
zur Zeit, SWINEMÜNDE, *den* 17 *August*, 1905.

DEAR SIR! We thank you very much for your kind answer and permission to translate your paper on Eugenics! Excuse, please, the delay of this letter, since I was travelling in the last time and lacking the necessary leisure. Regarding the permission by the Sociological

* Edward Wheler, Galton's nephew, who had succeeded Darwin Galton as squire at Claverdon.

† Darwin Galton's widow, Galton's sister-in-law.

‡ The lady of the house was a daughter of Francis Galton's old friend, Mr Justice Grove.

§ I suspect this stands for "calorifère," a heating apparatus, such as old ladies from the Midi sometimes placed when sitting down under their ample skirts to keep themselves warm.

|| Sister of Mary Phillips, Darwin Galton's first wife.

Society, I shall apply for it to the Secretary. We take the highest interest in your eminent and important Eugenics, which is so closely connected with the subject of our *Archiv*, and shall keep our readers acquainted with the further development of your ideas. That you will belong to our readers is, of course, a great satisfaction for us. We hope that an article in the now appearing number, "Die Familie Zero," the history of a family with its degenerating and regenerating branches, will be of interest for you. From your standpoint you perhaps take also some interest in a little book, which I published ten years ago and which I allow myself to send you with the same mail. I started from an English use of the word "race" and tried to investigate the conditions of preserving and developing a race—race-hygiene ("Rassen-Hygiene"). Afterwards, in the first introducing article of our *Archiv*, I tried to sharpen the meaning of the word "race," so as to make it suitable for the theoretical and practical needs of a man, who will seize the real *long* (beyond the individuals) lasting unities of life, their conditions of preservation and development. I should be very much indebted to you, the senior of the practical application of the principles of evolution on man, if you would in an hour of leisure read my essays and write me your cool judgment. My book is written mostly in a small town, where I practised as physician, absent from a good library and therefore without much knowledge of current literature. That, together with the haste, with which I was compelled by my editor to deliver my paper to the press, may declare many omissions in respect of modern authors.

Excuse, please, my bad English. I am sitting here at the sea without a dictionary, and have to feed on that little English fat which I have by and by gathered on my German body during my lifetime. Yours highly respectfully, ALFRED PLOETZ, M.D.

To SIR FRANCIS GALTON, LONDON.

THE RECTORY, OCKHAM, SURREY. *August* 20, 1905.

DEAREST MILLY, I am so glad the French Humane Society have done their belated duty to Guy, and I return the scrap of newspaper, which you will wish to keep. We came here on Thursday in beautiful weather, and had our tea on the lawn in a selected place by a big tree. But the bees began to buzz alarmingly; and well they might, for they had built a thriving hive in the hollow tree, and were flying in and out of a hole therein as fast as they do in an ordinary hive. No harm was done. We changed our place quickly enough. It is all green fields here, much timbered, chiefly with oaks, and very English. We are three or four miles from the Downs near Guildford, and go to-morrow afternoon to tea at a friend's house on the top of them—Sir H. Roscoe, the chemist's—600 feet above the sea; this is a not uncommon height hereabouts. Our "landlord," the clergyman, is our guest for last night and to-night for his Sunday duty. His wife, Mrs Ady, is a well-known writer, chiefly on Italian subjects. The book that first gave her her reputation is well worth reading, if you care about our Charles II and his sister "Madame" (the title of the book), who married the brother of Louis XIV, and did a world of sisterly good*. She had all the grace and not the faults of a Stuart. She died young, immensely regretted in France. What an adventurous drive, both for you and Patrick†, and then the sad Princetown. Why don't they use false webbed soles for swimming? They ought to get through the water much faster if they did. A neat patented design might bring in lots of money, if brought *well* out, just before the bathing season. George Darwin's Presidential Address at Cape Town (the first part) is first-rate. I am most curious to read the second part which will be delivered at your *favourite* Johannesburg. I suppose, as time goes on, that place will purify itself as the American gold-digging camps did. "Honesty, boys, is the best policy: *I tried them baith.*" Will the Japanese send missionaries to Exeter Hall? Their reception would be amusing to a cynic. My Lady Secretary begins work to-morrow. Ever affectionately, FRANCIS GALTON.

THE RECTORY, OCKHAM, SURREY. *August* 26, 1905.

DEAREST MILLY, Don't regard this magnificent address stamp. (That of the Royal Institution of Great Britain.) It is only that I am up in London to-day and am writing here. It is a quiet

* I have retained this paragraph, although it is repeated in the following letter. This is the first occasion on which I have come across a repetition in letters to the *same* person, a sure sign that Galton, strong as he remained mentally to the end, was still liable to one at least of the failings of old age—he was now 83.

† A much treasured horse of Mrs Lethbridge.

place, with lots of books, and my club is temporarily closed for repairs. I quite forgot this morning to put your letter in my pocket; I feel sure there was something in it I wanted to write about, and have forgotten it for the moment. All goes on happily with us. We have friends about and the country is delightful. Perhaps too many wasps; one stung Eva yesterday through a thin cuff and left ever so much poison in a stain on the cuff. I read somewhere that more people died—I forget where—of hydrophobia through bites of *wolves* than of dogs, the reason being that wolves fly at the face, but dogs bite through the clothes, wiping their teeth thereby, so their bites, as a rule, are far less dangerous than those of wolves, though much more numerous. You would enjoy seeing Upton Warren, where the widow of my old friend Charles Buxton still lives. Fondness for animals is the tradition of the house. They have parrots that fly loose in the woods and sometimes build nests, and there are very many artificial birds' nests, nailed against trees about five feet above the ground. I was told that they often opened the lids to see how the eggs or broods were getting on. Why don't you try a few at Edymead? They must be arranged (i) so that a cat can't get at them and (ii) so that a tom-tit's reasoning powers would be satisfied that it could not. The birds who used these artificial nests were principally tom-tits. There is an excellent library at the Rectory. The clergyman's wife, Mrs *Ady*, better known as Julia *Cartwright*, has written not a few important biographies, that of "Madame" (the sister of our Charles II, who was married to Monsieur the Duke of Orleans, brother of Louis XIV) is one of the best and well worth getting from Mudie. She was a far more interesting and good person than I had any idea of, and played an important sisterly part in politics. She died before thirty, immensely regretted*. One knows so little of the actors on the big stage of the world, so big that there is room for many important ones. Loves to Amy and to all with you.

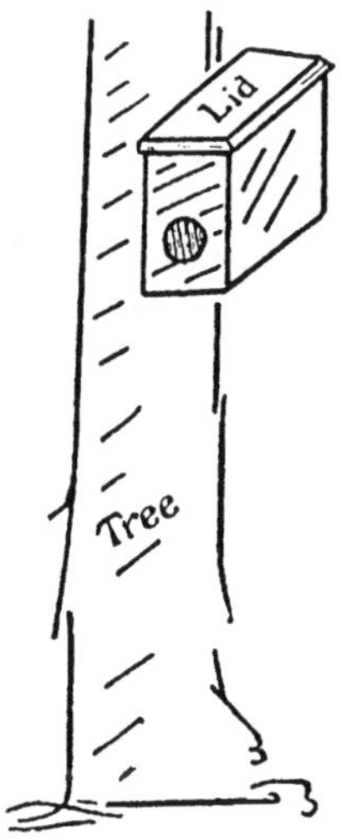

Affectionately yours, FRANCIS GALTON.

THE RECTORY, OCKHAM, SURREY. *September* 11, 1905.

DEAREST MILLY, Poor Patrick †, "hors de combat"; add an *e* and he would be a war-horse, a "horse de combat." You can gather my state of weakness of mind to attempt such a pun. I was bothered here to write a motto for a sundial, and after many attempts wrote this, "Love rules Man, Sunshine rules Me"—not wholly bad; anyhow it is a new one. "Vivent le roi d'Angleterre et M. le Capitaine Lethbridge"! Don't let King Edward hear of it, or he will be still more savage than he is said to have been when General Baden-Powell struck coins under his own name at Mafeking. What an excellent time the British Association has had in Africa. We are very well placed here and happy in a quiet way, which I like above all things, with a scrimmage now and then to stir us up. Tin-foil is a trouble to get good. They adulterate it with lead, it looks equally shiny but is not so good. I shall be in London to-morrow for a few hours and expect to pass near a trustworthy shop. If I do, I will get and send you some. Your garden must be very pretty, ours has lost its best flowers already. Excuse more, as there is much to do and little free time before post. Loves to you all. Ever affectionately, FRANCIS GALTON.

THE RECTORY, OCKHAM, SURREY. *September* 11, 1905.

MY DEAR BESSY, You will be glad to have Gussy back. How much did Edward and his colleagues fine the motorists? Eva was taken in a beautiful one, and said she felt her *disposition worsening*. Every minute she felt more careless of other people on the road and more superior to them, and it was doing bad to her morals! It was Lord Rendel's motor and probably made at the Armstrong works, in which he is a partner. Life passes very pleasantly and quietly here. Now and then an interesting luncheon or tea. There are very nice people hereabouts. Our Oxford friend Professor Weldon stayed with us last night and we went this morning to look at the "Swallows," or big pits made by rain in the chalk. One was as big as a small Coliseum, we did not go down to the bottom where there was probably a hole into the depths. Another was not

* See the footnote to the previous letter.

† See the last footnote p. 546.

a "Swallow" at all but an immense chalk-pit with vertical sides 100 feet high and grass and trees growing in it. Quite a charming place to spend hours in. It is melancholy how late the sun gets up now, and how early he goes to bed. Did you hear of Guy Lethbridge's reception at the banquet in France, and of the toast (in French) of "King of England and Captain Lethbridge"!!! He had better not communicate the news to King Edward, who might not like it! They gave him *both* the medal and the diploma. I am very glad of it; it will brighten him up and he richly deserved them. Loves to all. Ever affectionately, FRANCIS GALTON.

THE RECTORY, OCKHAM, SURREY. *September* 17, 1905.

MY DEAR BESSY, Remember me to Fanny Wilmot*; you will be glad to have her by you. I am very sorry about your continued rheumatism, etc., especially as cold weather is coming on gradually. Edward will I hope be able to give much help to Erasmus, who thoroughly appreciates and trusts him.

I have at last finished my small but troublesome book, and sent it yesterday to a friend to criticise and revise, before it goes to the printer. Its title is *Noteworthy Families* (*Modern Science*). I hope the rats at Claverdon do not mean drains out-of-order. There used to be a professional rat-catcher to be seen in London who has long since disappeared. He dressed in a sort of uniform, I think a greenish coat, with silver rats round the collar and a leather band crossways, also with silver rats. He was very picturesque, and reputed a great scamp.

All goes on well here. Frank Butler is with me for a couple of nights. We have many nice friends within reach and I shall be very sorry to go, as we must, on Thursday week, so I shall only get one more weekly letter from you here.

Only think of Mrs Gilson and others going as a matter of course to Khartoum and even to Gondokoro from the North, and the whole posse of the British Association going to the Zambezi from the South, and those places being actually undiscovered until lately. Bruce first wrote about Khartoum, but not much, and Livingstone of the Zambezi. It is much the same in N. America, where Fenimore Cooper's scenes of prairie and wild Indians are now big towns. It is a good story you send from Grant Duff about D'oyleys. The Sandwich Islands were of course called after Lord Sandwich, who I presume was then at the Admiralty, but how did the things we eat come to be called "Sandwiches"†? You know of course the old riddle: "How can sailors, wrecked on a barren coast, support life?" *Answer*, "By eating the sand-which-is there." The town Sandwich is an uncommonly interesting old place and so is the ancient Richborough which is near it, with its big fortifications, and which was the main landing-place when Kent was covered with thick forest through which were very few roads.

Ever affectionately, FRANCIS GALTON.

THE RECTORY, OCKHAM, SURREY. *September* 17, 1905.

DEAREST MILLY, Poby...stett‡ (I can't spell the name) vastly attracted Bishop Creighton when he went over to the Czar's what-was-it? There is much about him in Creighton's "Life." I wish that I could be sure whether it *was* he who was the ecclesiastic about whom Archbishop Benson told me in connection with the Jew persecution. I dare say Torquemada was an amiable man to some. Wasps, too, may be beloved in their own nests. So glad the Dartmoor pony is good and fit. We are dependent here on a dear old pony, sixteen years old at least. No, he is older than that—that is the age of the still dearer but less useful dog. My plaguy little book is finished! I sent it off yesterday, partly typed and partly in proof sheets, to my critical friend, Collins, to score over with corrections as he is sure to do. I always learn much from him. An amusing *measure* of *memory* about relations has cropped up. The paternal uncles (*fa bro*), and maternal uncles (*me bro*), are recorded in exactly equal numbers, so they are equally well recollected, but the different sorts of great-grandparents and great-uncles add up in impossibly different proportions. There are four sorts of each. One of the sorts *fa fa fa* and *fa fa bro* bears the writer's

* A daughter, Emma Elizabeth, of Sir Francis Sacheverel Darwin had married Edward Woollett Wilmot of Chaddesden.

† Said to be named from the fourth Earl of Sandwich (1718–1792), who had provisions brought to him in this form at the gaming table, so that he might not be compelled to leave it in search of food.

‡ Galton is probably referring to C. P. Pobedonosteff.

surname; call this group A. Another sort *me fa fa* and *me fa bro* bears the mother of the writer's maiden surname; call this group B. The other two sorts bear different names; call them C and D. Then the numbers recorded are in this proportion: $A\ 1 : B\ \frac{1}{2} : C\ \frac{1}{4} : D\ \frac{1}{4}$, showing how largely family recollections stick to the family surname. It all comes out very sharp and clear and consistently. The male first cousins follow the same rule, but less emphatically. I had somewhat tired myself about the book, but am *quite* rested and well now. I wish Bessy was more free from her rheumatism, etc. She seems at present to be quite confined to her sofa, but writes as cheerfully as ever. Frank Butler came here yesterday evening for two nights, quite well and not apparently overwhelmed by his three little daughters. I begin to count the days that remain to us here. We have to go on next Thursday week. There are very nice people and not a few old friends within pony-trap distance. I called on one, Mrs Archibald Smith, the mathematician's widow, whom I had not seen for many years. Her hall was hung round with African trophies. There was a beautifully strong and light iron chain with loops in it, which I thought was some kind of chain ladder, the loops being for the feet. But it was a *slave chain*. A gang of slaves was found by her son, the men were released and the chain kept. The loops went round their necks. Another thing was what looked like a big firescreen, with black leathern drapery. It was made of the two ears of an elephant.

Enclosed is some tin-foil. I had an amusing hunt after it in London and learnt much. It is only made at two or three factories, partly for druggists, partly for wine merchants to cover their bottle-mouths. Ever affectionately, FRANCIS GALTON.

42, RUTLAND GATE, S.W. *September* 25, 1905.
(This will be my address now.)

MY DEAR BESSY, You are a "bonne *écrevisse*" in the sense the gentleman meant. One never gets the big crayfish to eat in London, but I see them in shop windows. They are the most divine-right-of-King sort of fish. The biggest one in an aquarium sits as it were on a throne and the others gather round like courtiers in the most comically humble positions. I know they are good eating and must get one when we return. We pack up to-morrow and leave here on Wednesday, but not direct to London, which we reach on Saturday evening. We are sorry to go, but have a store of pleasant things to recollect. Evelyn Cunliffe* was to come to tea to-day, but it rains and we hardly expect her, it is a long drive. It gets cold too at nights. I have started winter underclothing to-day, and wanted it. I shall be interested to hear Edward's report of Erasmus. It seems so dreary for him to be practically alone in that wooden hut, but he has friends near and likes it.

Thank you for Miss Johnstone's address; I will write soon to her. All my things are in arrear now, that blessed book has thrown them all behind. A packet with the MS. of it, addressed to the publisher, is at this moment lying on the table by my side. It will go off by the same post as this.

What a disagreeable intruder upon her finger Gussy† seems to have had. Suppose it had come suddenly beyond her rings! There is some Arabian Night, or the like, story of a man who has a ring of mystic power, about which he knows nothing and is on the point of selling it to a wicked magician, when his guardian fairy takes the form of a wasp and stings the finger, which swells, so the ring cannot be removed. I wish some fairy would give me a better pen than this to write with. It scratches like a needle. Best loves. Ever affectionately, FRANCIS GALTON.

42, RUTLAND GATE, S.W. *September* 27, 1905.
(This will be my address now.)

DEAREST MILLY, The convicts must have been depressing. They are not however so homeless when set free, as big societies work in unison with Government to take care of them. But a broken-kneed horse and an ex-criminal are not favoured. It is all very sad. Government *can't* set up a factory, for all the trade unions are up in arms against competition by *state-aided* workers. We pack up to-morrow and leave on Wednesday, not directly for home but for three nights with friends near, and return to Rutland Gate on Saturday. It was amusing about your

* Sir Douglas Galton's elder daughter.

† Second wife of Herman Ernest Galton, Francis Galton's cousin: see our Vol. I, Pedigree Plate A.

dog and the looking-glass. Probably the little creature was terrified because the reflection did not *smell*. We shall be very sorry to leave; the people about are very nice and sociable and the quiet country is delightful. My little book is as troublesome as an ague, I thought it was off my hands but it has bothered me up to this instant, when I sealed up the MS. in a packet to go by post to Murray. And still there are odds and ends left and revises to come, etc., etc. But it is comparatively calm now. And it is such a small book after all. My friend F. H. Collins, who is a prince among proof correctors but cannot now leave his arm-chair, has been giving all his working time last week to putting Schuster's contribution into better shape. The material was good but the arrangement too higgledy-piggledy. I started winter underclothing this morning. Among the people we have met is that wonderful Arab-horsey lady, Lady Anne Blunt. She had a great deal to tell. She and her husband go to Arabia to buy horses. She lives by the Tombs of the Kings near Cairo where a stud is kept, and they have annual sales in England. She is apt to appear in marvellous dresses, of some outlandish cut and colour, not necessarily Arab. She came out on one occasion in bright scarlet from top to bottom, as I heard. She is grand-daughter of Lord Byron, so may do mad things with propriety. Best loves.

Ever affectionately, FRANCIS GALTON.

42, RUTLAND GATE, S.W. *October* 27, 1905.

DEAR MR CONSTABLE, I am flattered that you have thought my book worthy of attack, hip and thigh. You have chipped off many bits of paint but I am so incurably self-conceited that I do not yet feel any timber to be shaken. If I were to reply in print I should fix on the second paragraph of p.138 and follow out the conclusions to which it leads.

You will be scandalised at a forthcoming volume, *Noteworthy Families* (*Modern Science*), but if you see it, I think you will find the Chapter on "Success as a Statistical Measure of Ability" worth reading. Now I not only take your scourging with a smiling face but have the impudence to ask if you could get the enclosed forms suitably filled up for me? If you do, the reply will probably arrive after I have left London (for Pau) for the winter. Therefore the address at the bottom of the Circular is the best to use. Faithfully yours, FRANCIS GALTON.

I send this *via* your publisher, being not sure of your present address.

42, RUTLAND GATE, S.W. *October* 28, 1905.

MY DEAR BESSY, I more than fear that it would be very unwise for me to yield to the pleasure and wish of seeing you, before we start for *Pau* next *Thursday*. I had a sharp attack of shivering on Wednesday morning, and the doctor sent me to bed on fever diet all Thursday; yesterday the fever went, and to-day I may get downstairs a little while. He says I ought to be fit to start next Thursday, and the sooner I get away the better. So I must reserve every ounce of strength for the longish journey, and fear much that a long day to and from Leamington beforehand is more than I can stand. As soon as I cross the channel, as a rule, I feel better in breathing and general fitness. I am very sorry indeed. I wanted so much to see you and Erasmus before these many months of banishment. Louisa* will write her views and she must represent me in person.

You always take such interest in family matters, such as mine, that I send you a letter just received from the Master of Trinity College, Cambridge. It is about a copy of the portrait which Charles Furse painted of me. I heard unofficially that the Fellows of the College would be very glad to have one, so I got an excellent copy of it made by Frank Carter, and sent it with a suitable letter. You will see that they accept it both warmly and gratefully. It will be hung according to the recommendation of their "Memorials Committee," probably in the Great Hall alongside of many far more distinguished worthies. Anyhow, as a *picture* it would hold its own in any collection. Don't *destroy* the letter. It ought to be preserved somewhere. If it can be copied and returned it would be a good plan. I am very sorry that the rheumatism continues. Your news of Lucy and the Colonel is not quite as good as we could wish. You will have been hearing much of Lord Leigh's funeral. The death of a foremost man in a county must leave a large void for a time. Before we go, I shall certainly write again and send my address, which cannot be fixed until the reply of an hotel-keeper to my note arrives. It is due this evening or Monday morning. Ever very affectionately, FRANCIS GALTON.

* I believe this is a slip for Eva, Galton's great-niece and comrade. He used by accidental habit the name of his dead wife. See the following letter.

Enclosure in letter above:

TRINITY COLLEGE, CAMBRIDGE. *October* 27, 1905.

MY DEAR FRANK, The Council is over and I am desired in the name of the College to thank you warmly for this beautiful gift and to say that it is gratefully accepted. It is left to the Memorials Committee, of which I am Chairman, to consider the question of where it is to be hung, and to report to the Council. We are now arranging for a very early meeting of the Memorials Committee. You know how very earnestly I hope that this noble portrait will soon be on the wall of our Hall. Always affectionately yours, (signed) H. MONTAGU BUTLER.

42, RUTLAND GATE, S.W. *October* 28, 1905.

DEAREST MILLY, So glad that you would like to have James's book. It shall be ordered this morning, but, being Saturday, you will hardly get it before we are off to Pau on Thursday morning next. I have had a stern reminder not to delay, in the form of a sudden severe shivering for nearly a couple of hours on Wednesday morning. The *amplitude* of the shiver was remarkable and interesting; my hands shook through a range of fully 7 if not 8 inches. The doctor sent me to bed at once (two days before yesterday) on fever diet; yesterday I was much better and to-day I may leave my room a little. He promises that I shall be fit to go on Thursday and recommends it. So much for self. You recollect my picture by Charles Furse? The Master of Trinity saw it and wrote me a letter urging me to send a copy of it to Trinity College, Cambridge, where he felt morally sure it would be accepted by the Council of Fellows. Asking elsewhere, I heard the same thing, so I had an *excellent* copy of it made by Frank Carter, and sent it for acceptance. The Council met yesterday and accepted it "warmly" and "gratefully." The place where it is to be hung is referred to the Memorials Committee and will, as I am told, in all probability be in the *Great Hall*, alongside many of my betters. It is a great honour anyhow. I could never have dreamed in old times that they would elect me an Honorary Fellow and care to have my portrait. What a nuisance your Range must be. It might have occurred in bitter weather, so that trifling favour is something to be grateful for. You don't speak of the poor horse. If you want a nice *animal* book, get from Mudie *The Call of the Wild*, by Jack London. It has had an immense sale. I read it through yesterday from cover to cover, almost without stopping. They tell me that good as Hôtel Gassion is, a new one, H. du Palais et du Beau Séjour, is better. It is quieter, has a *large sunny balcony*, and the same south view. Moreover it is next door to the Winter Garden; so I have written for particulars which should arrive to-night. I will recollect about the "K" in Acland. Thanks for the introduction promised. I am living in hope that I may get the revise of my little book *in page* all sent to me to-night. If not, before I start, I must delegate the final look over and the index making to Schuster, but I should like to have a share in it. Always, at the very last, there is some difficulty to be settled. I think now, what with Schuster's willing help, Miss Elderton's business-like ways and the Advisory Committee, the Eugenics Office ought to run on its own legs while I am away. I will write again, at least a post-card, before we are off. But I *dare not* give a previous day to Leamington. Every ounce of strength must be reserved for the Pau journey, but Eva will go to Bessy for a day. Ever affectionately, FRANCIS GALTON.

HÔTEL GASSION, PAU, FRANCE. *November* 14, 1905.

DEAREST MILLY, At last, we are all again in normal condition and comfortably housed, with the splendid view (when we can see it) right in front of our windows in the very middle of the second story. I have not yet seen Mr Acland-Troite, on whom I called on Friday, leaving a letter explaining and saying we were about to change quarters. But I went to his church on Sunday. There were two clergymen and I do not yet know which was he. How the forms of Christendom do change! A Rip van Winkle, sent to sleep at the time when I was young, would have been bewildered at this service and have thought he must have mistaken the building. Do *clerks* still exist? The three decker arrangement of my youth has wholly disappeared, and one never sees the Royal Arms. It always used to be there with the White Horse of Hanover in its middle and often quartering the fleurs-de-lis of France. I recollect it gave me quite a shock, when I first went to Cambridge, hearing the choristers singing in their white surplices, but I had never, I think, at that time been at an English Cathedral Service. We

were brought up on such a Quakerish-Puritanical diet. The names of the Colleges also shocked me. Talking of Cambridge, I have now heard that the Council of Trinity College unanimously and warmly accepted the portrait, and that the Memorials Committee to whom the question was referred as to where it was to be hung unanimously recommended that it should be in the Hall (at a specified place), which was agreed to by the Council*. And there I presume it hangs at the present moment and may hang for an indefinitely long time. It is needless to say how pleased I am. Everything was done by the parties concerned in such a nice and kindly spirit. And Eva is equally pleased. We have had the whole gamut of Pau weather. At first it was wild and stormy, then perfectly beautiful; then more or less broken, and during the last two days a big thunderstorm, followed by swirls of rain with intervals of dry; now the sun is out and the weather promises to mend. What a picturesque place it, Pau, is in many parts, but I have not yet been able to get about much. The climate seems thus far to be something like that of Biarritz, damp soft air; perhaps like Rome too; without the dry, cold winds and piercing sun of the Riviera. It is quite a new experiment for me. This hotel is, as it was in your time, excellently managed and very clean, but rather dear. However I can stand that. We have two communicating bedrooms and Seabrooke's is just on the opposite side of the passage. A lift comes up whenever we ring for it. We have as yet made no friends here. The season is not yet begun. Those in the hotel are Russians, French and Americans, and one couple half-English and half-foreign (nice), and though the front rooms are full, those to either side of the big hotel are not. You recollect Charlotte Wood, afterwards Charlotte Batt, of old days? She died here. When Louisa and I were for a day at Pau we hunted out her gravestone, but I fear it will be difficult to identify it now after more than half a century has passed by.

My book, all except the index, has at last gone to Press, so you will get your copy about the end of this month, probably. I am so glad you like James's book. The criticism I would make on it is that he confines himself to *selected* cases. It would have been better if he had *also* given a *résumé* of *all* cases known to him, and of the experiences of doctors of the insane. George Fox must have been crazy when he went like a Jeremiah, and shoeless, into the heart of Lichfield. Best loves to you all. Ever affectionately, FRANCIS GALTON.

Wednesday, Nov. 15. I reopen the letter to add that Mr Acland-Troite made a long call yesterday evening, and was most pleasant. He told us ever so much, and has already undertaken to get me an introduction which I wanted to the Director of the great horse-breeding establishment here. Thank you so much for the introduction. He struck me as a cultured gentleman, full of interests, the chief of which was his church, which he called his *baby*.

There are two other English churches here, plus a Scotch Presbyterian: four in all!! His wife just now is a little unwell. So we are not to call just yet. F. G.

Fragment of a Letter to Mrs Wheler (Galton's sister Bessie) written in 1905 from Pau.

To go on with my broken off letter; I shall be glad to hear that Lucy's visit to Southampton did her no harm. It is very unfortunate for Col. Studdy that both his cough and his other malady continue to plague him. Please tell Erasmus when you see him that I feel I owe him a *full letter* in reply to the nice one that he sent me before I left England, but he must take what I write to you as partly to himself also.......

I was so *very* glad to read of George Darwin's K.C.B.ship. He thoroughly deserves it. His work in science has been of a kind that cannot be *popularly* appreciated, but is rated by experts as very high indeed. In every way it is a good and timely distinction. His wife will I am sure like it; though it is said that these titles always increase the charges of tradesmen!

Ever affectionately, with many thanks to Fanny Wilmot, whose letter I will keep,

FRANCIS GALTON.

P.S. The death of Edward Darwin† from *angina pectoris* is an interesting link between Dr Erasmus and Charles, both of whom died of that comparatively rare malady.

* This is a second instance of repetition to the same person.

† Son of Sir Francis Darwin, and grandson of Dr Erasmus, thus whole cousin to Francis Galton.

HÔTEL GASSION, PAU, FRANCE. *November* 14, 1905. (Post-card.)

B—— had a nephew, I believe; that is all I know about his family. I wish indeed that I knew more. Your letter reached me after a long round, hence the delay. The acceptance of the portrait by the Council, and its destination, have given me the greatest pleasure. Thank you all. I am so sorry to miss your fresh account of S. African experiences. We are probably here for the winter. F. GALTON.

Professor Sir G. H. Darwin, K.C.B., F.R.S.

HÔTEL GASSION, PAU. *November* 24, 1905.

MY DEAR BESSY, The blowing up into the air of the "pincushion" legend* is like the loss of a dear friend to me. I can only bow my head in grief, and submit. But it *ought* to have been true. Did not at all events our grandmother see a Doctor every week?—also with some ceremony? Where can I have got these notions from? Mrs Schim.'s† virtues, however, I will still stand up for. She had plenty of warm friends up to her death, and Douglas, to whom I mentioned her iniquities, rather laughed at the account with scepticism. I have latterly found a fourth admirer but only of her "Port-Royal" collections and enthusiasms.

We spent last Tuesday afternoon, which was beautifully fine, at *Lourdes*, and saw the place pretty thoroughly, including the going up a *funicular* railway to a famous mountain view. The place is wonderfully beautiful, and white, and clean, with abundance of smooth sward and a rushing river, which comes on here past Pau. I drank the holy water, of course, straight from the tap, and did *not* find it cold. Oh! the flare of wax candles in the Grotto, and the crutches and sticks fastened to its sides and roof, as votive offerings. There was no crowd of pilgrims, but many very devout-looking, praying people.

We can't get lively. The air is so unexhilarating even on the finest days. This place has, I find, that reputation, so I expect we shall soon make trial of Biarritz (three or four hours off).

I am very sorry about the Studdys' bad drains. I gave your messages to Eva. Tell us when you next write, how Mrs Skipwith progresses. You will be glad to see Penelope‡ again. It is of course a trial to every one to see alterations in old places, but what we two went through that day, in our search for Ladywood, Duddeston, and the Larches, can hardly be beaten by the experience of any one else.

I am news-less, day follows day monotonously with its meals and sleeps, newspapers, novels, and with sadly too little out of door exercise. The weather is usually so bad. Yesterday was execrable, and we have no GO in us just now. They want me to write a book on "Eugenics" and I am disposed to accept the offer. If I see my way to do it, it will give pleasant occupation for a year. But it will be a difficult job to do creditably. Many loves to all.

Ever affectionately yours, FRANCIS GALTON.

HÔTEL D'ANGLETERRE, BIARRITZ. *December* 10, 1905.

DEAREST MILLY, I will go through your letter in order, leaving the Tollemaches to the end, after I have seen them. Démolins *is* the man you mention. I have not yet read but have sent for his ...... *Anglo-Saxons.* His *French of To-day* is, to say the least, stimulating, but I find it raises many unsolved questions and criticisms, and especially as to whether his foundations are as solid as he believes. But I must read more before judging how far his methods would really help in "Eugenics" inquiries. So glad that you have Amy back, and a house full of sons and grandsons. The "Hilda" disaster must have come very home to you; all the more after your "Alliance" shock. I am so glad that you are again in correspondence with Mrs Benson, whom I myself knew only slightly, but whom I always heard so highly spoken

* The good lady was reported to have found it difficult to remember the names of the various parts of her frame and still more the locality of the pains she had experienced during the course of the week previous to the doctor's visit. So she caused a doll to be made and stuck a pin into the appropriate place as each pain troubled her. The doctor at his weekly visits gravely extracted pin after pin and discussed the corresponding pain and its cause.

† Mrs Schimmelpenninck, Galton's aunt, the well-known writer.

‡ Widow of Francis Galton's brother Darwin.

of; also that Amy's visit to Cambridge proved so interesting to her. The week here has passed pleasantly. There are interesting people here and very sociable ones. Mrs Tollemache is invaluable with her big collection of books and intelligent sympathy. He is greatly invalided and can work little, if at all, now. I sit with him and talk the philosophy he likes. He is quite blind, or rather can *just* distinguish light and dark out of the corner of one eye. Even that much is far better than pure darkness. You recollect (or if you do not my little book when it arrives will remind you) my nomenclature for kinship. It occurred to me that its *particularising* power would be greatly increased by foot-figures, thus bro_3 would mean the 3rd brother in the family. Taking yourself as the Subject of a pedigree, I am your $me_3\ bro_3$, Eva is your $me_3\ si_2\ da_1\ da_3$. In other words your mother was the third sister of her ("Geschwister") brothers and sisters, and her second sister's eldest daughter's third daughter is Eva. So a great deal of additional information can be given by these foot-figures without necessarily interfering with the general simplicity of the formulae. I sent a brief paragraph to *Nature* illustrating it by the more highly placed relatives of the newly elected King of Norway, and have just posted their proof of it, with corrections. It will be used in the next publication of the Eugenic Office, whenever that may occur, for which Schuster is now busily collecting materials. I wish I could get information about the principal Eugenic *centres* or *districts* in England. I mean those that are reputed to turn out the best sort of people, however the phrase "best sort" may be interpreted. The finest men come from Ballater in Scotland or thereabouts. I am trying to get an inquiry into this made. I suppose the "best sort" of persons are those who have so much energy that they are fresh after finishing their regular day's work to get their living, and who employ their after hours in some *creditable* way. The sun is at length out in fitful gleams. It has been foggy and rainy most days. The day before yesterday there was a marvellous sea and turmoil of waters at the Barre (the mouth of the Adour). I will add a scrap about the Tollemaches. Ever affectionately, FRANCIS GALTON.

Tollemache recollects well all about you and reminded me that I had suggested the book to be given to you.

HÔTEL D'ANGLETERRE, BIARRITZ. *December* 11, 1905.

DEAR SCHUSTER, You have indeed your hands full with Miss Elderton's batch of extracts. It is a big job and will be very interesting at the end. About my "Extension of the Nomenclature," you will see a letter of mine (I presume) in this week's *Nature*. But I heartily wish I had waited a bit and got the thing clearer in my head before writing it. Please—in the letter—imagine the F_1s to be replaced everywhere by F_1, and that this sentence had been inserted—"The foot-figure to every male, whether he appear under the title of *fa*, *bro* or *son*, refers to his rank among his fraternity; so the same person, who happens to be a third son, may appear as fa_3, bro_3 or son_3. Similarly as regards females, in respect to their sisterhoods: the same second daughter may appear, according to circumstances, as me_2, si_2 or da_2."

Of course, other things might be conveyed by foot-letters, but it would not be wise to encumber overmuch. Still, the phrase "only son" or "only daughter" seems to deserve a special sign, *u* (for unity) might do, as "fa_u," but *z* might do better—not *s*, which means son. This can stand over.

Of course footnotes would often be wanted. *wi* (1), *wi* (2) must stand for 1st or 2nd wife—then wi_2 (1) or wi_2 (2) would be easily understood.

I am sorry that the "Advisory Meeting" does not seem useful. If experience confirms this, have them less frequently and for special occasions only, or drop them altogether.

You do not mention whether *any* replies are coming in to the circular issued by the Sociological; when you next write, I should be interested to hear.

I am afraid that Branford is overworking himself dangerously.

When the corrections to Miss Elderton's papers come in, you will probably attack first some particular *class* of noteworthies and get them as far as may be off-hand before beginning another. However you will soon discover the lines of least resistance.

I find this place suits me much better than Pau did. It is rather too foggy and rainy, but this month in the Republican Calendar is called "Brumaire," the month of fogs. I hear the Riviera weather is far from good, so I am well where I am. The waves are sometimes magnificent.

Very faithfully yours, FRANCIS GALTON.

Enclosed. "Extension of the Nomenclature of Kinship." The method I adopted in your columns, August 11, 1904, of briefly expressing kinship has proved most convenient; it has been used in a forthcoming volume by Mr E. Schuster and myself on *Noteworthy Families.* I write now to show that it admits of being *particularised* by the use of foot-figures, as in the following example, which refers to the more highly placed relatives of the newly elected King of Norway.

Haakon VII, King of Norway (b. 1872).

fa_{1s}	Frederick, Crown Prince of Denmark (b. 1843).
$fa_{1s}.fa$	Christian IX, King of Denmark (b. 1818).
$fa_{1s}\ bro_2$	George I, King of the Hellenes (b. 1845).
$fa_{1s}\ si_2$	Dagmar, widow of Alexander III, Tsar of Russia, who d. 1894.
$fa_{1s}\ si_2\ son_1$	Nicholas II, Tsar of Russia (b. 1868).
$fa_{1s}\ si_1$	Alexandra, Queen of England (b. 1844).
$fa_{1s}\ si_1\ son_1$	George, Prince of Wales (b. 1865).
$fa_{1s}\ si_1\ da_3$	also *wife*, Princess Maud (b. 1869) of England.

The formulae are to be read thus: "His (the King of Norway's) father is the 1st (eldest) son, and is Frederick, Crown Prince of Denmark"; "his (the K. of Norway's) father's father is Christian IX"; ... "his father's 2nd sister's 1st son is Nicholas II"; ... "his father's 1st sister's 3rd daughter, who is also his (the K. of Norway's) wife, is the Princess Maud." These foot-figures need not interfere with the simplicity of the general effect, while they enable a great deal of additional information to be included. FRANCIS GALTON. *Nature*, December 14, 1905.

HÔTEL D'ANGLETERRE, BIARRITZ. *December* 17, 1905.

DEAREST MILLY, All goes on here steadily and well, but it is mostly overcast and foggy with only occasional sunny days, suitable for excursions to the battle-fields. It seems English ground. We possessed it for 299 years (from 1100) and occupied it in 1813–14. How Wellington's army did fight! His movements were so quick and sure. I see that one important success was due to the then newly invented Congreve rockets. He got with difficulty and at the close of a day a comparatively small body of his men across the Adour, upon whom a manyfold larger body of French swooped down from Bayonne. Each English soldier had two rockets in his knapsack and others in store. They launched these against the French, who had never seen the like before, and who were seized with panic and ran back. I find I was wrong in saying that bayonets got their name from Bayonne. It was from a neighbouring village, Bayonnette.

"Explain the relationship between (1) a gardener, (2) a billiard-player, (3) an actor, (4) a verger." The gardener attends to his p's (peas), the billiard player to his q's (cues), the actor to his p's and q's, and the verger to his keys and pews. If you want a short novel to read, try *The lost Napoleon*, by Sir Gilbert Parker (who is here). How happy you must be with your housefull. I suppose Amy's headache at the time you wrote is long since a thing of the past. The impending dissolution has stopped for a time the publication of books, as Murray told me it would some time ago. So I suppose mine is delayed for that reason. People are found to read little else than newspapers at the time of General Elections. Eva is particularly well and I ditto, with reasonable reservations due to getting older and less mobile by far. This is an excellent place for carriages, but driving is usually too cold now to be pleasant. I want vicarious exercise, like being tossed in a blanket (of course, not occasionally bumped on the floor as in school-boy days). Some mechanism ought to be devised for shaking elderly people in a healthful way, and in many directions*. Music might go on mechanically at the same time, with its rhythms and shakes all in harmony. Excuse nonsense. With much love.

Ever affectionately, FRANCIS GALTON.

HÔTEL D'ANGLETERRE, BIARRITZ. *December* 22, 1905.

DEAR SCHUSTER, I am glad you wrote; the point raised requires careful consideration. I had thought of it from time to time, and have done so more carefully now. The precise

* Cf. Sir Alfred Yarrow's electrically-driven rocking or rolling bed!

method that you suggest is not the best possible, for it would require the space of an extra half line, if not a whole line, in the printing (see fig.); neither does it contain all that is wanted, viz. number of brothers *and* sisters. Moreover it rather mars the simplicity of the notation. Lastly, there would be unnecessary repetition. After numerous trials it seems thus far (subject to discussion) best to leave all this to a separate paragraph in smaller type. In this nothing need appear that did not relate to an *entry in the pedigree.* Suppose it is a case of fa_2 bro_1 son_3 in which only the *son* is noteworthy. The paragraph would always contain entries corresponding to those below. In these the brackets mean "self, brothers and sisters"; and the first numeral is the number in it of males, the second that of females, so the "self" falls into the first or last of the numbers according to the sex.

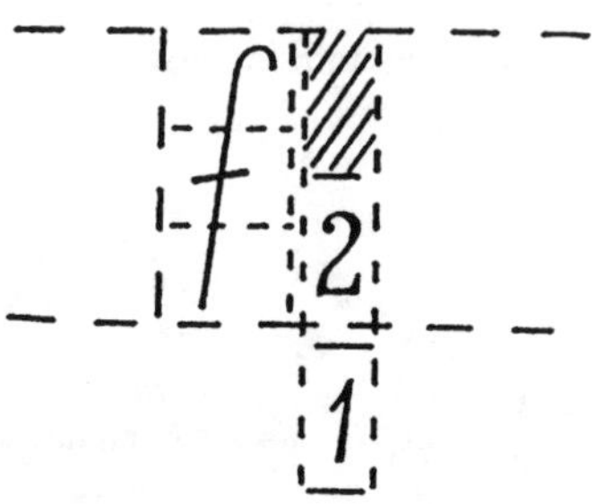

fa (5, 3); *me* (2, 2); *self* (3, 1); *sons* and *daus* (2, 4).

fa_2 John; bro_1 Edward; and so on for every non-noteworthy kinship in the pedigree.

Of course, this paragraph may contain *fa fa* ( ); *me fa* ( ); *fa* bro_n *sons* and *daus* ( ); *fa* bro_{n+1} *sons* and *daus* ( ); and the like, so far as data exist and it seems useful *in the case in question* to insert them.

Many complexities due to double marriages or to intermarriages could be made clear in the footnote, the object of which should be confined to explaining *the text*, not to bothering out *all* relationships. In brief the syllables with suffixes will *particularise* the persons concerned, the footnote will tell particulars concerning them which the text does not.

I hope this is clear enough for you to experiment with and perhaps improve on? Please try, and report.

I have been in correspondence with Mr Hartog about the destination of the Report to the Senate. I trust that it will be handed to you to send on to *Biometrika.* A *very* brief account of it—its title and an explanatory sentence—will be wanted for the Report of the Committee to the Senate, I suppose; but you will be advised by Hartog. I am very sorry about Miss Elderton's illness. I hope it is nothing bad.

Very faithfully, FRANCIS GALTON.

HÔTEL D'ANGLETERRE, BIARRITZ. *December* 23, 1905.

DEAR SCHUSTER, Since writing yesterday I have written out the enclosed as a *full* example. The *complete set of names* is more of a luxury than a necessity. Without them, the entries could go consecutively, thus:

fa_1 + fa_1 bros + fa_1 sis (3, iii); self + bros + sis (5, iii); sons + daus (1, 0); fa_1 bro_2 sons + fa_1 bro_2 daus (5, i); etc.

In this way they would take little room, especially if printed smaller than the text. The fault in this very concise form is that it fails to identify *by name* the non-noteworthy links. The *rated* order of birth may not be correct; one wants the Christian names as well, for certain identification. This difficulty could I am sure be got over. How would it do in the following to write the third and fifth lines thus:

self + bros + sis (5, iii) 2. Charles,
fa_1 bro_2 sons + *fa_1* bro_2 daus (5, i) 1. Constantine,

and so on, giving only the names of the persons who come *directly or indirectly* in the genealogical account.

It deserves a great deal of care to arrange once for all these and similar matters, to ensure uniformity, and to avoid costly printers' corrections hereafter. You would do well to prepare two or three typical genealogies for consideration. I would send them to Howard Collins whose advice on such matters is probably the very best to be had, and who is always ready to help me. Very faithfully, FRANCIS GALTON.

Example from King Haakon VII's Pedigree, which might be added to the Table in *Nature*, Dec. 14, 1905, p. 151. The data are taken from a pedigree in the *Graphic*, Nov. 25, 1905.

fa_1 + fa bros + fa sis	(3; iii)	1. *Frederick*, 2. *George*, 3. Waldemar; i. *Dagmar*, ii. Thyra, iii. *Alexandra*.
me + me bros + me sis	(?; ?)	Data wanting.
self + bros + sis	(5; iii)	1. Christian, 2. *Charles*, 3. Harold, 4. Ingeborg, 5. Gustav; i. Louise, ii. Thyra, iii. Dagmar.
sons + daus	(1; 0)	1. *Alexander*, b. 1903.
fa_1 bro_2 sons + fa bro_2 daus	(5; i)	1. *Constantine*, 2. George, 3. Nicholas, 4. Andrew, 5. Christopher; i. Marie.
fa_1 si_1 sons + fa si_1 daus	(2; ii)	1. *Nicholas*, 2. Michael; i. Xenia, ii. Olga.
fa_1 si_3 sons + fa si_3 daus	(1; iii)	1. *George*; i. Louise, ii. Victoria, iii. *Maud*.
fa_1 si_1 son_1 sons + fa si_1 son_1 daus	(1; iv)	1. *Alexis*; i. Olga, ii. Tatiana, iii. Marie, iv. Anastasia.

To the Table in *Nature*, Dec. 14, 1905, make the following additions:

son_1 *Alexander*, Crown Prince of Norway (b. 1903).
fa_1 si_1 son_1 son_1 *Alexis*, Tsarevitch (b. 1904).

HÔTEL D'ANGLETERRE, BIARRITZ. *December* 23, 1905.

MY DEAR MILLY, A most happy Xmas and New Year to you all. Your little Xmas card reached me to-day, your letter yesterday. The suggestion of a cob to ride is rather like—I hardly know what to resemble it to. I was somewhat less incapable four years ago when in Egypt than I am now, yet even then it required engineering before I could mount an ass, much more a pony. The muscles that first fail in age are those that enable one to use a stirrup for mounting and those that throw a leg over the saddle. I can do neither of these things now. With a crane like this, I might get on a donkey's back, or even a quiet cob's. Otherwise, it is

nearly hopeless. The past week has gone pleasantly, but rather uselessly. The cold freezes my wits, but the weather improves. We have some idea of trying St Jean de Luz for a week, but nothing is fixed yet. What an electioneering turmoil there will be. One M.P. has just left here to look after his constituency. I get quasi-philosophical talks on most days with Tollemache, and there is a quasi-resident here, Col. (with Mrs) Hill-James, who is an excellent authority on Wellington's campaigns. It seems he was a yearly guest of Darwin's* at Claverdon, and being a tandem-driver they had many common points of interest. It was very pleasant to hear how much he appreciated Darwin. We tea-ed there to-day. They are to have

* Darwin Galton, Francis Galton's eldest brother.

a big Xmas function at this hotel, dinner, champagne (probably bad), speech-making, holly and a dance,—"sauterette." This house is full of Russians. The Duke of Oldenburg, whose wife is some near relation of the Tsar, is among them. The wife is rather maid-servantish looking, and reminds me of Temple in a way. Oddly enough Mrs Hill-James' name was Fanny Arkwright. She is a cousin of her namesake, Darwin's (first) wife. Other news is wholly local; you would not care for it. Over again a happy Yule-time to you all.

Ever affectionately, FRANCIS GALTON.

Mrs Robb's son is made a general, the youngest in the army. He did excellent work at the Intelligence Department. Sir George Darwin's son (Charles Galton) has just won a major scholarship for mathematics at Trinity College, Cambridge, while still at school (Marlborough). Montagu Butler's brilliant son, James, has done the same for classics, while at Eton, at only $16\frac{1}{2}$!! Both of these are promises of great University success. I *think* each is about £100 a year for 5 or 6 years.

HÔTEL D'ANGLETERRE, BIARRITZ. *December* 27, 1905.

DEAR SCHUSTER, The enclosed "Note," to follow your typed account and to be printed in *smaller type*, is what I mean. Perhaps you will type it and submit it to some "Devil's advocate" to hear the worst that can be argued against it, and to get hints for improvement. It certainly contains much information in a form well suited for statistics. There is some repetition of what appears in the text, but that does not seem objectionable. Please supply the number of Sir Edward's children.

I have pencilled some small suggested corrections in the letterpress.

It would be well to use the typed "Note," after such revision as you think well, as a guide; then, after doing a few more families on the same plan, to seriously reconsider the form and finally to decide. Recollect that all *complexities* of kinship should be unravelled in this "Note."

Very faithfully, FRANCIS GALTON.

HÔTEL D'ANGLETERRE, BIARRITZ. *January* 5, 1906.

MY DEAR EDWARD*, Our morning post goes out a few minutes after the incoming post arrives, so I had not time to finish Eva's letter. Give my dear love to your mother. She knows, and so do you, that I realise the discomforts of asthma and bronchitis. I had a brief bout only yesterday, which has passed off happily. It is a queer thing. I wish I knew how to cure it as well as I do how to bring it on *myself*, viz. by a cosy, well-warmed and carpeted room, and good feeding. But what is bad for me may not be bad for others,—who knows? Please thank Lucy† much for her long letter. I grieve at her family misfortunes and own uncertain health. Do not think of bothering her or yourself overmuch, but I should dearly love a frequent post-card until your mother is convalescent.

Bugs for ever! I am delighted that your brochure is to be published. It will be timely. *Snails* are the interest here according to Professor Weldon. There is one sort that in *ancient forest times up to the present* does not, or hardly does, cross the Garonne (the river at Bayonne). Why, no one yet knows. It is very common on the south side.

We are going in the middle of next week to St Jean de Luz for a week or so, but the above will be our address till I write again. I have no news you would care about. Three sets of people here know Claverdon well: 1. Miss Hodgson, 2. Col. and Mrs Hill-James (she an Arkwright and a *Fanny* Arkwright too), 3. (I forget at this moment). I am busy finishing off an outline account of occasional experiments during the past few years on the *Measurement of Resemblance*. It is possible, I find, to give a brief account of the essentials, but the subject with its many side issues is big, and would require a book full of illustrations to treat properly. I have not now time nor "go" enough for such a task, I fear.

Do you know young Sir G. Skipwith? Our friend Mr Townsend, who died very lately, has left him the Honington Estates. He was one of the nearest though a distant relative.

Ever affectionately, FRANCIS GALTON.

* Galton's nephew, Edward Wheler, son of his sister Bessy.

† Lucy, Mrs Studdy, sister of Edward Wheler.

January 6, 1906.

I had written the enclosed when your grave telegram arrived of "Mother much weaker." We are very sad. I realise only too vividly what is probably going on to-day, which is even worse to the onlooker than to the sufferer. My father constantly repeated this in respect to his violent asthma. He seemed to suffer terribly, but did not suffer so much as we used to fear. F. G.

Address next letter please to Hôtel Terminus, St Jean de Luz, Basses Pyrénées, France.

January 7, 1906.

MY DEAR EDWARD, I am so glad that the end was peaceful and not preceded by long suffering. You and M. L. will be conscious of having been an infinite support and help to your Mother, and will look back to her even more than motherly affection to you with continued remembrance. I lose in her the only remaining person who knew our family and family friends in the days of my boyhood. All her store of memories is now irrevocably gone. You, together with all your sorrow, will doubtless feel a dearly bought sense of liberty, for all your movements have been guided by the thoughts of her convenience and happiness. Still it is something gained; also the pecuniary gain to yourself and to Lucy. It is all in the order of nature. I wish you all well through the sad ceremonies previous to and at the churchyard. I would have asked to share in them had I been within easy reach.

Ever very affectionately, FRANCIS GALTON.

HÔTEL D'ANGLETERRE, BIARRITZ. *January* 7, 1906.

DEAREST MILLY, "The end came this morning, so peacefully. Wheler." Such is the telegram just received about dear Bessy. It is the last link with my own boyhood, for Erasmus was at sea, etc., and knew little about me then. So much of interest to myself is now gone irrecoverably. But it was time, according to the order of nature, and I feel sure it will give longed-for liberty to Edward and M. L. to see distant parts. They were so devoted to Bessy and made their arrangements so subservient to hers, that the liberty *must* be welcome. But how they will feel the loss. Bessy's was a stoical life for a long time, not only after her widowhood but long before when her and her husband's income was very small. She battled bravely then. We go on Wednesday to *St Jean de Luz*, to the *Hôtel Terminus*, for a week or perhaps more, for a change. Please address your *next* note there, but only the next one. We shall probably return here afterwards. Count Russell was staying at this hotel, and we had pleasant talks with him, and kind invitations to his caves which Eva burns to accept, but I could not walk up to them. He has 500 acres of snow, 8000 and more feet above the sea, with rocks around it, as his property. Here are some of his caves*. He has been sleeping through the summer in sleeping-bags, not in beds, for the last 40 years! Something in the food, or what not, has somewhat upset us and we shall be glad of a change for a little while. It is still very warm on the whole, but variable. I doubt whether it will be fit weather for San Sebastian yet, but we could so easily spend a day there from St Jean de Luz. Mr Webster, the Basque scholar, lives like a Basque in the hills nine miles from there, where I hope to see him. Excuse more, I have had to physic myself and to keep upstairs to-day, and am in addition a little upset by the sad news. Ever affectionately, with best loves, FRANCIS GALTON.

I am very sorry that your eyesight still gives some trouble.

HÔTEL TERMINUS, ST JEAN DE LUZ, BASSES PYRÉNÉES, FRANCE. *January* 17, 1906.

DEAREST MILLY, I have delayed replying until everything had been received bearing on Bessy's death and funeral. She has indeed had much to be thankful for, especially in her closely reciprocated affection for Edward and Lucy, and the painlessness of the end. It must leave a terrible void in their affections and interests, judging even from what my own loss is to me, and of course far, far greater to them. But there is little use in talking of these things, even sympathy does not much help when wounds are deep and yawning. You must have been much distressed, and I look much for your next report on Guy and yourself and Amy. Sorrows come in battalions. They certainly are doing so to Lucy Studdy. Erasmus *cannot* attend long functions, he is medically unfit to do so. Edward and M. L. sat with Erasmus every evening in his room

* Presumably at Biarritz.

at the Regent, which was a solace to all. We are healthfully situated here. Did I tell you that we spent a pleasant afternoon with Mr Webster, the Basque scholar, in his home among the hills at Sarre, nine miles off? He is growing very old and feeble, but is full of interests. It was a great pleasure to make Count Russell's acquaintance. He has sent his charming little book, full of genius, called *Pyrenaica* to Eva (it is in French). But at my age, I don't take kindly to the thoughts of a sleeping-bag in a big hole in a rock some 10,000 feet high, with the chance of sluices of rain and tempests and a most disagreeable descent afterwards. "Peace is of the valley"; Valkyries were not peaceful ladies, and are not at all to my fancy. We drove yesterday to Fontarabia; two hours there, two hours at the place, and two back. But I sat still and left Eva and a lady friend, to whom I gave a lift, to do the sight-seeing (which I *had* seen forty and more years ago). Lucy wants me to write recollections of her mother to put into her mother's book of recollections which you probably know, or know of. She never liked talking about it, but I had once a good read at it. It is all very nice and interesting and well deserves being typed, which is being done by Lucy's niece. The only thing I could do, would be to give my own recollections of the family, my father, mother, brothers and sisters, *as a whole*; and I shall try, but fear making inaccurate and one-sided remarks, also I should be deficient in dates. The family is a curious one, from consisting of very heterogeneous elements; my father and his three brothers and three sisters, Theodore, Hubert, Howard, Mrs Schim., Aunts Adèle and Sophia, having totally different temperaments and characters, and each very decided in its way.

Ever affectionately, FRANCIS GALTON.

HÔTEL TERMINUS, ST JEAN DE LUZ, BASSES PYRÉNÉES, FRANCE. *January* 19, 1906.

MY DEAR EDWARD, I was glad to hear from you, though letter-writing is more than I can expect now, in this miserable time for you. The yawning gap left by the loss of a mother, and all the interests connected with her, and the extremely painful business of going through her things, which is a repetition of what you went through a year ago on dear Emma's account, are grievous to think of. Your Mother was very thoughtful and you are very good, to suggest my having some memento of her, but I really do not know what to ask for, for I want nothing. The many things I had on Emma's death fulfil the present purpose of family memorials. Don't let any Darwin or Wedgwood things, or anything referring to my Grandfather, or even to Mrs Schim., be *lost*. They are all family mementoes, but I cannot say either that any of them would be suitably bestowed on me, or that I should really care to have them. There is so little spare room in my house and I am perforce so large a part of the year away from home. Any *trifle*, such as a bit of tape, if characteristic, would quite serve my purposes.

We are staying on here, which suits us, and Biarritz seems about to be over-crowded with Royalty and their suites, and therefore not attractive to return to. When tired of this I shall probably try San Sebastian for a bit, also Sarre, a thoroughly Basque village where there is a clean Inn and where Mr Webster, the Basque scholar, lives with his family—but the present address holds good for a while.

I shall be curious to learn in time the fate of No. 5, Bertie Terrace and other particulars resulting from the great change. I was very glad on all accounts that you were both of you able to see so much of Erasmus during the sad week. For my own part, I feel that almost all interest in Leamington is gone; it lay so predominantly in No. 5, Bertie Terrace. I had not only the personal affection to it, but some of the mere house-affection, like a cat's, also.

What a strange political change! Everything seems going topsy-turvy in England. We shall soon see some results, and can only hope they will not be dangerously bad.

Best love to M. L. Also please to Erasmus when you see him next.

Ever affectionately, FRANCIS GALTON.

HÔTEL TERMINUS, ST JEAN DE LUZ, BASSES PYRÉNÉES, FRANCE. *January* 22, 1906.

(We stay on here till I write to the contrary.)

DEAREST MILLY, My letter will have crossed yours, and explained why I had delayed writing. Erasmus tells me by letter this morning that Edward and M. L. propose going soon to the Mediterranean for a complete change and rest. It must have been a most sad and trying work to look over all the old papers and things, and to arrange about them, a repetition with additions of what he went through a year and a half ago. The sad event has brought them and

Erasmus still closer together. He, Erasmus, suggests that I and he should interchange periodical letters, say once fortnightly, to which I cordially agree. There are great merits in this place and I like it. We spent a long day a little since at Fuentarabia in Spain close by, but it is rather too cold and the days are still rather short for excursions. Your letter reached me yesterday as wet as if it had come out of Guy's trousers' pocket after his "humane" feat by St Malo. The story was that Seabrooke* bicycled to Biarritz, being invited to dine there by her friends. On returning, the weather there was at first rather blowy but dry. Here it was a gale with squalls of rain and we were anxious about her return. She ran into the gale about half way, and had to walk with her machine four or five miles, arriving here late with your letter in her pocket and every stitch of clothing on her wet through. A dose of hot tea and brandy, followed by dinner and early bed, has put her quite right. It is remarkable how popular she becomes wherever she goes. There was, and is, a Russian Archduke at the Hôtel d'Angleterre at Biarritz with his suite, and there are other Russians also. On *their* Xmas Day (while we were still there) a big servants' dinner was given and according to Russian custom they chose a queen for the evening, and that queen was Seabrooke. She was crowned with and wore a handsome *dish-cover*.

I feel just like you about Leamington. All special interest in the place is gone for me, which was for a long time so close and grateful. I greatly miss Bessy's weekly letters, too. In fact it is a big loss to me, that time cannot now go far towards supplying. I am glad to know that your affectionate heart feels it deeply too. What anxiety you must have had about Guy's very sharp attack and fever. Is not the present form of *bad* chickenpox a special type recently imported from Germany? I fancy that I have heard so. Free Trade in microbes and diseases! Hurrah for Free Trade! We get good things however as well as bad through Free Trade. This cataclysm in the political world is ludicrous as well as terrible. Most likely it will be a refresher and turn to good in the end, but there are many wrong ways to one right one, according to Bunyan's *Pilgrim's Progress*. Eva went to the Basque Cathedral last Sunday—most imposing—then a long, long procession through streets sparsely strewn with sweet-smelling rushes, which with Basque orderliness were all swept away that same evening.

Loves to Amy and all, ever affectionately, FRANCIS GALTON.

Address for next two or three days: Hôtel de la Rhune, Ascain, Basses Pyrénées, France.
February 1, 1906.

MY DEAR WELDON, I owe you thanks for your kind sympathy and was indeed about to write when your letter to Miss Biggs arrived yesterday. As yet we have not noticed any arbutus but shall be to-day in less cultivated districts for we go to the above picturesque Basque village for a week certain and will take care to look out and to ask. Probably we shall go on four miles to Sarre where Mr Webster, the Basque authority, lives with his well-informed wife and daughter, and I will put them on the arbutus inquiry. We should have gone there instead of to Ascain, but some unexpected bother arose about the contemplated rooms. I should be grateful for a few lines about the horse-colour question and the Royal Society discussion, where I have heard from Pearson that Bateson drew a red herring across the track of the discussion. He also sent me a slasher in the *Chronicle* about X. and tuberculosis, well-deserved. X., with his fluent pen and Oriental character, strikes me as a precarious combination, not to be depended on over-much. I heard from Schuster a few days ago but having been much preoccupied his letter is still unanswered. How does the *book*, the magnum opus, get on? You can hardly believe how much I thirst for its appearance, for your zoological facts are just those I am most deficient in.

We have had a quiet pleasant three weeks here, at St Jean de Luz, and feel the Spring in the air and the good time coming. The Royal Lover whirled through the town in his motor, to and from Biarritz, but I did not see him. One ought to "cast" a future (like a horoscope) for the prospective children. A queer medley of good and bad breed will run in their veins.

Ascain, and the inn there, where we go to-day, is where Pierre Loti wrote *Ramuntcho*. His ship was somewhere near and he got leave to stay on shore. The Bay is now wonderfully calm, such a contrast to when we arrived, when the waves ran wonderfully high and a newly wrecked ship lay on the shore. The sea reminds me of a gorged cannibal, sleeping with his stomach full. Ever very sincerely, FRANCIS GALTON.

* Evelyne Biggs' maid.

[In Evelyne Biggs' handwriting.]

I should like to read the *Fogazzaro*, also the book on Naples, but I find as a rule I can't read Serao's writings. Just begun *Pêcheur d'Islande*, it is very charming in spite of being French!

This place is most paintable, the Basque buildings are delightfully irregular and no street is at right angles, or rather the houses in it aren't and that makes it so interesting. I think we've got the tunnel murderer in this hotel! Lucky we are just moving! E. B.

Address 42, Rutland Gate, "please forward." *February* 3, 1906.

DEAR SCHUSTER, At length, after a scarcely pardonable delay, I have had a good go at your paper. (1) Take great pains to describe the Subjects' doings in terse and forcible language. It is a most difficult task, so it would be well to be in touch with some classical or literary friends to criticise helpfully. An epitaph is a work of art; the late Lord Houghton was frequently appealed to to compose them for public characters, and these are like epitaphs. I have pencilled suggestions of my own.

(2) About the appendix to each family, such as that to the Freres, which please look at for explanation, it will of course be printed in smaller type. I *think* that the *Subjects*, as (fa fa + bro sis), had better be the bracketed entries, and their brothers and sisters or sons and daughters be separate, thus:

(*fa fa* + *bro sis*) 2 bros, 3 sis || (subjects + *bro sis*) 3 bros, 0 sis || etc.

Think this over and do what *you* then think best, for it will be *your* book.

(3) In the Butler family you have tried bro_a, bro_b, etc. instead of bro_1, bro_2, etc. I like the numerals best. It would hardly do to *combine* the notations as bro_c son_3, because, however well it might look in one pedigree, the term bro_c might appear as bro_3 in another, as applied to the same person, which would puzzle. You have taken great pains with these families.

I have been twice in correspondence with Murray, first in regard to whether the book was to be one of the University of London Series, he replying that he understood *not*. I referred this and him to Hartog, to whom I also wrote, fearing to make some technical blunder; Hartog suggested at least the University arms. The second was yesterday in reply to half a dozen sample covers of diffused hues, all printed alike and *with* the arms. I suggested the addition of the words "modern science," which no doubt he will put in if he gets my answer in time. Otherwise there would be nothing on the cover of the book (though there is in the title-page) to distinguish it from forthcoming volumes of the same kind. The cover looks uncommonly well and suitable to attract attention favourably, as it lies on a table.

I have just received an offprint of a German translation of all my Eugenics papers, inserted in that excellent periodical *Archiv für Rassen- und Gesellschafts-Biologie* of which I know Professor Pearson has a high opinion. I have not had time yet to look at it but am sure it will be done well, as the co-editor who translated it writes in excellent English.

We are in a funny and very comfortable Basque Inn, in a village, Ascain, four or five miles from St Jean de Luz. It depends upon procurable rooms whether or no we move hence to Sarre, another Basque village, or possibly even go a little way into Spain. So you had better write to 42, Rutland Gate, "to be forwarded," if you have occasion to do so before you hear from me again. Very faithfully yours, FRANCIS GALTON.

HÔTEL DE LA RHUNE, ASCAIN, BASSES PYRÉNÉES, FRANCE. *February* 8, 1906.

DEAREST MILLY, Your last letter, that you wrote of on a post-card, has miscarried. We get all letters forwarded from St Jean de Luz, but yours has not been among them. I am so sorry. We like this place, but having been house-bound by bad weather I have as yet seen little of the neighbourhood. There is a pet here that even you have no experience of, viz. a wild boar, 10 months old, as high as my knee. He is kept mostly shut up, but was let out yesterday for a run. It was the funniest sight conceivable to see his twists and turns and gallops about the field and garden. His tusks are fully 3 inches long, not sharp but formidable looking, and he shakes his head continually as though ripping up at something. He will be dangerous soon. The landlord picked him up quite young on the hills. There are no *events* here to tell you of. As regards personal matters, I packed off my paper to-day on the "Measurement of Resemblance" to the typist, who is to send one copy to Karl Pearson for his criticism, which

I await with no ordinary anxiety. Another is that I have received a German translation of my Eugenics papers, printed in a first class periodical, which reads and looks extremely well. My "Resemblance" dodge *may* turn out very useful in inquiries bearing on Eugenics, for it measures among other things family likenesses, racial likenesses, etc., and is especially adapted to measure those between composite photographs, respectively representing the features of different races. But it has to be criticised, well tried, and then developed. You will have received one picture card from Eva. We are collecting them by degrees, but are far from the parts you are likely to go to. Argelès sounds promising. So Edward and M. L. go on March 1 for a five weeks' cruise. It is sad for Guy not to be with them. But it is rather a blowy and cold time of the year, for Constantinople especially. Still they are sure to have many delightful days and to see delightful places. I wonder if we shall by chance return *via* the Simplon tunnel when the time comes. I know it is open, but do not know when trains will run. It is however a good deal out of our way. The King of Spain has driven frequently in his motor through St Jean de Luz, waving his handkerchief and looking very happy, as a friend who walked over from there this afternoon told us. Ever very affectionately, FRANCIS GALTON.

[In Evelyne Biggs' handwriting.]

This is a duck of a place, so very simple and picturesque, and St Jean de Luz being four miles off one can go there for books, shops, etc. The Basque churches are beautiful, quite unique, and the people are very devout, the church here being quite full at every service; every man and boy seem to go. I believe this would be quite cheap in the summer but I will inquire here and at Sarre and all the little places; Argelès would be dear I am sure. How very sad that Guy can't go with the Whelers. I do call it the most disappointing thing. They would all have enjoyed it so. Much love to all from E. BIGGS.

HÔTEL DE LA RHUNE, ASCAIN, BASSES PYRÉNÉES, FRANCE. *February* 9, 1906.

MY DEAR EDWARD, We are staying on at this cosy picturesque inn, for the weather has been too bad of late for gadding about, so the above continues to be my address.

Thank you much about the things that you have offered me. I endorse the list with a "yes" in pencil on those I would gladly have, my Father's portrait especially. Few relatives now living recollect him, none as I do. The trifles about my own early life I should be glad to keep, with a few others of the same quality that I possess already.

When you return, the garden and trees will have begun to be green, and you will appreciate the result of the clearings and improvements. I am so glad you are going, but it will not be all fair weather. *If*, when you are at Porto Empedocle, a party is made to go to the *town* and *cathedral* of Girgenti, do go with them and manage to hear the wonderful acoustic properties of the building.

You sit at A where the confessional used to stand (before these properties were discovered) and the slightest whispers are heard by a man at B who stands in a gallery hidden by a perforated

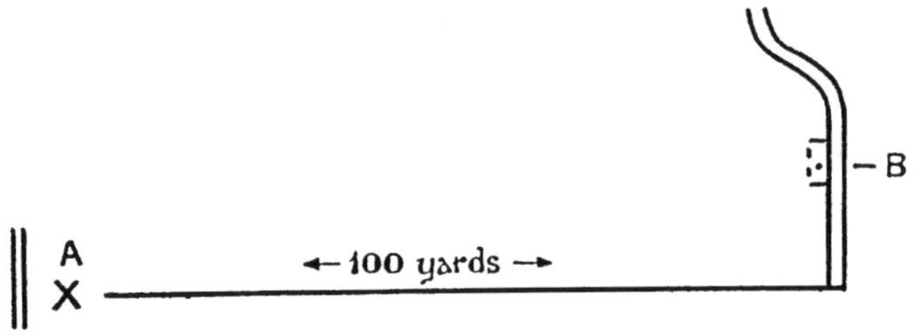

screen of wood and who repeats them. Eva and I sat on the same bench placed for the purpose. She whispered numerals "venti tre," etc., so low that I myself, through my deafness, could not properly hear them, and back came the loud repetition from the man at B.

The feature of this hotel is a pet wild boar, 10 months old, with formidable tusks already. He is kept in a pen and allowed an occasional run and frolic with friendly dogs. It is very funny to watch his short gallops, sudden stops and twists, but above all to see the instinctive way in which he twists his head as though to strike upwards with his tusk. I don't feel quite easy when the animal runs to or past me, for he is as high as my knee and could do mischief.

He will soon be dangerous, and have to be converted into ham. He does not smell a bit in the open. His hair is thick and bristly and of a rich brown, and his head and mane ("*hure*" is I think the technical word) are grand for his size. I like these Basque folk much, they are so quiet and orderly and substantial. But as for their language, it is impossible to a stranger.

Eva asks me on her part to say that if one of the *steel spectacle cases* that your mother wore, which were very characteristic, *happens* to be available, she would prize it much as a souvenir.

I am glad the *Report* is so nearly ready. What a long time always intervenes between the time when a book is apparently ready and that at which it actually appears. I hope that the new Ministry will go in for research. Best love to M. L. You both need a complete change of scene and a rest. Ever affectionately, FRANCIS GALTON.

HÔTEL DE LA RHUNE, ASCAIN, BASSES PYRÉNÉES, FRANCE.
February 16, 10 a.m., 1906; aet. 84 yrs. 8 hrs.

DEAREST MILLY, I had not realised before receiving your letter of a week ago how anxious you have been about the eyesight. The Doctor's favourable verdict must have given great relief. A change of spectacles may do much. I must write in a much clearer and bolder hand, like this*. We are in the midst of bad weather, February being the worst month in these parts, and I have been house-bound for days together, but very cosy and very happy with plenty to do. They feed us so well and the cooking is so juicy and good. The place is said to be beautiful in summer, now of course it is bare but some fields are very green. The typed copy of my "Resemblance" paper arrived yesterday, so I hope to hear from Karl Pearson in two or three days. Enclosed I send a pencilled *résumé* of the chief points in it, in case you care to read it. It is not worth keeping. It does one good to have to try to explain oneself in a clear way and briefly, so this pencilled scrap was a self-discipline. The late John Murray, the publisher, advised those who were about to write for the first time each to keep some one friend in constant view, and to fancy he was writing to that friend. You ask about that German translation so I send the only copy I as yet have, but more are promised; in all probability I shall not want this again, so pray keep it until I write. (See postscript.) That blessed book on *Noteworthy Families* is not even yet published, but *covers* were sent to me to choose from, which I did. It looked quite nice. The report of the Louping-ill, etc. Committee, of which Edward Wheler is an active member, is on the verge of publication, and is an admirable piece of scientific work. Part III mainly written by himself is a summary of the rest. It is most instructive. I think the results will form an epoch in the progress of knowledge of disease and how to cure it. The strangest part of all is that the blood of sheep differs notably in its *quality* at two different seasons of the year. In one it kills a particular sort of microbe, in the other it does not. It is equally the case whether the experiment be made on the live sheep, or in a test-tube with cultured microbes. I fondly believe that the time will come when doctors, after feeling pulse and taking temperature, will ram a sharp tube into the patient and take from him a drachm or so of blood to experiment with. I ought to have begun by thanking you for your kind birthday letter. I am now four times as old as when legally a *man*, viz. 21 years, and cannot in retrospect make up my mind which of these four spaces has left most impression. They all seem very long and very different. I don't quite catch the point of the following remark, which has been sent me by letter; perhaps you or Amy can. It is that there are three sorts of religion: Religion, Irreligion and *Bi*-religion. It was sent me by a shrewd person as containing a shrewd meaning, which I however cannot discover. Best loves to you all. Ever affectionately, FRANCIS GALTON.

P.S. Alas, I can't send the German article by book-post because I have pasted the writer's letter inside it.

Enclosed with letter of February 16th.

Measurement of Visual Resemblance.

When a person is walking towards you the first thing you notice in *his face* is its *general shape*, which may be the same as that of many people (I leave all the rest of his body now out of account). When he comes nearer, the *general markings* of the face are seen, but not enough

* Change of handwriting, but it was not maintained.

to identify him surely. When closer still, you see the *individual features* clearly. I ought to have begun a stage earlier, by saying that when first he is seen at all, his face is little more than a dot and *cannot be distinguished* from that of any one else. You can see all these grades of resemblance in the faces of a group-photograph of any crowd. Each grade of resemblance is connected with a "critical distance." Further off it ceases. But it is not simple distance that we are concerned with, it is with distance *and* size, in order that what is true for a big picture shall be equally true for a miniature. Therefore the unit is the *angle*. The size at a distance is expressed by the *angular size*; the distance *and* area by the *angular area*. The particular angle I use is approximately that subtended by the disc of the sun (paled by a cloud). It is that of a breadth of 1 seen at a distance of 100, as one-tenth of an inch at 10 inches, one yard at 100 yards, and so on. It is only $\frac{1}{9}$th wider than a sun-breadth, so, wanting a word for it, I call this angle a *Sol*, and the square whose side is a Sol, a *Square-Sol*. My *measure* of any of the above grades of Resemblance is the number of Square-Sols at the *critical distance*, this being proportional to the number of *just-distinguishable* plots. The number of square-sols is easily determined by a low-power telescope with appropriate cross-lines in its focus, such that each little square is exactly a square-sol, and one counts the squares that cover the image. I have quite another page-full about this with which I need not bother you now. The above is a mere outline of what I am at. F. G. February 16, 1906 (aet. 84 years and 8 hours).

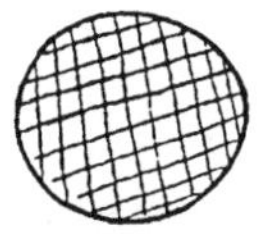

HÔTEL DE LA RHUNE, ASCAIN, BASSES PYRÉNÉES, FRANCE. *February* 18, 1906.

MY DEAR EDWARD, Part III is most clear and pleasant reading, and the results both practical and theoretical will strike everyone as first-rate. You have vastly improved Part III. Somebody will be down on "*Endemic*—confined to a particular *district*," but let that be.

I am very glad all is getting ready for your Mediterranean holiday and sincerely hope you may have none of the abominable Bay of Biscay weather that has plagued us during the past ten days, up to yesterday morning.

As regards some of your dear Mother's things,—on or about whose birthday you may receive this—you mention the engravings of *Hodgson*. Yes, I should be very glad to have one of them, but *without* the frame, as its destination will be in a portfolio. Hodgson brought me into the world 84 years ago, he advised my father on my education, I worked under him at the Birmingham Hospital, travelled, on his recommendation, with one of his pupils, afterwards Sir Wm. Bowman, lived in the house of another pupil, Professor Partridge (father of the caricaturist), when medicalising in London, and saw much of him up to his death. So he fills a large part of my recollections and I should be very glad to preserve his portrait.

The photo you send of Claverdon garden from the verandah gives a capital idea of part of the changes. I shall be most interested in the full result when I return. We stay on here, having only partly seen the neighbourhood yet owing to vile weather, but this is a land of surprises in that respect. The sun may burst out at any moment, as it has done while I was writing this; then, squalls of hail and cold and all that is unpleasant.

I wrote to Erasmus two days ago and hope to hear that he is well again. Milly and I write weekly. She told me of your kindness in asking Guy to join you, and of his very great regret that his lost arm unfitted him for sea voyages. I suppose the least bit of an arm is of much help, and he has none left, which makes the difference between what he and, say, Lord Nelson, could do on board ship. With best loves to you and M. L.

Ever affectionately, FRANCIS GALTON.

P.S. Eva would of course join, but she is at the Basque Church.

Address now: Hôtel Terminus, St Jean de Luz, Basses Pyrénées, France.
February 24, 1906, Saturday.

DEAREST MILLY, Yours is just come. Yes, it is evidently Birrelligion. I had forgotten that he was the new Minister of Education. We leave here on Monday and stay loosely on at St Jean de Luz, so as to get a few days at San Sebastian. The weather with the exception of two beautiful days has been execrable, and is so now. Eva will send you a post-card about this place. I fear it is liable to inrushes of noisy French, who go up the Rhune (3000 feet) and have a grand dinner, sleep here and return to Biarritz, etc. on the morrow. One noisy party of six men in two motors appeared here three days ago. They drank like Britons and sang the Marseillaise like Frenchmen and danced in rhythm to the chatter of the motors in the *place* in front of the hotel. Much of this would be a nuisance. The inn does not possess a third story, only two of them. I would send you the final proof of the *Louping-ill, etc., Report*, had I not already sent it on to Erasmus. The scientific part of the inquiry is by Dr Hamilton. Edward Wheler's part was making a readable summary. Parts I and II are technical and confused, but they contain the facts on which the summary, Part III, is based. I have no doubt that Edward Wheler gave much help all along. I *think* that they have good men now inquiring into S. African cattle plague, but these inquiries take much time. It is long before a true clue is found. Louping-ill was at first ascribed to *ticks*, but it was proved that they had nothing to do with it. The malaria mosquito, and the poison *carried by*, not emanating from, the tsetse, are instructive instances. When the Blue Book is out, you will like to read Part III, which I presume will cost under 6*d*. Parliamentary Reports are always issued so cheaply, at little more than cost of paper and printing. I am glad you could make something out of my brief summary *re* "Resemblance." Karl Pearson approves, but I do not, of the paper I have had typed. The subject has many side issues, and I must publish nothing without examples, but I see my way now pretty clearly. Karl Pearson helpfully suggests that I should work out *fraternal* resemblance and compare the numerical results so obtained with those already derived from measurements, and which are now certainly determined to within less than $\frac{1}{10}$th of the stated value. So I shall take steps towards doing this. Arthur Galton has been staying at Claverdon. He has many staunch friends. I wish I could appreciate him more, but his ideals in every way differ much from my own. I am always delighted to hear good words of him, being a relation. Eva gave to a young lady friend an introduction to his brother Ralph Galton in Ceylon, and she wrote from there a few days ago charmed with him. The young lady is a Miss Riardon, a Canadian, who travels far afield with her aunt. We met them in Sicily. How you are all marrying. Eva knows well about Mr Cope's merits. I shall be half sorry and half glad to leave this restful place, but it is becoming too restful in this weather, that keeps me for many days at a time wholly indoors. Novels are a great resource in the evening—good ones and big type. There is a very good circulating library at St Jean de Luz. I described our wild boar as having tusks. He has none; I mistook tufts of light-coloured hair for them. When I first saw him he had been, I suppose, shut up over long, for he rushed about hither and thither with short turns like a lunatic snipe and I could not see clearly. Since then he has been very quiet and the boys scratch him as he lies on his side to his great enjoyment. He is a funny and a handsome beast. Best loves to Amy and you all. Ever affectionately, FRANCIS GALTON.

Post-card. Hôtel d'Angleterre, Biarritz, France, will be our address. *February* 26, 1906.

Such a disaster to our best clothes. They were all left at the Hôtel Terminus, St Jean de Luz, which was totally burnt on Saturday night. Such a panic, we hear. All the tenants out in the streets, in a gale, in their chemises! We were to have gone there for a week, to-day. As it is we go to Biarritz to refit. All my papers and valuables were fortunately with me, so none of them are destroyed. Only a holocaust of good clothes. It was the fire in a *Frenchman's* chimney that caused the mischief. We leave Ascain to-morrow morning. FRANCIS GALTON.

Portion of a letter addressed to Edward Wheler.

SAN SEBASTIAN. *March* 12, 1906.

Your Marseilles card of March 8 just come. So glad that all goes well. We are tripping in Spain for a very little while. Our chief news is already a week and more old, namely, that all our smart things, which had been left in charge of a smart hotel, while we roughed it in the country, were utterly burnt, hotel and all. It is funny being clotheless, but the natives in these smart districts all wear clothes and have tailors and linendrapers where they can be bought. I had all my papers with me.

Lots of Royalty here, and Grand Dukeries, and I presume Royal Courtship too. The King of Spain looked older and more set-up than I had expected. His profile is pronounced. It is a very different face from that which is printed on the front of this card. You fly *with* the mail and I doubt whether Alexandria would still be a feasible address, so I send this to Malta. I hope you liked what you saw of the "unspeakable" Turks. You must tell me which of the various sea-port rascalities you come across strikes you as the worst. I should back those at the Piraeus. I have tried to make a tinted map of European knavery, marking the most knavish parts with the darkest tints. English public schools were the whitest, and shades thickened about the Levant. A friend who in his youth was appointed (?) "Judge of Appeal" to the Ionian Islands, told me there were more cases of *Appeal* (not of Law Suits) in one of the Islands than there were adult male inhabitants. I am in for law now, to try to get some compensation for my burnt clothes. I don't expect any but shall certainly have to pay the lawyer. Best love to M. L.

Ever affectionately, FRANCIS GALTON.

SAN SEBASTIAN (we leave to-morrow). *March* 25, 1906 (one quarter of the year gone).

DEAREST MILLY, Your dates will apparently suit us well and accordingly I will arrange generally, leaving details for later on. We have had some abominable weather, cold and snow storms, just like you have had according to the papers. So expeditions have been nil. I wonder if the following got into the English papers. A week ago, a party of friends, including the French Consul and a charming Spanish lady, a quarter and more English in her ways, drove some 12 miles from here to visit a famous grotto and cave (prehistoric remains, etc.). It is a long way underground; the party go with lighted candles; at one place the path crosses a deep crevasse, over which a wooden bridge had been put four years ago, for the King of Spain. Nobody had visited the cave since. The wood had rotted, the bridge gave way and down went the two. The lady stuck 10 metres down against some débris of the bridge. The Consul fell as far again down, upon a ridge. Fancy the alarm! It was two hours before ropes and help could be procured and they were pulled up; the lady in blood and dishevelment, the Consul barely conscious, having been stunned. They are, I believe, not seriously the worse for it all.

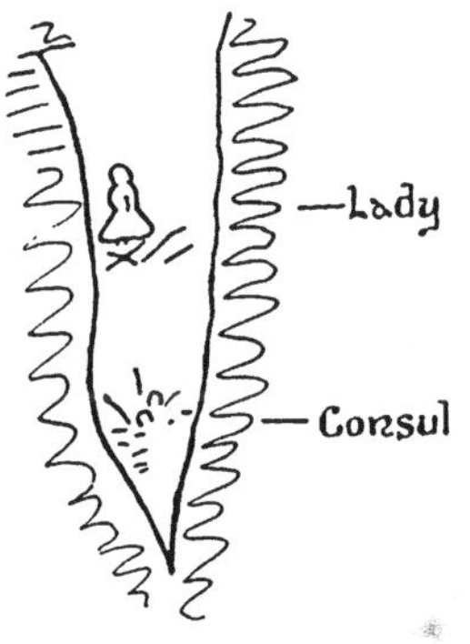

A narrow shave like this suggests epitaphs. I heard the following lately, and the last two lines "obsess" me. I don't know to what careless, vicious young genius they referred.

"He revelled 'neath the moon,
He slept beneath the sun,
He lived a life of *going-to-do*
And he died with *nothing done*."

So the *Noteworthy Families* is at length published and you have received your copy. I am glad to have made much of Schuster. He is a good, gentlemanly fellow and feebly protested against it, but it has encouraged him and he is working hard at families now. I think I told you of the speciality of these parts of inlaying iron with gold thread and making ornaments. They do it very cleverly and prettily. The pattern is engraved, the very thin wire is punched in with a fine punch, the whole is heated which somehow solders the two metals, and then it is polished up. This exceedingly fine work is done by the naked eye. The man I saw at work is a fine big fellow, but his sight must be such that he could see as much detail in the eye of a needle as you and I could in—what shall I say—not exactly a "barn-door."

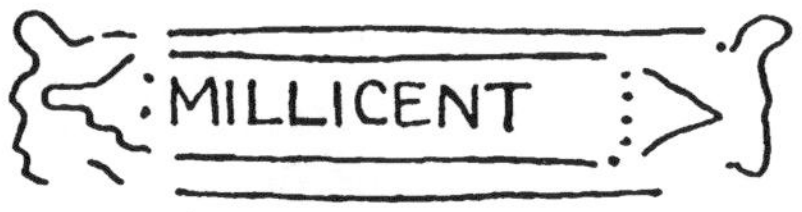

I got him to do a brooch like the above (Eva sketched the outline), which I will send as a memento, when we return within reach of easy and honest posts. After beginning this letter by abusing the weather it has suddenly changed into calm sunshine and we *are* going at once for a good expedition, so I will finish this later.

Monday morning. We had a grand day yesterday to Loyola's place—perfect weather, two hours train, four and a half in carriage in all. The place itself seemed hardly what it should be. I had hoped to find a record in portraits, pictures and maps, of the progress and misfortunes of the Order, and a good library, besides some idea of Loyola's own surroundings. But there is nothing of that. Plenty of gorgeousness in marble and gold, small taste, and nothing of graphic historical value. However the drive, etc., was most picturesque and to-day the cold rain has recommenced. I got a ring of no value, which may do for Amy, with the Spanish equivalent to SOUVENIR OF LOYOLA upon it. It shall be sent with the brooch. I am writing on my knee in a bad light, the morning is so dark and dreary. We sleep to-night at St Jean de Luz.

Ever affectionately, FRANCIS GALTON.

I don't know where to tell you to send your next letter to me, so don't write at all!

42, RUTLAND GATE, S.W. *April* 7, 1906.

(I enclose the German pamphlet of which I wrote and promised to send.)

DEAREST MILLY, It is so pleasant feeling in one's own clean home again and receiving smiling welcome. We had an excellent passage cross-channel and all is well, except that Eva has a rather bad head cold and keeps half this day in bed. I send the brooch and ring; also, I have ordered to be sent to you, from me, Skeat's *Concise Etymological Dictionary* which is the book you refer to and is I find a capital book to lie about, at hand when wanted. You praise me too much in your letters, please don't any more. This reminds me of a true story of the present Lord Thring when he was Parliamentary draughtsman and had in consequence to discuss familiarly the terms of proposed Government bills with the Cabinet Ministers who introduced them. He was very outspoken and uncourteous-like and talked of everybody as d...d fools. Bob Lowe one day said to him "Now, Thring, we will understand once for all that everybody except ourselves are d...d fools, so you need not trouble to repeat it, and let us stick to business." By the way, Lord Thring told me in answer to a question, that of all the many Cabinet Ministers with whom he had worked, he rated Gladstone and Disraeli as quite the first. He said they were of different "clay" from other men. On the first occasion, he put Ayrton third, but in later years when I asked him to verify my recollections, he did not particularly dwell on Ayrton. We did not do much sight-seeing in Paris, only Notre-Dame, Sainte Chapelle, Louvre twice, etc. Eva quite thinks our unpretentious hotel would suit you. You have to give three days' notice before leaving and would of course have to arrange before coming, and not take your chance on arrival. I have seen but few friends yet, being busy. One was William Darwin just now, when for the *first time* I saw the *Noteworthy Families* book, Murray having omitted to send me a copy, knowing I was abroad. I tea-ed yesterday with Miss Baden-Powell (who does the honours of the house), having just found a card asking me to come. There is a wonderful collection of curiosities, Ashanti as well as S. African. She had a wire gauze thing like this

with a cross-bar to hold it by to put tea in, and to lower it into the teapot. It is taken out after standing long enough. She got hers at the Army and Navy Stores. It seems a capital plan. I have just got such a pretty card of invitation to a golden wedding, with portraits of the pair 50 years ago and now. It is from Sir William and Lady Crookes. Also, a still more ornate and grand card, engrossed in black and red letters, of invitation to a 400th centenary of Aberdeen University in September. It is all in Latin and drawn up in a very complimentary form. But I can't accept, it is too far and bustly. Sir George Darwin has gone to Philadelphia, as the representative of England at the forthcoming Benjamin Franklin commemoration. His wife is a descendant, great-grand-daughter, of a fellow-worker of Franklin. So it is very appropriate. Ever affectionately, with many loves, FRANCIS GALTON.

42, RUTLAND GATE, S.W. *April* 17, 1906.

DEAREST MILLY, I am glad you like the brooch and Skeat, and Amy her very unpretentious but characteristic ring. The sudden death of my—I might almost say colleague and—friend, Professor Weldon, has been a great grief and will be a serious scientific loss to my other colleague and friend, Professor Karl Pearson. Weldon came to London alone for a night or two, while suffering from incipient pneumonia. He became rapidly worse and was put by friends in a nursing-home, his wife having no previous idea of it, and being then happy with the Pearsons, and he was dead in a few hours, his wife reaching him while he was still conscious but very ill indeed. He was one of the strongest of men constitutionally, but took liberties with his strength. It has cast a gloom over this house. We go to Lucy Studdy to-morrow, Wednesday, and the plans are to stay there over Friday night and to go to Claverdon Leys from Saturday to the ensuing Thursday. The next Saturday–Monday I go to the Frank Butlers who have a charming little house at Witham. He is now full Inspector of Schools and will probably before long be promoted to London work. One effect of the fire at St Jean de Luz has been to show how much fatter I have grown of late years. Certain clothes, left in my wardrobe of recent years, have been tried on and found too tight, and are being sent to the tailors to alter up to date. I heard of a man who said to his tailor, "I am now forty and never had occasion to be re-measured by you." The tailor smiled and said, "We generally ease the fit a little when our customers seem growing stout, without troubling them about it." I can't now take enough exercise to keep muscles fit; it is no good trying. It only fatigues and I have capital digestive health as it is. Chamberlain never takes exercise, neither did the late Lord Salisbury. I have now got back with proper appliances to my "Resemblance," but am less confident than I was of getting *useful* results; the theory is all right, which is something though not enough. You will have enjoyed this weather. I hope that Dartmoor won't be set alight. Many moors are burning, I see. We have had long sits in the Park, which is growing beautiful. Dear old England. She *has* merits. Best loves. Ever affectionately yours, FRANCIS GALTON.

CLAVERDON LEYS, WARWICK. Saturday, *April* 21, 1906.
(I go to Rutland Gate on Thursday.)

DEAREST MILLY, On arriving here about lunch-time I found your letter. Guy's chill, Amy's influenza, and the combined want of Cook and Parlour-Maid are a large tale of mishaps. The bitterly cold weather of the last few days and the blackened moor fill your cup almost to overflowing. Claverdon gardens and shrubbery are greatly improved. All the former stuffiness from overgrowth of trees is gone without any sign of bareness. The ground too is judiciously levelled here and banked there, so it is becoming both pretty and interesting. Edward and M. L. look very well. The voyage and change of scene had become a necessity, for they were overworked. Lucy and Col. Studdy *seemed* quite well, though he is not so really, but he mends slowly. Their house is very pretty. Lucy's embroideries, framed and hung on the walls, make a brave show. I read through the typewritten copy of Bessy's memoir, which is very readable by any one and full of interest to her own family. It wants "perspective," treating all occurrences too much on the same scale. We are discussing how to treat it to the best advantage, by adding notes and illustrations, and probably printing it for private circulation. Eva is quite done up, I fear, and fit only for quiet at present. She has Gwen Chafy with her, otherwise the house for the moment is almost shut up.

The loss of Weldon is a severe one, from many different points of view. I attended the funeral service at Merton College on Wednesday, but the weather was far too bitter for me to go to the Cemetery. All was very sad, and through change of address I am temporarily out of touch with the Pearsons, and through them with Mrs Weldon. His death will modify many of our future plans and movements. It is very sad for us, and almost desolation to the Pearsons.

I did not go yesterday to see Eva's stained window given over to the church at Ettington. It was too cold, but Lucy went and brought back Constance Pearson for the night. I left her with them. What a large scale she is on! I must leave off now as tea is coming in and it is nearly post-time. Very best loves. Ever affectionately, FRANCIS GALTON.

42, RUTLAND GATE, S.W. *April* 28, 1906.

DEAREST MILLY, Yes, the window in Ettington new Church was put up by Eva and her brothers and sisters to the memory of their Father, Eva taking all the trouble and bearing nearly all the cost. She has lent Count Russell's book; you shall have it in time. There is little in it bearing on the picturesqueness of the lower heights, but great lamentation over the want of enterprise in not building hotels, etc., upon them, as in Switzerland. I will read the

Bishop's Apron as soon as I can get a sight of it. The rain had not reached Claverdon when I left on Thursday, but there had been some snow. Edward is greatly improving the place by cutting down a great amount of overgrowth and re-forming much of the garden. He is also evidently becoming an important man in the county, being so familiar with county duties, and eminently useful and kindly. His Land-Agents' Society is quietly growing into a great institution. The Studdys have arranged their house very prettily. He *looks* well, but is not yet quite well. I go this afternoon to the Frank Butlers for Saturday to Monday. He is now *full* Inspector of Schools and lives at Witham, in what I hear is a very pretty house, with his wife and three little daughters. His eldest brother, Cyril, will be there. Cyril married a rich Miss Pears many years ago, bought latterly a country place near Shrivenham and is High Sheriff this year for Berks (or is it Bucks?). Anyhow it is the county in which Reading is. I can't think of the preceding lines to Canning's "Buck-, Buck-, Buckinghamshire dragoon." How Canning must have bubbled over with fun! Last week or ten days have been in great part melancholy. Weldon's funeral on Wednesday week in bitter weather, and the cold weather subsequently, had given me a sort of chill, which all the warmth and hospitality of the Studdys and Edward Whelers did not wholly overcome. You would have laughed to see how I was covered up at night, and fired—big fires, I mean, in bed-room—just like a decrepit nonagenarian. I drove over with Edward (shut up) to see the last of 5, Bertie Terrace, which has been a second home to me for more than fifty years. It was very painful. Bessy's old house, which with the garden went to Lucy, *has* been sold. We lunched with Gussie and had news of Grace from Athens. Eva would have added a line, but is just now upstairs. She is still far from strong, I am sorry to say. With loves to you all. Ever affectionately, FRANCIS GALTON.

42, RUTLAND GATE, S.W. *May* 6, 1906.

DEAREST MILLY, You will be growing restless like a migratory bird. I shall be curious to learn your plans and dates generally; by the 27th the weather ought to be warm in the south, but early June is full early for the Pyrenees near the bigger mountains. You will find flowers at all events. Your son Frank will have a busy and responsible time in Natal, such as young men love and parents fear. Eva and Walter Biggs* (who is up here for two or three days) went last night to the great meeting of Roman Catholics in the Albert Hall. That big building was full to overflowing, and vast crowds gathered in the streets. It was very impressive, I hear, most enthusiastic but well-ordered, and the speaking both good and temperate. The whole eleven thousand sang a hymn in unison. Frank Butler (as School Inspector) tells me that he thinks the bill would be quite *workable*, independently of its merits, I mean. I am going with Eva this morning to hear, or rather to try to hear, a sermon in its favour. All the same, I don't profess to really understand it, and have not fairly tried as yet to do so. Ethel (Galton) Marshall Smith† lunches here to-day and Violet looks in after. Then Eva and Walter go to St Paul's, and he is to hear the Education Debate in the House of Commons to-morrow. It is "history in the making."

Lunch is over. The sermon was, alas, almost inaudible to me, to my great regret. I have no news. Eva and I went to Hampton Court yesterday afternoon. The morning was brilliant, but clouds and cold wind came, and the expedition was a failure. So many things of Bessy's and Emma's have been offered very kindly to me, by both Edward and Lucy. One of these is the original picture of my Father, signed by Oakley, which I have put up in the place of the *copy, also by Oakley*, which I had. The latter is good, though not equal to the original, and it is now of no use to me. Would you like to have it? It ought to be in the family. I will send it at once, frame and all, in quite good order, if you like. Family matters remind me of Mrs Schim., and she, of Bristol Cathedral where she is buried, and that of the Bishop of Bristol, with whom I was talking two or three days ago at the Club. He is delighted at being asked to be President of the Alpine Club and has gladly accepted. He was a great climber in old days and more especially an explorer of caverns! Ever affectionately, FRANCIS GALTON.

* The Rev. Walter Bree Hesketh Biggs, Evelyne Biggs' brother.

† Ethel, daughter of Cameron Galton, married a Marshall Smith.

42, RUTLAND GATE, S.W. *May* 14, 1906.

MY DEAR LUCY*,

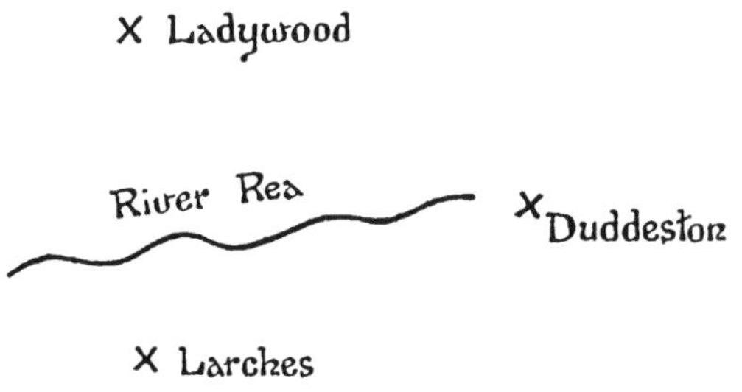

My *impression* is that the three places† are the corners of an equilateral triangle, three miles to the side—but I have no map of Birmingham whereby to verify. *Ladywood* is by the "Crescent," to the right of the road from the Town Hall to the Five-Ways. *Duddeston* is located by St Anne's Church, and the *Larches* by Sparkbrook; I can give no more exact reference to the latter. The River Rea, once sparkling, subsequently filthy beyond compare and finally diverted into a sewer, fed the Duddeston ponds. One was called the Mill Pool and, I presume, not only *had* acted but did act during my grandfather's life-time, as such, to the Duddeston Water Mill, which subsequently was partly if not wholly replaced by steam power.

I am very glad that Arthur takes kindly to the idea. He is not handicapped, as I am, by crowds of ancient recollections, which had my Father and Mother, Uncles and Aunts, as their focus, and are with difficulty adjustable to the focus in which you are concerned, namely, your Mother.

I feel as if I did not deserve to be forgiven for my blunder about the paper of dates. It confirms a strong impression I have long had, that the way to mislay a document is to put it in some peculiarly safe place. Ever affectionately, FRANCIS GALTON.

42, RUTLAND GATE, S.W. *May* 15, 1906.

DEAREST MILLY, It is a relief to hear that the picture arrived safely. Glass, when cracked over a water-colour, does or may do great harm. Our letters will get again into order this week. To-morrow we go to Cambridge for the day to see Montagu Butler and my portrait. Also, some of the Darwins, *not* George I am sorry to say, who will be away on business. Eva went yesterday with Gwen Chafy to see both Lucy Studdy and the memorial window at Ettington which they both liked greatly, I was glad to learn. Eva is much better. Guy must be glad of a fortnight of his old work, which he does so well. What an account you send of Johannesburg rascality. I have arranged to have a look shortly at the Identification Department in Scotland Yard. The Chief Commissioner, Mr Henry, was, as you may remember, *lent* by the India Office to the Colonial Office, in order to get the Johannesburg Police into order, before taking up his present appointment under the Home Office. He told me that for Kaffir police purposes, a great desideratum was that each man should have, and be always compelled to use, the same readable *name*. It would be all the more necessary with the Chinese, whose names have less variety than those of Scotchmen (Highlanders). We have been very quiet at home. Last week there was a "gentlemen's soirée" at the Royal Society, where one of the most beautiful exhibits was a set of four large maps including only a small part of the Milky Way. The multitude of small stars that photography reveals far exceeds what could have been imagined, and the brilliancy of these multitudes of specks is astonishing. Edward Wheler comes to us on Thursday for two or three nights. He has much business to get through—the Land Agents' Society, and so on. I am going to subscribe to the *Times* library and shall put down the *Bishop's Apron* on my first list.

* Mrs Studdy, daughter of "Sister Bessy." It may interest the reader to know, that on the death of her mother, Mrs Wheler, she came into possession of several Darwin relics, and of these she left, on her death, the armchair of Dr Erasmus Darwin and silhouettes of his second wife and him to the Galton Laboratory.

† With regard to these three homes of the Galtons, closely associated with Francis Galton's boyhood, see our Vol. I, pp. 50–51 and Plates XXIX, XLV.

I am quite enthralled by one of Renan's books, the *Antichrist.* (I am reading it in an excellent translation with an excellent introduction by W. Hutchinson.) He makes out that Nero is the Beast of the Apocalypse, and brings in an enormous amount of the history of those times, most of which was quite unknown to me. It is a book well worth reading. Best loves to you all.

Ever affectionately, FRANCIS GALTON.

42, RUTLAND GATE, S.W. *May* 20, 1906.

DEAREST MILLY, The paragraph about Guy is pleasant reading. Edward Wheler, who has been staying here three nights, heard to the same effect from —— Smith (son of the publisher), who is (?) Colonel of Guy's regiment, a few days ago.

Eva and I had a most pleasant 24 hours at Cambridge, lunching and spending the bulk of the day at Trinity Lodge, and sleeping at the George (*Sir* George) Darwins. The portrait looks particularly well in the Hall. The background being much lighter than those of the other pictures, and all being surrounded by dark oak, gives a welcome light to the general effect. Nothing could be better all round. Eva is also quite pleased with the memorial glass window in Ettington Church. She went down to see it on Monday last and to lunch with the Studdys. She has got Count Russell's book back and proposed to, perhaps she already *has*, post(ed) it to you. I can quite fancy Biarritz becoming enormously expensive. This is the beginning of its summer season, when wealthy French and Spanish grandees visit it in large numbers, and ordinary French and Spanish go in shoals and sleep six in a room, as we were assured often occurred. I shall be eager to know where you yourselves finally go to. We went last night to Stephen Phillips's play of *Nero*, having read it first. It is very "spectacular," but the acting was on the whole not quite first-rate. Still it was extremely interesting and apparently a just rendering of Roman Court life in those days. What villains they were! Talking of villains, I spent an hour in the morning yesterday seeing the finger-prints in Scotland Yard. Mr Henry (the Chief Commissioner of the Metropolitan Police) has got them into good order. The methodical arrangements are excellent. He has about 84,000 sets of prints and thinks he could deal with 150,000 without straining the method. There are more than 500 identifications a month, now; in the old days, there were not so many in a whole year. The burglars begin to use gloves, and now and then they destroy the skin of their finger-tips, but this grows again.

Ever affectionately, with loves to you all, FRANCIS GALTON.

42, RUTLAND GATE, S.W. *June* 2, 1906.

DEAREST MILLY, I wonder where and how you are, after your very hot start. To think of your having been so near Rutland Gate on Monday! It was fortunate for *you* that you did not call, for on that day we had arranged to go to Claverdon. But in the morning I was suddenly seized with a strange ague-fit like I had last autumn; the doctor sent me at once to bed, and Claverdon had to be telegraphed to. I was rather bad for a few hours, and could not have seen you. The thing is gradually working itself off with bronchitis, and I get out, but am still over-weak. .... When inquiring about interesting places, did you ever hear anything of the Mediterranean side of the Pyrenees—Port-Bou, etc.? The places are said to be very picturesque, but the accommodation is not *smart.* You reach them *via* Perpignan. Louisa and I once spent some time at Vernet-les-Bains, at the foot of the Canigou. It had merits but is probably now a noisy bath place. Ibrahim Pacha was sent there in his old age to recuperate from his excesses. The doctor said that he must stop wine. On a second visit Ibrahim was worse and the doctor rebuked him for not obeying orders. Ibrahim flew into a fury and said, "Oh, pig-brained son of a she-ass, I have not touched wine, only two bottles of Cognac each day." Ever affectionately, FRANCIS GALTON.

42, RUTLAND GATE, S.W. *June* 8, 1906.

DEAREST MILLY, If you are having splendid weather, like us, you will indeed be joyful. Eva will enclose her letter to Count Russell. She saw more of him than I did and has corresponded with him already. What an ovation you have had at Montauban. Amy must have rejoiced in the Bishop, and you both have been delighted at the happy ways of Jeannie Ronsell and her kindred. We English *are* a nation of natural snobs, which Southerners rarely are. We do however bear some polish, though it is costly and laborious to rub it on. The servility to persons of high social rank seems an expression of a conscious want of the polish that those have acquired and that they have not. If so, it is a pardonable feature, so far. I am glad you recall the *zinc* figure in the garden at Royat. It has left a deep impression on myself, not unlike that of Millet's "Angelus"—very sad, very brave, very noble. I had no idea that one of your sons had the honour of having played hockey with the present Queen of Spain! A lady who was here had joined in *eating a bun* with her, some years ago, at a pastry cook's. What a deal she has gone through already. Among the *minor* Spanish events is, I see, a resignation of the Premier, followed by a reconstituted cabinet. I am getting straight again and have driven out the last three days, and to-morrow we go for the week-end to friends at Haslemere. Next week I (and Gifi) go for three or four nights to Oxford, to the Arthur Butlers, which I think can be now safely effected. It is always such a great pleasure to see him. I am pitching into "Eugenics" again, seriously discussing the possibility and advisability of offering *certificates*, that must be trustworthy in reality as well as in popular appreciation, and that must be inexpensive and yet self-supporting. Though the thing is full of difficulty, I now think I see my way, so have just sent a paper to be typed, and to be submitted to a few critical friends before taking the next step. Lucy Studdy is in town and dines here to-night. Her embroidery won two prizes at a recent exhibition at Oxford. If you come across a Pyrenean sun-dial, such as the shepherds

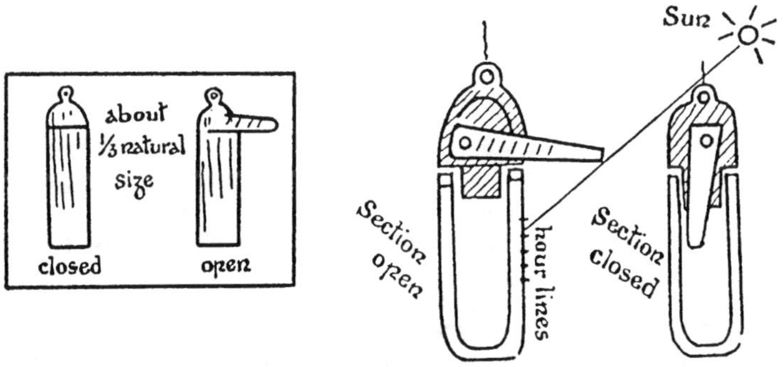

always carried with them, I wish you would invest in one for me. They can hardly cost more than 1 franc. I gave mine to the Pitt-Rivers Museum. The principle is to find the time by the altitude of the sun at *any given* season. The head of the dial is turned to the right place (month and day). The gnomon sticks out and casts a shadow. The cylinder is marked with proper curves, and is dangled at the end of a string, and the hour is read off. I have drawn the top badly here. Ever affectionately, FRANCIS GALTON.

42, RUTLAND GATE, S.W. *June* 16, 1906.

DEAREST MILLY, Your letter is very interesting, but I grieve at Edward's rheumatism. We have had three cold and rheumatic days here but the bad spell seems just over. Thank you much about the Shepherd's sun-dials. The rougher and more every-day order that they are of the better. I will even ask you to get me *two* of them. I had neat box-wood ones made for me in England some fifty years ago, and I calculated the curves, and had them cut in them, for the latitude of London, but I liked the rough native ones better as objects of interest. I hope you may come across Count Russell after all. Amy is well out of Montauban hospitality. Your account of her reminded me vaguely of Vathek, who, absolute and incomparably learned monarch that he was, was so upset by his inability to decipher the magical letters on the sword given him by the magician, that of the 163 dishes presented to him at dinner he had so lost appetite as to be only able to taste 35!

We had a most successful week-end visit last Sunday, when I was well "molly-coddled" under the surveillance of Eva, and three other nights at Oxford with the Arthur Butlers—all most pleasant. George G. Butler (whom you know) and his boy are with us now. They all go to the theatre to-night, with others who dine here. I shall smoke the cigarette of peace and quiet in great comfort at home. Fred's account of the Chinese would have been most welcome to the Unionist newspapers a few days ago, but after Mr Churchill's confession on the part of the Government that only twelve Chinese in all had asked to be repatriated, the case is closed.

The Ministry seem learning to blunder less, but have a difficulty in carrying out their party programme, etc., as stated at the Elections. I heard a good story of, let us say, Lady A., a great lady who lives in Grosvenor Square. She told her friend, say Mrs B., "I have asked all the new Ministers to my reception in July." Mrs B. said, "What, all of them? Have you asked John Burns and his wife?" Lady A. answered, "No, not them: they are impossible. Besides, I have never called there." Mrs B. said, "But you *must* ask them, or it will be a slight and they and many others of their party will be angry." So Lady A. went home and wrote, "Dear Mrs Burns, I hope you and Mr Burns will give me the pleasure of your company at my reception on .... Pray excuse my not having called, but the distance is so great from Grosvenor Square to Battersea." The answer came: "Dear Lady A., I fear that Mr Burns and I shall be unable to avail ourselves of your kind invitation, for I have *studied the map*, and find the distance from Battersea to Grosvenor Square to be *just as great* as that from Grosvenor Square to Battersea." Neat, wasn't it? Mrs B. told Lady Galton who told me. I am getting answers and suggestions to my typewritten circular about the Eugenics Certificates, which were sent to about half-a-dozen experts. We shall see the final results, probably in the first instance in a paper published somewhere. Best loves, ever affectionately, FRANCIS GALTON.

42, RUTLAND GATE, S.W. *June* 24, 1906.

DEAREST MILLY, July 12th or 13th, as you propose, will suit excellently; so come here at once on your arrival. Eva will stay two or three days to overlap you and to have the pleasure of seeing you. What about Amy? The little dressing-room will be the only room available for her. It would be heartily at her service if she came with you. As the time approaches, you will tell me more particulars?—day and train, etc. To continue business, you will, of course, stay a full fortnight. Later on *I* should greatly like a week with you and I would arrange about Gifi in the way you describe, but I can't say more just now, as Eva's plans are not certain—cannot be certain—just yet, and mine would be somewhat governed by hers. The plan in outline is that she should go with her artist friend, Mary Savile, to some picturesque place, yet to be decided on, in conformity with Miss Savile's portrait-painting arrangements. Eva writes to-day to fix more particularly, but cannot hear for some three days, I expect. One idea was to go to Polperro, which would be very convenient to her and to myself. But this must stand over for a few days. She is most obliged for your very kind invitation, but she wants a bit of artistic Bohemianism badly. Miss Savile, too, who is coming into high vogue with great people, wants the same. Lucky for Bob, not to have been blown up! So glad the Pyrenees have been a climatic success, though not a social one. Hugh and Fred will, I trust, enjoy it all thoroughly. I have been in what is now for me a whirl of doings. There was a big dinner at Trinity College given to us old fogeys, once undergraduates at Trinity. I was the oldest fogey but one, but it was very interesting meeting many scattered friends. Llewelyn Davies was one, who sat next to me. Lord Macnaghten (the Judge of Appeal) was one of the guests at the Lodge and talked to me very pleasantly about R. Cameron Galton, who was his contemporary. They both won rowing prizes and were great friends. Macnaghten was a Senior Classic of his year. Then I had a good deal of talk with Sir Fowell Buxton, who told me he had a genealogy of the descendants of the Gurneys of Earlham. There are upwards of 1000 now living, but of these some 200 must be subtracted owing to cousin marriages, which include duplicate entries of the same name. He has sent me some figures and asked me to suggest how to work the thing to the best advantage. I had some ideas and have written them out fully and sent them. There were many others of great interest to myself, but tedious to narrate. Then, one day, I went with George Butler and his boy, with Eva, into the country to hunt up family portraits of the boy's family, contained in an old house, whose representatives welcomed us warmly. In the evening at 11.15 I went to a big affair in the offices of the *Daily Telegraph*, to which the German Editors now visiting London were invited, and a lot of English to meet them. We saw the set-up of the paper in all its details and the beginning of the printing of it at 12.15. The scale of the whole thing is enormously costly. One sees that home industries, in producing things in *wide* demand, have no chance against big machinery. There are eight big machines, all fed from duplicates cast from the same type. Each machine is fed from a roll of paper four *miles* in length and drops out *Daily Telegraphs*, ready folded and dried, faster than it is possible to count—certainly at least five in each *second*. It was a wonderful display.

Ever affectionately yours, FRANCIS GALTON.

BRIDGE END, OCKHAM, SURREY. *August* 12 (St Grouse of the Philistines), 1906.

DEAREST MILLY, The four letters which I return are like the opening of an Etruscan tomb, where all the contents appear just as they were deposited 2000 and more years back. How human we all are! I can quite understand your having felt just the same about your child and grandchildren as your mother did about me. Erasmus might, or might not, greatly like to see his own letter and others about Loxton. I have to write to him, and will mention their existence, especially that of July 19, 1839, so he can apply to you if he wishes. I recollect so clearly coming home—to Leamington—in 1840, and my sisters all in mourning array for my grandfather; Eva's *great-great-grand*father. It is pleasant to read of the strong affection that your mother then had—even at that early date, I mean—for Aunt Brewin, or Aunt Sophia as she then was. We were rated by outsiders as a most united family and the letters show that we were so then. But after my Father's death the hoops that bound together the staves of the family cask seem to have given way, and with independence we mostly flew off in different directions. What good paper and ink they used in 1839–40. It is the bleaching and the shoddy (short fibres) and other material than linen, that cause modern paper to be so weak and perishable. But it is marvellously cheap. I suppose that paper is "pulped" over and over again until it serves no other purpose than to give bulk. The strength, such as it is, being due to a scanty intermixture of proper fibre. It is worth while to scrutinise paper through a strong lens and to notice its curious structure. I have been busy with my machine, out of doors, I sitting under an awning and the machine projecting out into the open, and now that I can test the plan experimentally and for the first time with proper appliances, am more doubtful than ever as to its real usefulness. But there are still some tests to be applied and some variations of method.

Yes, this August is a sad month to me, or rather a month that brings sad and solemn recollections. Dear Emma, I feel the want of her more and more, but she fully lived out her life. There were grounds for fear that her faculties would noticeably weaken before long. There is a Greek phrase, I think, "he was still young and his tomb was not yet in sight." In my fancies, I don't see a tomb but a greenery with some cypresses in it showing over a bit of old brick wall on a hill about a mile off, where the peaceable cemetery lies. There are many small, nice, old-fashioned churches hereabouts and, as Sir Lucius O'Trigger expressed himself, nice quiet lying in the attached churchyards. We feel much at home here, having made many friends last year. The weather continues lovely—no rain; the trees don't show the want of it, though the gardens do. Your rain is wanted here by the farmers. Best loves.

Ever affectionately, FRANCIS GALTON.

Try to excuse smears and blots.

Lucy Cameron Galton comes here next Saturday for a week. I much look forward to going to you on or about September 4. We leave this on Thursday, August 30, thence to town to refit and to settle matters, and I should be free to come to you on September 3 or 4 (Monday or Tuesday). Please let the exact date stand over for a bit. (You must of course consider *your own* convenience *first.*)

ROYAL VICTORIA YACHT CLUB, RYDE, I. OF W. *August* 14, 1906.

MY DEAR FRANK, Our friend Collins is quite wrong about the compass points as used by seamen. No sailor would dream of saying N.N. West by West half North; he would say N.N. West half West. No doubt Collins had picked the term up among yachtsmen, who do make ridiculous mistakes among themselves, and there is no one to correct them. In mercy to Collins pray tell him. Many thanks about sister Adèle's letter and F. Miller's address. I am glad you like your quarters so well. What a cheerful companion Jenny would make for me; how we would converse and understand each other—like yourself and the camel did some 40 years since at the Zoo, when the camel flopped down on its knees and toppled over a lot of children and two ancient parties. I simply made a run for it—yourself ditto.

I get afloat in steamers often, but not in sailing vessels. How curious it is how people keep on using old terms, now meaningless. The newspapers and others constantly say such and such a company's steamers *Sail* on such a day in place of saying *Leave.* They carry no sails and have no masts except for signal uses.

I am as well as possible, but the toss by a cow three years since has spoilt my walking powers and my sea legs. Ever very affectionately yours, ERAS. GALTON.

P.S. Kindest remembrances to Eva.

BRIDGE END, OCKHAM, SURREY. *August* 18, 1906.

DEAREST MILLY, I am glad that the proposed time for my visit suits you. The progress of the drains must be interesting to watch! We, like you, have at length had much rain, sorely needed here, and it is cold in the evenings, so we have begun fires and sit but little out of doors. The hammock had been put up and the tent peg seemed as firm as iron when I lay in it, but after the rain had soaked the ground, Eva tried it and out came the peg and down she came "Hammock and all" like the nursery song of the bird in its nest on the tree-top. We had tea yesterday with an interesting man, Mr Stokes, the iron contractor for the big Egyptian dams. His wife, an Ionides, is half-Greek, and grand-daughter of the Ionides who gave the collection of pictures to our government. Everything about his cottage by the side of a rushing mill-stream is thought out, home-made for the most part, and interesting. The stream is of considerable width and he built with his own hands two bridges across it; each is on a peculiar principle, and a water-wheel pumps water high up to his garden. Lucy Cameron Galton is with us for some days. The weather has been much against her sketching, but she thoroughly appreciates the paintability of the place. Cameron is taking a walking tour in Wales with Violet.

You recollect my pinched thumb-nail. It happened about July 26, more than three weeks ago, during which time the black has travelled only 7 millimetres forwards, say at the rate of 2 millimetres a week. It is now, as well as I can draw it, like this—

and there is more black to an as yet unexplored distance below the flesh. In fact, it gets blacker and the nail seems more rotten nearer its root. I wish the surgeons would make bioscopes of healing wounds. In fact everything that grows might be bioscoped—humans, trees, etc. For a landscape, either a stone, or three smaller stones or bricks, would have to be fixed permanently in the ground, with holes in them for the legs of the camera, that it might be always in exactly the same place, then the photographs would have to be taken at the same hour on different days*. My machine† is now worked in my little so-called dressing-room here—sometimes still out of doors. All it wants is a common (it can't be too common) table to put it on, and it does not hurt a good one. A small table, so long as it is not less than 2 ft. 8 in. in length and 9 in. in breadth, is handier than a larger one. If you have not such, I could easily buy one in Bovey, or if none are to be bought, I could get a carpenter to nail up something thus:—

So I will bring the machine down on *spec.*, unless I shall have done enough with it before then. But the inquiry is very troublesome. I am still uncertain as to the real utility of such results as I am likely to get. Best loves. Ever affectionately, FRANCIS GALTON.

BRIDGE END, OCKHAM, SURREY. *August* 26, 1906.

DEAREST MILLY, If all goes well and the May Bradshaw is to be trusted, Gifi and I will reach Bovey Station on Saturday next at 3.9, and will drive straight thence to Edymead, with the awning and with the machine. I have only just got it all into *fluent* working order, though it is so simple. Best congratulations to Amy on the Pope's autograph. May it convey a blessing! I feel quite interested about the progress of your drain. About the black on my nail, it is now 10 millimetres long and has five more to accomplish to reach the end. By pushing the skin back at the base I fancy I see the root of it (the black part), and its very rough commencement. This too is a subject of present interest and amusement. Nothing out of the common way has happened here since I wrote, and being at this moment rather hand-weary with much writing, and as we shall meet so soon, I will not write more. Lucy Cameron Galton has been a most pleasant visitor. She left two or three days ago. I take such a sound sleep in the middle of the day that each one seems to be two days and my dates are apt to be mixed. Best loves. Ever affectionately, FRANCIS GALTON.

* The coming of spring, or the passing of summer, might be effectively represented in this way.

† The "Measurement of Resemblance" apparatus, still preserved in the Galton Laboratory.

Francis Galton and his Great-Niece Eva,—Evelyne Biggs, at Bridge End, Ockham, in August, 1906.

Address now to 42, Rutland Gate, S.W. *August* 29, 1906.

DEAR SCHUSTER, The advantage of Doubleday's work is its direct way of meeting the question whether in the long run such and such classes prevail. The fact that one class is more fertile than another in one particular generation cannot be trusted altogether. It may well be the case that the marriage rate is different in small and large families or again (for which some grounds exist) that the tendency to fertility is more or less periodic. It is astonishing how often large families have few descendants in the next generation owing to causes that may be partly social, partly physiological, whose existence we may suspect but of which we as yet know next to nothing. "The fate of large families" would be an interesting inquiry, in its way.

Tell me more precisely the use that you think might be made of Burke's *Landed Gentry*. I leave Ockham to-day for three nights in London and thence, first to Devonshire and afterwards to the Midlands, but letters to 42, Rutland Gate, London, will always be forwarded on.

Very sincerely yours, FRANCIS GALTON.

Galton's Reply to a request on the part of Mr Frederic Whyte to know what he thought of Phrenology, September, 1906.

The localisation in quite modern times of the functions of the brain lends so far as I am aware no corroboration whatever, but quite the reverse, to the divisions of the phrenologist. Why capable observers should have come to such strange conclusions has to be accounted for—most easily on the supposition of unconscious bias in collecting data.

Whoever may seriously re-examine the question must procure a collection of appropriate cases by persons who know well the characters of the person named, and who, if possible, should be wholly unacquainted with the purposes of the collection.

A good way would be by fixing after much consideration on some strongly contrasted characters appropriate to the inquiry, and then to obtain returns from masters, etc., of large schools of the names of those boys in whom they were notably in excess or deficiency and to photograph the heads of those boys on a uniform method for subsequent comparison.

Trustworthy conclusions might be reached; but what qualified persons will undertake the labour of what will probably end in showing phrenological bumps to be meaningless?

42, RUTLAND GATE, S.W. *September* 12, 1906.

MY DEAR EVA, Will you very kindly do the following job for me, and send the results to me to Claverdon Leys, Warwick? I have traced a few of Miss Baden-Powell's silhouettes and send them herewith. Also I send a sheet made up of 15 smaller ones pasted together, all numbered. I want two or three of the silhouettes enlarged according to the instructions by the side of the sheet of numbers. All I want in the end is something like this on a very large scale. You can mess the paper with crayon as much as you please but in each case draw a thin, firm, equally thick line through all the mess, to indicate clearly the outline. Choose whichever of the silhouettes you prefer, Marconi and myself being comparatively beardless would do for two of them. Generalise the hair and beard as much as you can, fancying it has to be worked in tapestry. I shall be so much obliged for this.

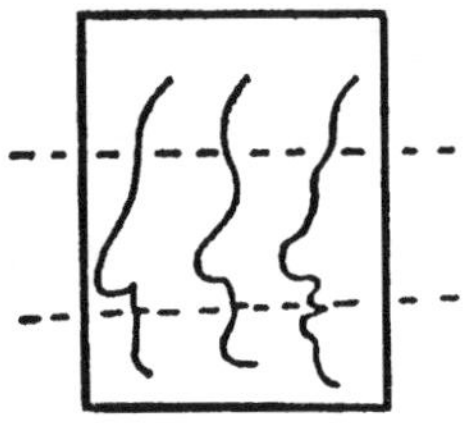

I have been working hard and getting on. The Athenaeum is shut up, and I don't care to go to its temporary substitute, but tea and lunch and dine here. There are various little things to tell, hardly worth writing about. I got a big strong kit-bag yesterday, to replace the burnt one. Kindest remembrances to your fellow-lodger. Ever affectionately, FRANCIS GALTON.

Tell me your plans when you write; I have somewhat forgotten what we arranged.

Address Malthouse, Bibury, Fairford (Glos.).

[CLAVERDON LEYS.] Sunday, *September* 16, 1906.

DEAREST MILLY, The past week seems to me an age through change of scene, though there have been no notable events. You are barricaded, I suppose, more or less. Here at Claverdon, where I am at this moment, there are big waterworks going on. In a spare spot between the garden and the stables a cistern 20 × 10 × 10 feet is just dug out and built round, and in an ingenious way all the water that falls on the many roofs connected with the farm-yard is collected into a pipe to feed it. Ever so much is going on besides. I had a quick but stuffy journey from Newton Abbot. The fast train was easily caught there, but there was overcrowding in it. The three nights in London were very profitable, for I finished my little paper and did various jobs, and I reached here, as arranged, on Thursday. Eva is at Bibury. I join her there to-morrow, for the week; then I go home, and she for a few days to Warwickshire, and we converge in London afterwards. I wonder if the drought has continued with you. There has been a little rain here, and yesterday a big threatening cloud, with apparently waterspouts of rain, hung over Leamington and elsewhere, but only a few drops touched us. The newcomers at Gannoway Gate (where Darwin lived as a bachelor, and the Torres till lately) were here while getting their furniture in. It transpired that the male could whistle through his fingers and after moderate persuasion he did. He gave us lessons in that musical and very useful art, but although I blow with his diagram by my side, in front of the looking-glass, for five minutes at a stretch, I have not yet caught the trick. Edward occasionally succeeds. I shall go on night and morning till I can. How many useful accomplishments are neglected in our youth!—this of making "cat-calls" among the number. I want it every day to get a cab in London. It beats all whistles hollow, but confessedly is not elegant to the eyes. I do not suggest Amy's acquiring it. 5, Bertie Terrace is not yet sold, several things that they all are glad to have stored are still there. I had not the heart to look at it. Gifi cycled over to Leamington and saw Temple, who had been here for a little while in Claverdon, and learnt that since then she had been somewhat seriously ill, a doctor-three-times-a-day business; I don't know more. She is convalescent now, but weak. Yeales is, I hear, losing her memory. Everything ages, and is extruded when of no further use. Among others, I am glad to reckon my pinched thumb nail, only one half of the old one is left now. Good-bye, loves to you all. Fred was very patient. We *were* ½ hour too early at the station and the train was late as well! Ever affectionately, FRANCIS GALTON.

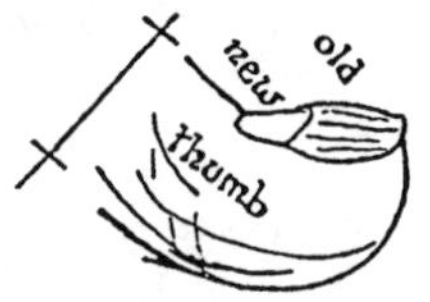

I am writing before breakfast so have no message to send.

42, RUTLAND GATE, S.W. *September* 23, 1906.

DEAREST MILLY, I am really at this moment still in Bibury, but go home for good to-morrow. We, like you, have much sunshine and warmth, but I hear dismal stories from the Cumberland Lake country of what the weather is and has been there. Thank Amy ever so much for Mrs Benson's letter which I keep and which confirms essentials. Amy seems to have told her that I said it WAS *Pob* etc.*, my point was that I thought it could *not* have been him but that I quite forgot *who* it was and wanted to learn. This, Mrs Benson supplies. The object of the visit to Lambeth was to see some papers in the library there which bore on the history of the Greek Church. For all the rest, I can trust my own memory. The interview was described by the Archbishop most graphically and forcibly. I have now been a week here at Bibury in extreme cottage-comfort. Eva has a lady artist friend, Miss Savile, who has to sleep out but takes her meals here. The post has just brought me the photo-process reduction of the diagram in my forthcoming paper (composed at Edymead); the proof will doubtless be at Rutland Gate. I shall be glad to have this preliminary off my hands. You shall have a copy of the thing when *Nature* has published it. I am receiving *excellent* tracings of profiles, full of character, from Dance's big works. They are large enough to fill (allowing a full margin) one page of this note-paper, and are all of well-known contemporaries, sketched from life. They are making a most interesting subject for study and comparison. The caster of the British Museum coins is on his holiday, but undertakes to cast them all when I come back. I mean all in the list of 100 or so that I sent him. To cast all in the British Museum would indeed be a large order. To-morrow we all separate. Eva goes for a week to Warwickshire and then she rejoins me for good in London. We have not yet absolutely, but approximately, decided against wintering partly in England. The probable event will be that of going slowly Rome-wards early in November. I look back with ever so much pleasure to Edymead. Pray give suitable remembrances all round, not omitting the Signora.

Ever affectionately (from the awning as usual), FRANCIS GALTON.

* This probably refers to C. P. Pobedonosteff. See p. 548 above.

EDYMEAD, BOVEY TRACEY. *October* 19, 1906.

MY DEAR EDWARD, At length a newspaper notice has appeared of your Report which I enclose. It is by Prof. R. T. Hewlett, as I see from the table of contents of this number of *Nature*, Oct. 18, 1906. They have spelt your name with two *e*'s. I am surprised that nothing was said about it at the British Association. It should have been referred to by the President of the Biological Section, but he (Lister) gave an Address by no means up to the mark, in at least certain particulars, and which is at this moment undergoing scathing criticism by Prof. Karl Pearson.

I have had a trying 12 days of Rheumatics and Bronchitis and though much better, am not yet sound. I funk now foreign travel and probably shall try *Plymouth* for November and December. Eva went down for a night to prospect, and reports favourably. Milly and I are to go down on Monday and conclude. London in November would help to, or quite, kill me. It is sad being banished. There are great offsets however to the discomforts of invalidism, in the care and affection one gets, the fires in one's bedroom, and the lots of sleep. Guy has been from home, but returns to-day. His renewed adjutancy hopes are now finally disposed of, by the appointment of Captain Weston as his successor. Sidmouth was a haven of rest to me for a week in bed. Thence I came here on Tuesday last in two easy stages, sleeping at Exeter. I was far too ill to see *Beer**, but I read about it and saw a picture of it. Neither could I make an excursion anywhere. I have learnt nothing whatever during the last fortnight except the virtues of a new (to me) *pill*—Podophyllum, with a little colocynth and hyosciamus. I shall adopt them in the place of compound rhubarb, of which in average health I take about one in two months. Of these, I have already had to swallow three. Erasmus wrote me such a nice affectionate letter in reply to mine; so also did Grace Moilliet. How beautiful Devonshire is, and how varied! Two seas, two (?) moors, lots of harbours and rich pastures, besides red earth and red cattle. Best love to M. L. I trust your Agents' meeting went off as usefully as hitherto. Eva goes to-day to London to look after winter clothing and the house. Ever affectionately, FRANCIS GALTON.

7, WINDSOR TERRACE, THE HOE, PLYMOUTH. *November* 7, 1906.

DEAREST MILLY, I must write my first letter from this charmingly placed club (Royal Western Yacht Club of England) to you, to ask you to thank Guy again for procuring me admission to it. We get on quietly and happily. I have one friend at the Aquarium, Bidder, grandson of the "calculating boy," and have made acquaintance with the others there. They do excellent work. Their steam trawler, the "Huxley," is just back from the north seas, and is off westward for a few days, and I am to go over her when she returns. Bidder has been making prolonged experiments on the drift currents of the North Sea, by sinking closed bottles with a paper inside and with a very legible label asking the finder to break the bottle, take out the enclosed small roll of card, fill up the spaces with date and place of recovery, and to post the card to him. It appears that fishing with trawls is so *searching*, that 77 per cent. of the bottles are recovered by the fishermen, some after drifting up to even 140 miles. My other important experience regards cutting a cake. The object aimed at is to make the *arcs* of the two slices equal, without regard to the part of the circumference, so (1) does as well as (2). I have tried both, and rather incline to (1). It is an excellent plan for keeping the cake moist. An indiarubber band keeps the halves together. Cross cutting is not necessary, (3) being as good as (4).

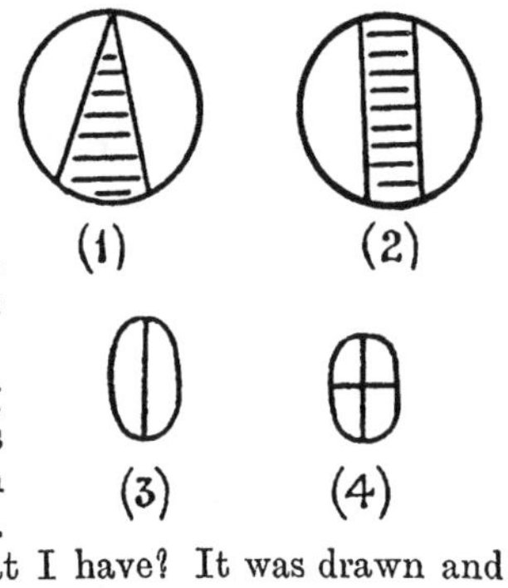

We were grieved at your bad cold. I hope it is disappearing at a normal rate. It has been too cloudy and rainy to tempt us out much, but the day before yesterday we had a grand forenoon seeing the dockyards. The Nasmyth hammer worked beautifully. Do you happen to recollect the crayon picture of a *meteor*† that I have? It was drawn and given to me by Nasmyth, who saw the meteor at his own place in Kent, I having seen it (and

* "Beer" or "High Beer," near Winterbourne-Kingston, one of the homes of the original Galton yeomen; see our Vol. I, p. 40, ftn. 2.

† This drawing is deposited in one of the drawers of Galton's writing table at present in the *Galtoniana* in the Galton Laboratory.

published a brief account of it) at Boulogne. What a noise it made! People thought a magazine had exploded somewhere, and the trail of white that it left behind lasted for a long while. With many loves. Ever affectionately, FRANCIS GALTON.

7, WINDSOR TERRACE, THE HOE, PLYMOUTH. *December* 20, 1906.

MY DEAR EDWARD, Best wishes of the season. It will be the shortest day when you get this, and then the year will turn—Hurrah! I am particularly glad you will be on the Advisory Council *re* Agricultural Biology. It will be just the thing you could help so well in, especially when the stage is reached of the Agricultural Farm.

Very amusing and pleasant your Gloucester host's account of Erasmus at the Regent. I heard of his luncheon party there a few days since, from Lucy, and how happy he seemed to be. She too seems at last to be getting strong and happy, and her husband as well.

A friend of mine, Pryor by name, has a collection of old silhouettes, mainly of certain Quaker relatives and their friends including my Grandmother of Duddeston (whose pastel portrait you have) and one of *dear* Mrs *Schim.**, made at Bath in 1809, according to my Father's note in pencil upon it. She was then an uncommonly handsome woman of 30 odd years with a profile greatly like that of her very promising brother, Uncle Theodore*, who died young of plague at Malta. You naturally do not share my (reserved) admiration for Mrs *Schim.*, for your Mother certainly did not, but she interests me on family grounds, so when I return home I think I shall frame her.

So James Keir Moilliet is buried to-day. Poor Lewis with his twin brother gone and himself blind. Amy Lethbridge is quite well again, after a bad sore throat to begin with. Then she was taken to Weston and got quite well. Eva saw her at Edymead House two days ago, just returned.

Plymouth atmosphere is not enlivening, but I get on well enough by leading an invalid life. Driving is no good, for the ground is very hilly and the ugly suburbs stretch far.

You mentioned that you read *Nature*. Look in to-day's issue at a paragraph, with small diagrams, on how to *cut a cake scientifically*, signed by a certain *F.G.* We have used the plan regularly for at least a fortnight. It suits our modest wants. So you have two bulls! Claverdon Leys will become "Bashan" (I have however no conception what the Biblical "Bulls of Bashan" refer to). I am delighted that you are so fit, so busy and so happy. Loves to you both.

Ever affectionately, FRANCIS GALTON.

7, WINDSOR TERRACE, THE HOE, PLYMOUTH. *January* 17, 1907.

MY DEAR EDWARD, So glad to hear of your doings, of the house "bursting full of boys and girls" and of the six calves.—Also of the forthcoming wild geese in Wales.—The poor old bank in Steelhouse Lane!!† Nothing endures. One of Bewick's vignettes is of a churchyard on the edge of a cliff that is crumbling into the sea. The havoc has reached so far as to cut a monument in two. The part that remains is inscribed "To the immortal memory of..."; all the rest is gone. I was more sentimental about the little Slaney Street, where there was an Office connected with the bank, which my Father kept up till his death, I think. It had an old copying machine, given him I believe by James Watt its inventor, and which looked not unlike a mangle. A huge thing worked by cross arms. I went with him there on not a few occasions, but never into the big bank house. I wish I could get rid on fair terms of the small remainder of my Duddeston property, for the reason you mention. But after all there is not enough of it left to be risky overmuch. The cistern must now be a pleasure, also the pond.

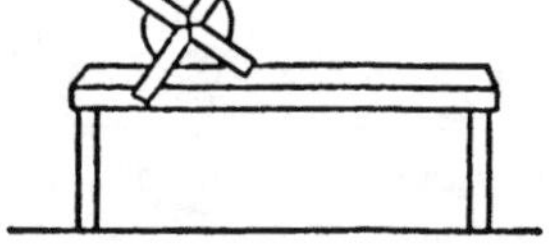

I am not yet by any means *fit*, having had a week ago another shiver with bed and doctor, but I feel now well cleared out and particularly comfortable in myself, leading at present an invalid life, which I hope will not last for many days longer. I am to take regularly every morning a purgative fizz, and strychnine after meals as a nerve tonic. The prescription seems reasonable. I should greatly like to accept your kind invitation later on, but dare not make any plans yet. I suppose I must stick here till spring sets in. The doctors strongly urge it.

* See our Vol. I, Plate XXXV.
† See our Vol. I, Plate XXXII.

Poor Milly Lethbridge has had nearly a fortnight in bed with influenza; Dim* was again sent to bed, I hope only for a short time. Their principal maid has been quite ill, etc., and is gone away for a while to get strong again. I expect Eva back to-morrow, but have urged her to stay on, if good for her head.

This is a season of sad recollections for you. I can hardly think that it is only one year since your Mother's death. Best love to M. L. Tell her that I have learnt one good cooking receipt—viz. not to serve Whitings *boiled* with their tails through their eyes, but to spitchcock, take out their bones, and fry them. They are quite good eating in this way. Very like soles and, if possible, better. Ever affectionately, FRANCIS GALTON.

3, HOE PARK TERRACE, PLYMOUTH. *February* 4, 1907.

MY DEAR EDWARD, What an escape! Don't let the Egyptian sun get into the head, which *may* be tender for a while. I hope you will be able to go, and to enjoy and learn. Also that you may *not* get too much of the March *Khamsin* hot winds. "Khamsin" means "50," = the number of days during which that sometimes detestable wind is apt to blow. Thank you for the newspaper slip which seems to give a fair account so far as it goes.

The news that compound drenches are being well tried is good. In some future time, babies will undergo "suction" at the same time as their baptism, to preserve them from all microbic ills, and will repeat the same at about the age of confirmation.

I am just now at some statistics that might interest you. They are those of a weight-judging competition of the West of England Agricultural Society—800 returns. They show the sort of value possessed by the *Vox Populi.* The distribution of error is curious. Half of those who judged below the average of the whole lot were more than 46 lbs. lower than that average; on the other hand, half of those who judged above the average were more than 28 lbs. above it. So the distribution of error is *skew.* Why it is so, and what the correction should be for skewness, I cannot yet make out, but am busy at it. The average was 11 lbs. wrong.

My "Eugenics" has started on a revised scheme very hopefully. The laboratory is now attached to Karl Pearson's department in University College, and will be well looked after by him and become in all probability important. The staff consists of a Fellow, a Scholar and a Computer, and all statistical work will be rigorous and of the most recent kind. It, in fact, constitutes a new department of Professor K. Pearson's excellent Biometric Laboratory.

All goes on here comfortably though rather monotonously. Presumably you will start (if you go) for Egypt from here. It would be nice if you were to stop a night or so at Plymouth *en route*, but I am sure that you are unlikely to spare time for the purpose. Pray tell me the date of your start that Eva may have a chance of accompanying you on board. *I* am wholly confined to the house for most days. I expect Eva to go away (for a week) and Milly to take her place, on the 14th. Best love to M. L. Ever affectionately, FRANCIS GALTON.

Archdeacon Bree was very much better two days ago, and out of danger (at Bournemouth). Edward Lethbridge's girl has been very dangerously ill with typhoid. The last news is cheerful.

42, RUTLAND GATE, S.W. *March* 30, 1907.

DEAREST MILLY, Being alone, I was doubly glad of your letter. Karl Pearson simultaneously sent me a copy of the paper. Seabrooke has written to me, with an added postscript from Eva, to say that a longer letter from her is coming. Probably it will not arrive till after the last outgoing post of to-day. All seems *going on* favourably, but to what *end* who can foretell? Face to face as I now am with solitariness, it seems more endurable, even during illness, than I had feared, so long as servants work happily together. Also it draws me back to old friends, which is a moral gain. I have been busy with an old method of mine, adopted only at long intervals, of *stock-taking* of my own *character*, and grieve to find it has somewhat deteriorated in two particulars. The process may interest you, and if on this occasion I can elaborate it further, it may be worth publishing. Its essence is (1) to catch oneself unawares and to consider carefully the thoughts and moods that were at that moment in the mind, and (2) to note them. The (1) is not difficult at *first*, but after a while it becomes very difficult without independent aid such as a person calling out or a machine striking. (2) requires a good deal

* Pet-name of Miss Amy Lethbridge.

of thought and experiment to make a *logical* classification, and yet a *brief* one, of moods and subjects of thought. I based mine originally on the Ten Commandments (leaving out the 2nd, 3rd and 4th as archaic), but find this division can be much improved on for the present purpose. Thus, it is convenient to have a preliminary division into *vigorous* virtues and vices, and to *subservient* ones. Almost any reasonable system will work fairly well. I use generally two letters; one for the class, the other for the subdivision, with a dash (v′) to signify "faint" and an underscore ($\underline{v}$) to signify "strong." One does not like to put too much down with pencil or pen. I have hitherto burnt my notes (though they were mostly hieroglyphics), but memorised the results. Where I have deteriorated is firstly in a general *weakening* of the moods—perhaps this is merely the result of age. The second failure is more easily remedied; it is the want of frequent *withdrawal into one's self* and of looking at and directing one's own conduct as if it were that of an alien, together with all that action connotes, such as communion with a higher power. The fact is, I used to *overdo* this, and feeling myself becoming priggish, thought that simple naturalness, for a bout, would be good. But I have overdone this phase too, and must revert to the old one, which it will be grateful now to do. If you have ever attempted anything of this kind, or heard of any one doing so *briefly* (not by gushing out-pourings and self-revelations), do tell me. Ever affectionately, FRANCIS GALTON.

I shall be most interested in your and Fred's Swiss plans. I have been laid up during most of the week with my inflamed and eczematous ear. It is practically well.

42, RUTLAND GATE, S.W. *April* 11, 1907.

DEAREST MILLY, You will all be most welcome at luncheon (1 h.) to-morrow. I shall be particularly glad to make Miss Trail's acquaintance. Your tidings concerning Bob are very encouraging and will lighten the skirts of the hitherto terribly gloomy sky. Enclosed I return the beautiful letter of your monastic friend. Thank you much for letting me see it. No creed can compare with Christianity in its conviction of an all-pervading *love*.

Stoicism and most pantheisms are cold and cheerless for the want of it.

Ever most affectionately, FRANCIS GALTON.

I have not *yet* heard from Eva, but probably shall do so either by a late post to-night or early to-morrow.

42, RUTLAND GATE, S.W. *May* 29, 1907.

DEAREST MILLY, Eva's address is Moor House, Ringmer, Sussex. I had a letter from her yesterday—short, but pleasant. George Butler leaves to-morrow. Lucy Studdy does not come, as her maid's "shingles" has now attacked the leg and made stairs impracticable for her. She goes into lodgings. I am steadily recovering from the effects of an awkward fall on to the floor of my bedroom Saturday–Sunday night. It was about midnight, and getting up I rested on the edge of a three-legged table with "invalid comforts" on it. It tipped over and came down with a clatter of crockery, and I fell with it, heavily, on to the floor. I was so bruised and battered that I had not strength to lift myself up, so there I lay helpless till 6½ a.m. when the united forces of the awakened household lifted me, in no small pain, on to my bed. Things are mending one by one, and I can already almost get out of or into bed unaided. Hibbert, the nurse-housekeeper, sleeps in my dressing-room, and Gifi and she are most anxious to help.

So my Oxford lecture, "to be delivered by *myself*," is abandoned. I have sent everything prepared for printing to the authorities there. Excuse bad writing, my hand is still sprained. Here is an *Art of Travel* experience. It has twice occurred to me for want of better accommodation to sleep on a billiard table. I now find that an oak floor is less hard, also that it carries off the body heat less quickly. I dare not make any plans yet, but if improvement continues to-morrow, the doctor thinks I may. As it stands I should go to Claverdon for a few nights on the 7th. Love to Amy. Ever affectionately, FRANCIS GALTON.

42, RUTLAND GATE, S.W. *June* 6, 1907.

DEAREST MILLY, I do trust that you and Amy are now in a fit state for the calm and serenity that Italy can give. All my earliest and pleasantest recollections of the Italian lakes are associated with Baveno. How I used to watch the boatmen manipulating the heavy slabs of granite and getting them on rollers into boats, and there used to be simple merry-makings at night, all very picturesque and very Italian. My lecture* went off *well* yesterday. Arthur Galton delivered it effectively, as I am assured, and there was a large and attentive audience. I was utterly unfit for the exertion even of going to Oxford. All pains from the fall have wholly gone, but bronchitis remains, ever on the watch to become bad on the slightest imprudence. Cameron Galton made a brief call this morning. Lucy Studdy goes to-morrow to stay a few days with Eva. I shall be quite sorry to lose Lucy, she has exerted herself in every way to be pleasant and helpful, and allowed me to be quiet as long and often as I wished. Eva, according to Seabrooke, is better physically than she has ever known her, but complains of the headaches. I must whip up friends to keep me company occasionally, when Lucy is gone. I am so much stronger that I hope to be able now and then to get to the club all by myself, or at all events with Gifi to help me in. Mrs Hibbert seems to do very well and the cook is excellent. Gifi highly approves. I saw your post-card of Baveno, sent last Monday. Eva sent it to Lucy Studdy who gave it me. Is all that *white* on the hill behind and on those in front, snow or bared granite? I wonder if they spear fish by night at this time of year? The lights in the boats are so pretty when they do. Writing rather tires me, so I will leave off here. With most affectionate good wishes to you all. Ever yours, FRANCIS GALTON.

When you next write, tell me how Bob goes on. Lucy would, I am sure, send her best love if she were in now. You shall have a copy of the Lecture as soon as I receive any.

THE YAFFLES, HINDHEAD, HASLEMERE, S.O. *August* 25, 1907.

DEAREST MILLY, You are going through a sad and trying time and I greatly sympathise with you. It will be difficult for you and Amy to get as much rest as you want, the home duties being so many and so various and the terrible memories so obsessing. All goes well here and promises well for the future, thus far. The house and grounds are singularly agreeable and we have old friends within reach. Karl Pearson came on a bicycle (2½ hours each way!) to lunch yesterday and we had much pleasant talk together. Violet Macintyre† leaves us on Tuesday, I am sorry to say; she sails for America on Wednesday to see her husband's relations there. After that, she returns to England to sail by steamer to Penang. It is as short and cheaper and pleasanter than going there by way of San Francisco. I shall be very sorry to leave this place and may perhaps take another house somewhere hereabouts for the end of September and early October. Our tenancy is out on September 12. We had a most interesting afternoon with Mrs Watts, the widow of the great artist. She has a large collection of his works in a studio to which the public are admitted, and there is a beautiful memorial chapel designed by herself. The spirit of his works is so lofty that one feels the studio to be a chapel. Longfellow's introduction to his translation of Dante quite expresses my feelings and rang in my ears all the time. As she wrote me a very nice letter, I have ventured to transcribe it from memory and to send it you. With Eva's best love as well as mine to you all.

Ever affectionately, FRANCIS GALTON.

THE YAFFLES, HINDHEAD, HASLEMERE, S.O. *August* 30, 1907.

DEAR SCHUSTER, Part III of the Eugenic publications has just reached me and I have read your excellent memoir in it with great interest. Also I have heard good news of you from Professor Pearson who bicycled to this pretty place last Saturday. I am here till September 12, and then the owner of the house returns, and I must go, with much regret. Miss Elderton seems to be doing an immense deal of good work at the Laboratory. What a nice and capable lady she is. Very sincerely yours, FRANCIS GALTON.

WILLSHAM, BRENDON, NEAR LYNTON, N. DEVON. *September* 2, 1907.

DEAR MR GALTON, Very many thanks for your kind letter; it gives me very great satisfaction that you approved of the memoir. I seem to have been very lucky in the time of its appearance, since the University has come to the fore through its educational and pecuniary deficiencies and there is no parliament sitting to fill the papers.

* The Herbert Spencer Lecture: see our Vol. III^{A}, p. 317 *et seq.*

† Evelyne Biggs' half-sister.

Miss Elderton has certainly been a remarkable success at the Eugenics Office; but I think her marvellous energy and quickness to learn anything new would have enabled her to succeed at anything she undertook.

Hoping that you are in good health, and have not been too much troubled with bronchitis lately. Believe me, yours very sincerely, EDGAR SCHUSTER.

QUEDLEY, HASLEMERE. *September* 30, 1907.

DEAREST MILLY, I was remiss yesterday in letting the Sunday post-time pass, without writing to you. A lady who says that she knows you, Miss Bennett (? as to number of *n*'s and *t*'s), has been staying with our friends here, the Lionel Tollemaches, and returns to Bovey to-morrow. She will tell you about these surroundings and ourselves. I continue to think the choice of this place a wise one. The neighbourhood is rich in nice people and there are numerous drives, each different from the other. The house too is convenient in itself, very much so in its position, and is growing pretty inside under Eva's artistic touch. I have been occupying all my novel-reading hours with reading *Sir Charles Grandison*, and am ashamed rather to say how much I am carried on with it. Richardson has a remarkable power of keeping his characters distinct and vivid before the reader. What an enormous length his novels are! My edition of *Sir Charles Grandison* is in four closely-printed, large 8vo volumes, and *Clarissa Harlowe* is I believe about the same length. Violet Macintyre arrives in England to-day from America. Her baby is with Walter Biggs. She goes straight to Constance Pearson. The baby vastly improved while here, hardly any of her fits of *yelling*, of which she had many at first when with her former nurse. Poor little thing; her look-out in life is not a happy one, to all appearance. If Violet finds a good ayah to take her back, it will be a great gain to the child. I trust your own many domestic troubles are dispersing. Has Guy actually begun his new work? How is Amy? Where is Hugh? Is Bob better? We had a pleasant visit from my old and rather invalided friend, Lady Welby, who motored here for lunch all the way from Harrow and back again. She is a wonderful woman in many ways, and of wide experience in life, beginning as a pet godchild of Queen Victoria, and for the last ten or more years steeped in metaphysics! It is so pleasant to meet Mrs Tyndall and to talk of old times, as for the most part: "All, all are gone, the old familiar faces." Best loves to all. Ever affectionately, FRANCIS GALTON.

I have not yet found out the meaning of Quedley.

QUEDLEY, HASLEMERE. *November* 2, 1907.

MY DEAR GEORGE DARWIN, I fully sympathise with H. M. Taylor's [blind Fellow of Trinity*] proposal [for the blind*] and gladly send £2 to help it.

But my strongest sympathy is with the deaf. Had I a fairy godmother, I would petition that every experimental physicist should be made as deaf as I am, until they had discovered a good ear trumpet, and then that as many fairy-gifts should be heaped on the discoverer as should exceed all he could desire, as well as the thanks and gratitude of all whom he had relieved!

I am spending most of the winter here in hopes of evading much bronchitis and asthma. The place promises well.

Miss Biggs is not *quite* recovered. But now she is in a healthy position, among old friends who love and break-in horses, and she is busy and hard working all day, with little time to worry herself. You will be particularly interested just now at Charles' *début* and progress. All good luck to him. Affectionately yours, FRANCIS GALTON.

To SIR GEORGE DARWIN, K.C.B.

QUEDLEY, HASLEMERE. *November* 25, 1907.

DEAREST MILLY, You will be most welcome here on or about January 7, and for as long as you like. Eva will be pleased too, very pleased, to see you. She does not now seem to care about going clean away, but I am glad she should have variety, for I unaided can be but a tedious companion, and next to no companion at all out-of-doors. What you say about not requiring Charlotte, removes the only possible difficulty. I fear *she* would be impossible. Matters go on as smoothly now, though hardly so securely, as in old times. I have had a little bronchitic warning but nothing more, no fever at all, and sleep like a baby and eat like a boy. Methuen,

* Interpolations by Sir George Darwin.

the publisher, or rather his man of business, has written me a "fetching" letter asking if he might have my autobiography for publication. A curious double coincidence occurred, (1) Methuen himself, who has been seriously ill after some operation, lives here, though I do not yet know him, and (2) Frank Carter, the artist who copied my picture for Trinity College, was staying here for the week-end, and was engaged to lunch with Methuen (a connection of his) last Sunday. So I made him a sort of go-between. Briefly, I am disposed to attempt the job, making no further terms than the usual half-profits and an assurance that the book will be handsomely brought out and that I am liberally allowed to correct proofs. Also to have some simple illustrations and perhaps Furse's portrait. This will keep my hands very full indeed for months to come. Have you any old diaries or letters or documents that would help as to ancient dates? Now that Bessy and Emma are gone I feel singularly at sea about much. I *have* Louisa's diaries, but they refer little to myself; however, they should be very helpful. What a curious account you send of Guy's "dowsing." Edward Wheler had a like experience, but *his* dowser proved unsuccessful. There is a firm of dowsers. If I belonged to it and believed in it, I should construct a paved yard with waterpipes below and stop-cocks (×), any of which could be turned off or on, and should test people by it.

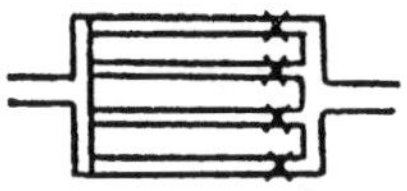

Ever affectionately, FRANCIS GALTON.

QUEDLEY, HASLEMERE. *December* 2, 1907.

DEAREST MILLY, The Bogatzki*, which I return, has given just the events I wanted at this moment. I got between 70 and 80 dates from it, many of which help me much. The search into one's memory opens so many doors of the past that are usually passed by unregarded. A strange bygone experience (which I published) testified to the same thing. It was that when capturing, as it were, the first associations connected with any word the moment it was presented, they were often connected with some long past and habitually forgotten experience. I am working at different periods of my life in turn and have done a lot already about my *medical epoch*. How the ghosts arise! What touching mementoes there are in Bogatzki's pages. So many by Aunt Brewin referring to 1700 odd. I can't, of course, decipher most of her initials, but some of them I can.

A man with a much more horrid name, which I can't venture to reproduce from memory, wrote to me yesterday asking permission to translate my recent "Herbert Spencer Lecture" into *Hungarian*, for his *Sociological Review*, of which he enclosed a prospectus. They do these things well in Buda-Pest. An old friend of mine, Kőrösi, lately dead, was the head of the Statistical Department there and wrote valuable memoirs. The numerous accents they use are to me unintelligible. I hope I have put them right in Kőrösi's name. It was pronounced Keresi. We have at last been visited by a "Yaffle," a green woodpecker. The old gardener had never seen one in this garden before, though they are common (they say) among the woods higher up. There were plenty about when our previous house "Yaffles" was built, but they disappeared *after* it was built and named. Two starlings are on the lawn now, picking up the crumbs I threw out for the Yaffle. How quickly they gobble them up. Our next neighbour is a famous *etcher*. Some beautiful specimens of his doing are now on exhibition in London. It has been much pleasure to make his acquaintance. They are Mr and Mrs Fritton, with an uncommonly attractive 16-year-old daughter, still at school. All goes on well here. You must be much grieved about Mrs Northy. How does Guy get on with his motor? Any further news from Africa? Many loves. Ever affectionately, FRANCIS GALTON.

Address now: 42, Rutland Gate, S.W. *February* 9, 1908.

DEAR MISS ELDERTON, The tidings in your letter about the Eugenics Education Society † pain me much. Thank you greatly for sending them. I have written to Dr Slaughter withdrawing an offer of help that I made in response to an exceedingly sober and well-written letter from him, and said that I cannot consent to be connected with it at present. It is very sad. We are turned out of this house, "Quedley," for a fortnight by a damaged kitchen boiler, but letters will be forwarded either from here or from London. I hope when the spring is advanced and the place around grows beautiful, to tempt you down for a week-end. I think you would enjoy it then. Very faithfully yours, FRANCIS GALTON.

* A work by the well-known pietist, used by the originally Quaker Galtons like a family bible for personal records.

† See Note at the end of this Chapter.

Address now: 42, Rutland Gate, S.W. *August* 29, 1908.

DEAREST MILLY, This will reach you in, or *via*, your new home. I look forward to your next letter, anxious to hear that you are all at length settled at Shirrell House, but we go home to-morrow for a bit. Hubert Galton had asked us and we had accepted to go to Hadzor, but his wife is unhappily ill again, so that is off. We shall have some house-hunting to do from London on fair days. Otherwise I think I shall be chiefly in London *all* September. The recent storms and chilly wave of air make me less adventurous-minded, and a study of Bradshaw reminds me what a long journey it is from London to Minehead, so I fear that running down there is and will remain a dream till winter is overpast. How well and cheaply the Germans illustrate books and newspapers! I post you one—don't return it please—in which I come in on page 178. I don't know why on earth they include me, for I take no part in the Geographical Congress, but the shape of the little photo was convenient to them. Proof revising and index making is tedious, but I am nearing the end of my book at last. It cannot, I should think, be brought out before mid-October, but that is wholly in the hands of my publisher, who has first to bring out a new book by Marie Corelli! Eva went yesterday to the Isle of Wight, and came back disillusioned as to Ventnor and the like being suitable for us next winter, as I felt sure she would. I enclose one of the new programmes of the Eugenics Education Society, which may possibly interest you. If you can sow it (like a seed) in any likely place to meet with a favourable response, please do so. I am busy on a paper wherewith to open its proceedings next October, and find it very hard to steer between the *Scylla* of mere platitudes and the *Charybdis* of disputable details. If there proves to be time enough, I will venture to send you a typed copy for suggestions, if I may? We have had squally weather with fine intervals. To-day it is as calm as a cat sleeping in a comfortable arm-chair after a night of fighting and caterwaulings. Ever affectionately, FRANCIS GALTON.

September 14, 1908. 42, RUTLAND GATE, S.W., but please address next letter to me at Claverdon Leys, Warwick, where we go on Wednesday; I for a week.

DEAREST MILLY, Your painful attack is grievous. One of my very few quasi-superstitions is that change into a new house spells illness for someone. In this case, you are the sufferer and Amy has escaped. What pleasant news you give of Guy's appointment. How many years does it last? and what pay does he get? I am so extremely ignorant about army matters. I suppose the "Brigade" is one of the new territorial army?

Adèle Bree is going on rightly but though the operation was not a serious one, the healing, as I understand, is a little delayed. Eva saw her for a few minutes one day last week, going down for the purpose, and returned quite happy about her. The Archdeacon too is quite well. So the house proves quite a success. I am so glad—also, that you do not feel at all cramped in it. The desideratum in life is to have all that you really want and as few superfluities as may be, and your house appears to fulfil that desideratum. The Roman Catholic Congress seems to have been uncommonly well managed by its officials. Eva has been to two or three services, and we both went together to see the school-children's long procession. I don't care much for great length in one. A sample is to me quite as good. Did you ever go to an oil-cloth shop, where they drop a box with reflectors on to a pattern, say one foot square, and at once the pattern is reflected and re-reflected into a great surface. One or two of these children, or of Eton boys, who outwardly are as much alike as peas, might be put under one of these boxes and, *hey presto*, they would grow into a multitude. But what a blunder the Home Secretary made in first permitting the procession for yesterday and then retracting it. The Premier shares the blame. I should have thought the question quite deserving of having been made a Cabinet one. The papers will shortly come in and we shall see what they say. They have just come, and say what I thought they would say. My personal news of this week is largely connected with dentists, tailors and hosiers, of no interest to others. Eva and I went to Methuen's and arranged about the cover for the book. Smooth green cloth with a flat gold band.

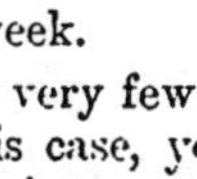

I have been very busy over a small matter which requires care, viz. a brief opening address to the Eugenics Education Society. It has been typed and then much cut up by the skilled hand of Crackanthorpe, and is now being re-typed in a shrunken and disembowelled form, but made much more suitable thereby. It is a delight to me to put myself again to school, as it were, under a competent critic. Generally my friends are diffident and won't *slash*, but I have two excellent friends who happily feel no compunction in performing that operation, and I learn much thereby. Best loves. Ever affectionately, FRANCIS GALTON.

42, RUTLAND GATE, S.W. *September* 27, 1908.

DEAREST MILLY, This is a prompt answer; Ravenscourt seems quite a success. Enclosed is one of the prospectuses of my book, which I *hope* may be published next Saturday. A small misunderstanding of the printer threw it back for a while. Adèle Bree is rapidly getting well and has no dread, I believe. The removal was an easy matter, though the healing was prolonged. —So you have to do with one of the "Feeble-Minded" of whom so much has been brought to light by the Royal Commission. In these border-line cases it is *most* difficult to know how to act. I know the Porlock Hill, perhaps it was then not quite so steep as you drew it, but was perilous-looking, and there were beautiful views. Motors are certainly great comforts, and bring far-off places near to one. The London taxi-cabs have a rare time of it in the afternoons; every one of them in the neighbouring stand being always taken. Eva is off to-day to Malvern to stay with Mrs Keir Moilliet and to bicycle on Monday to Lewis M.'s. She returns on Tuesday. Give my love to all of yours. You will be a large party now, if Guy has returned from Exeter. I am about now to be trundled in my sister Bessy's bath-chair into the park, which I find very pleasant. Sometimes Mrs Simmonds, sometimes Gifi, pushes it, and I have lost all sense of oddity in the matter and enjoy it without drawback. Ever affectionately, FRANCIS GALTON.

42, RUTLAND GATE, S.W. *October* 10, 1908.

DEAREST MILLY, You are indeed enthusiastic. The book seems successful, as a second edition of it is being printed; but I find that the first edition was only 750 copies. Still, it shows that the book has already paid its way, and my publisher writes prettily and congratulatorily (is there such a word?).

The idea of your troubling to join the Eugenics Education Society! I never meant to cajole you into it. Still, it is not a bad thing to do, and a few of us are taking pains about it. I shall understand "the ropes" better after next Wednesday's meeting. The absurd part of it is that the proper President of it, Sir James Crichton-Browne, has wholly absented himself for ever so long, and won't answer the letters of the Secretary to him. It was this that *obliged* me to take the lead, which I did not at all want to do. It is a funny thing that none of us can comprehend; Sir J.C.-B. is quite a pleasant man and seemed originally keen for the work. Personally I like him much. He sent much of value to Charles Darwin, who appreciated it. It did seem extraordinary in those far back days, that Crichton-Browne, then quite a young man and looking still younger, should have the control and mastery over the biggest lunatic asylum in England. He looked more like a man whom the hostess of a ball would introduce to partners lest he should be too diffident to ask them. Your gardening must be a great pleasure and matters of storage room must be difficult to solve. The Gibbons have built a cheap studio by their cottage. I sent them a perambulator and now the poor child is *dead*! How they will hate the sight of the little carriage! I pity them much. Best loves. Ever affectionately, FRANCIS GALTON.

42, RUTLAND GATE, S.W. *October* 17, 1908.

DEAREST MILLY, It would be amusing if the next year's camp in the neighbourhood of Dorchester should be placed on "Galton Heath." There is a wide extent of open land there and the high Downs are within marching distance. But how the midges bit me there, one summer! You must be full of gardening and hopes of flowers in the Spring; I now see flowers in shop windows here, that is all. My book is well reviewed thus far by most of the leading papers, but not yet by the *Times*, who kindly gave half a column to my paper at the Eugenics Education Society. It will be printed in full in the next (?) *Nature*, for they have sent it in proof to me to correct. You shall have a copy when it comes out. I contrived to read it myself and got through it creditably to a rather large audience, but was tired and bronchitic in consequence. All right now. Next Monday (to-morrow) week, *26th*, we go to "The Meadows, Brockham Green, Dorking," so I shall get *here* your usual Friday letter, but thenceforward the address will be as above. How I *hate* the thoughts of the coming winter. Eva went yesterday to see Mrs Gibbon, who is very sad. My old friend, Lady Pelly, has just undergone a very serious operation, I know not what. She is doing fairly well.

I had to break off, owing to the earlier-than-expected arrival of a Bordighera friend, Mr Bicknell, to stay with us. Such an interesting man. He is the scientific and literary soul of Bordighera and a good botanist and artist. He gave a small museum with a good sized meeting room and a beautiful little garden round it, to the place. His religious life has been in rough waters. At first a clergyman, then throwing off what he felt to be the trammels of orthodoxy, and now calm and sympathetic to all creeds. Just after my own heart! Best loves to you all.

Ever affectionately, FRANCIS GALTON.

I am rejoiced to find that my book pleases the Butler family. It was a difficult task to write about them without *gush* and yet appreciatively.

42, RUTLAND GATE, S.W. *October* 24, 1908.

DEAREST MILLY, Henceforth please address to me at "The Meadows, Brockham Green, Betchworth S.O., Surrey." This is the correct address of Brockham Green according to the Postal Guide; "Dorking" does, but it involves delay. We are packed up ready to go on Monday, taking such a lot of things, but four months is a long absence. How wonderfully well you get on with birds. I take bread with me in my arm-chair into the park, and feed pigeons and sparrows, but they are tame and easily attracted. I send a *Nature* herewith. The Address begins on p. 645. The *book* continues to be reviewed very favourably. The *Times* had a careful review in its *Literary Supplement* last Friday. Possibly some of the weeklies may be down on it to-day. I must go to the Club to see. As soon as we get to the "Meadows" I shall recommence the work I was at when with you at Edymead, about "Resemblance," and expect to be at it a good deal this winter, which, with some Eugenics, will keep me busy most of the time. How very bright much of Fräulein Schmidt is. It is rather too protracted, so I skipped a good deal and read the end. How those of this German middle-class who read her must hate her. What a handsome gentlemanly-looking man, judging from his portrait in the *Graphic*, the fourth son of the Kaiser must be, who is just married. This coming cold is formidable. Fortunately the "Meadows" is particularly well supplied with warm water and various cosinesses. Did you happen to hear of the military mayor of some small town in France and his interpretation of the order from his Governor to make all ready for an impending attack of cholera? The Sanitary Inspector called to see what he had done. It was confined to digging a big trench for the expected dead people. He was highly pleased with his work, saying of the cholera, "Je l'attends, pied ferme." I am sorry and glad, both at the same time, at leaving London. Love to Amy and to Guy, if he is with you. Ever affectionately, FRANCIS GALTON.

Eva sends her love also. She is very well, occupied and happy.

MEADOW COTTAGE, BROCKHAM GREEN, BETCHWORTH, SURREY. *November* 1, 1908.

MY DEAR LEONARD DARWIN, Your letter was grateful. We are planted here for the winter. In my *Memories*, p. 204, I say something about Speke's memorial in Kensington Gardens, and I am now sending a letter to Keltie which I have asked him to lay before Council, asking that a Committee may be appointed to consider a report on the question. If you think well of the idea, perhaps you will help its furtherance. Will you?

Re Sven Hedin. He arrived at Simla in *tatters* and was made the guest there of my wife's nephew, Monty Butler, who clothed him out of his own wardrobe and made him presentable at once! Very sincerely yours, FRANCIS GALTON.

MEADOW COTTAGE, BROCKHAM GREEN, BETCHWORTH, SURREY. *November* 6, 1908.

MY DEAR LEONARD DARWIN, You have indeed appointed a powerful Committee in which the artistic taste of the Harry Johnstons will be most helpful. As regards funds, I am prepared to give quite £100, as the object has long been a hobby of mine, if the design seems to me appropriate, and not too grand, and if I can do it anonymously, without ostentation. Your suasion about my giving a copy of my book to the R.G.S. is irresistible and I have instructed Methuen (the publisher) to do so, as soon as the 2nd edition is published.

Ever sincerely yours, FRANCIS GALTON.

MEADOW COTTAGE, BROCKHAM GREEN, BETCHWORTH, SURREY. *December* 7, 1908.

DEAREST MILLY, I trust that by now Edymead is *quite* off your hands for at least 3½ years. You will all be desirous to hear the doctor's report of Bob. Lady Galton is again downstairs, but I should fear not permanently recovered. We heard from Evelyn Cunliffe, who returns to her home near here to-day.

I have a donkey-cart and donkey lent me for two months and am just returned from a four-mile, in all, expedition. The donkey is an aged pet, much accustomed to have her own way. Still she pulls. I have no news. Life goes on monotonously and pleasantly and novels, etc., are read. A good deal in the Eugenics line is going on this week. Miss Elderton, the very capable Research Scholar, reads a memoir on Cousin Marriages. She has been working at 2000 of them for some months with the usual result that their ill-effects are statistically insignificant. When observed, they seem due to both cousins having the *same* bad quality. But I have not seen her paper yet. She is such a zealous, capable, nice girl, and is now familiar with the higher statistics. Her brother is a first-rate actuary too, which is all in the same way. I take it that the actuaries are, as a class, the hardest headed men in the community. The problems they have to deal with are sometimes very stiff ones. Tell me of any good book you know of, to get from the *Times* library to read. Best loves to you all from us both.

Ever affectionately yours, FRANCIS GALTON.

MEADOW COTTAGE, BROCKHAM GREEN, BETCHWORTH, SURREY. *December* 14, 1908.

DEAREST MILLY, I have nothing to tell. My life is largely taken up with donkey drives and novel readings. At this moment with *Waverley*, just previously with *Guy Mannering*. Eva went up to some Eugenics gatherings last week and reports enthusiasm in many quarters. Also some good work is being done. I have just got (from the *Times* library) Waldstein's new book about Herculaneum, which gives, to most persons, a new view. It is not embedded in lava. No lava came near the place until long after its burial, and then only in patches which afford useful covering to excavators. He, Waldstein, is very sanguine, and has been pushing forward international help with rather too much zeal, so that the Italians are made jealous. However, they are going to begin and have voted money. We, I in the donkey-chair, called to-day on some people. By a strange coincidence the daughter-in-law, Mrs A'Court, of the (blind) owner, Mrs A'Court senior, of the chair and of the donkey, Jemima, was staying there. Now Jemima has been petted all her life and the meeting of the two old friends was touching. Jemima is turned out into our meadow when the weather is suitable. She follows Eva like a dog through the garden on the way to it, and comes up and does the same conversely, when wanted for the carriage. Odd creatures donkeys are,—so near to perfection and yet short of it. With best loves. Ever affectionately, FRANCIS GALTON.

I am too late for the post to-day.

MEADOW COTTAGE, BROCKHAM GREEN, BETCHWORTH, SURREY. *December* 19, 1908.

DEAREST MILLY, This will be my Xmas letter to you, with all good wishes to all of you. You tell me many things, showing how occupied you must be at this time. On the contrary, I am sadly un-Xmaslike in arrangements. Lucy and Cameron Galton come down to us on Friday for a few nights; that is all the family gathering possible to me here. We asked Alice Corbett for Xmas but she was engaged. Also, three days ago we lunched and spent some time at Henrietta Litchfield's ($1\frac{3}{4}$ hours' drive off) to meet Frank Darwin, his daughter and Mr Cornford to whom she is engaged. It was all very pleasant. She (the daughter) managed the Comus masque at Cambridge, but did not act in it. He did. He is a Fellow of Trinity College. You say I have a kindly heart towards donkeys. You recollect perhaps Coleridge's not very wise ode to a *young ass* and Byron's comment on it: "A *fellow-feeling* makes us wondrous kind." An ass is certainly a mysterious animal, and the continual and usually independent movements of his long ears testify to the busy thoughts or perceptions of the beast. But its obstinacy! What a martyr an ass would make to any cause that it pleased to favour. I write this by Saturday's evening post and wonder whether it will reach you Sunday, Monday or even Tuesday. All depends on the route it has to take. I am puzzling all day, day after day, over an apparently simple problem in my favourite statistics, but can't wholly satisfy myself even yet in explaining it on paper.

Ever affectionately, with love to you all in which Eva joins, FRANCIS GALTON.

MEADOW COTTAGE, BROCKHAM GREEN, BETCHWORTH, SURREY. *December* 27, 1908.

MY DEAR SIR GEORGE DARWIN, Thanks for your letter—I am so glad you like the book, and am grateful for the corrections.

It is, alas, impossible for me to attend the Darwin Celebration. I could not do it with safety, if at all, even in mid-summer. I get about partly in a donkey-chair. The movements of the animal's ears in connection with his presumed perceptions and thoughts are an unfailing object of interest.

My brother (simply Erasmus Galton, Leamington; he has a post-bag there) would I am sure be highly flattered by an invitation, but I am still more sure that he would be unable to accept it. He suffers from an old man's ailment that keeps him always in the immediate neighbourhood of his home. But he *reverences* your father's memory,—if possible, as much as I do.

I am pulling through the winter fairly well thus far, thanks to the pure air of these parts. I see in a *Times* article, that they seem to have discovered an anti-toxin to bronchitis. It would be indeed a blessing to me.

I grieve to hear of your bad knee, one limb out of two cannot be easily spared. A centipede would not mind it.

Love to you all—not least to Charles. I would have sent him the book, had I foreseen that it might have been liked. One hates so to intrude. I hope his mathematics continue to prosper.

Ever affectionately yours, FRANCIS GALTON.

MEADOW COTTAGE, BROCKHAM GREEN, BETCHWORTH, SURREY. *January* 7, 1909.

DEAR COUSIN GEORGE DARWIN, This is I think the correctest commencement of a letter!

Thanks for your letter in the *Times*, standing up for me. I only found it out this morning by reading the replies by Sir H. Cotton and Pollaky.

In my *Finger-Prints* I *translated* Purkenje, having got his exceedingly scarce pamphlet with great difficulty, and through a curious coincidence. As to Sir Wm. Herschel I have acknowledged my debt to him in print over and over again, and dedicated my *Finger-Print Directory* to him. He however did none of the three things that (as you quote from me) are essential preliminaries. Sir H. Cotton is I think mistaken in saying that Sir Edward Henry had organised the method in India before he had visited my laboratory. He *had* then organised the *Bertillon* system in India with great care, but found it a failure there. But I shall not bother to write to the *Times* unless Sir E. Henry himself should write what seems to require an answer.

Your son Charles lunched here to-day. It was *very* pleasant seeing him, so bright and capable-looking. Few will be more interested on his behalf next June than myself. He tells me you have that painful malady, a gouty knee. I once had one which ultimately got quite well, and speedily, though it hurt badly at the time; I never felt better and happier than when ill with it. I suppose the gouty humours drained away mischief. That active man, Sir John Evans, had it also, badly, and I think more than once, but got quite well. Lord Avebury, who as you know is of the goutiest stock, told me that he had tried all diets, but the advice that suited him best was "Eat whatever you like but only a *little of it*"! Small quantity rather than good quality. This will reach you *via* Cambridge. Affectionately yours, FRANCIS GALTON.

To Professor SIR GEORGE DARWIN, K.C.B.

MEADOW COTTAGE, BROCKHAM GREEN, BETCHWORTH, SURREY. *January* 12, 1909.

DEAREST MILLY, I have been somewhat bothered and busy and cannot recollect whether I did or did not write to you on Saturday. If I did, excuse some repetition. You must feel quiet after the departure of two such restless though amusing guests, besides that of your own son. Quietude prevails here. Violet came two or three days ago for a week. My bother lay in newspaper letters declaring that my share in the Finger-Print System was very small, and it was indeed disregarded in a *Times* notice. Thereupon G. Darwin wrote a letter on my behalf, which led authorities (from India), on the other side, to write. It seemed at last necessary that I should say my say, which I have done in a longish letter to the *Times* which (if they insert it) will probably appear to-morrow. It gave me trouble to refer to past things, and to write in a way that shall not irritate but be conclusive. I have done my best, and I hate newspaper controversy. There is really some spring now in the air, and a snowdrop in the garden, but much that is nasty may happen before spring comes. I suppose your home will be in much beauty even before May. How you will all enjoy it. The account of Guy's motor expenses is very interesting. I see that much effort is being now made to produce small motors at small cost, that will travel at a moderate pace and be good machines. Being one's own chauffeur greatly facilitates matters. My loaned donkey grows lazier and lazier, and more caressing at the same time. If she was not so old, and so prized by her owner, and if neither Eva nor other humanitarian persons saw me, I should make her "taste stick." Do you know that "walloping" is derived from the names of the two (?) Generals or Admirals (?) who were ancestors of Lord Portsmouth, and who *walloped* the enemies of England? Best loves.

Ever affectionately, FRANCIS GALTON.

MEADOW COTTAGE, BROCKHAM GREEN, BETCHWORTH, SURREY. *January* 17, 1909.

DEAREST MILLY, As to that newspaper correspondence, I enclose my reply of which I have *one* duplicate; please therefore return it when you next write. It is in answer to very positive assertions by two men of Anglo-Indian weight, who ought to have informed themselves more exactly when they wrote. I purposely wrote as civilly as possible. Whether more will follow, I know not. Also, I enclose a short letter of mine in this week's *Nature*, on quite another subject, "Sequestrated Church Property," which may interest Amy. It arose through Eva's inclination to believe in the supposed curse. Please let me have this back too, when next you write. What interests me the most in this little inquiry is that the average tenure of landed property in England is between 25 and 26 years. Yesterday I had a long letter from Harcourt Butler, from India, enthusiastic about the finger-print system. He has indeed succeeded in life, being now Foreign Secretary to the Government of India, though still young. How responsibility brings out character. It seems to have done so with Signorina Cotta. I pity you with your fire-places. You may recollect my own troubles in Rutland Gate many years ago about the kitchen chimney, when I called in an advertising expert who spoke like an oracle: "Sir, I am a practical man and can assure you that all that is wanted is to enlarge the opening of the chimney pot." I allowed him to try, and the chimney smoked as badly, if not worse than before. Then I called in a still greater expert and he began just as the other, "Sir, I am a practical man and can assure you that all that is wanted is to constrict the aperture of the chimney pot." I think that plan also was tried. Anyhow a much more intelligible cause of the fault suggested itself and *that* was remedied and all went well. If you could remember, as I do, my dear Sister Adèle, your mother, long before you were born, you would probably have associated her as I often do with the fire-place of her bedroom in Lansdowne Place, which had two hobs, on one of which a kettle always stood most conveniently at hand.

Your ants must be a great interest. Do you yet know the features of any one of them? I see that Guy's motor account works out at a trifle under 3*d.* a mile, *exclusive* of *depreciation* of the value of the motor. That ought to be included, but I have not a notion of what it is—somewhere between 15 and 20 per cent. of its original cost, I suppose, but *quere*. Eva is gone for two nights to London. Lady Galton is very ill but not worse, and with no hope of ultimate recovery, for it is senile gangrene. Ever affectionately, FRANCIS GALTON.

MEADOW COTTAGE, BROCKHAM GREEN, BETCHWORTH, SURREY. *January* 31, 1909.

DEAREST MILLY, Poor Erasmus! He is so very stoical. When he felt "something give" as he was about to enter the tram and fell on the road, the first thing he said to those who picked him up was, "It's all arranged, and mind I'm to be cremated"! I hear that he is as free from pain and as comfortable as may be, but that the broken bone can never heal, so all his habitual walks and independencies must end. I am extremely sorry for him. He somehow seems to me to have failed to get as much interest and "go" in life as his circumstances might have given him. Thanks for returning the newspaper cutting. Sir W. Herschel wrote subsequently a very nice letter to the *Times*, which I was very glad of, for he of all men can speak out best on the early stages of finger-prints in India.

Why don't you try Charles Darwin's perfectly successful plan of warming your room? The air enters through the wall, behind the fire-grate, into a compartment closed in front and top, at the back of the grate, and issues warmed underneath the mantel-shelf, *on either side of it*, above its jambs. Campbell (of Italy) did the same to a summer-house in Niddry Lodge. Both were perfectly successful. Some fellow took out a patent, but spoilt the idea. He made it "decorative" and it acted badly. You, of course, can have anything you like set up for your

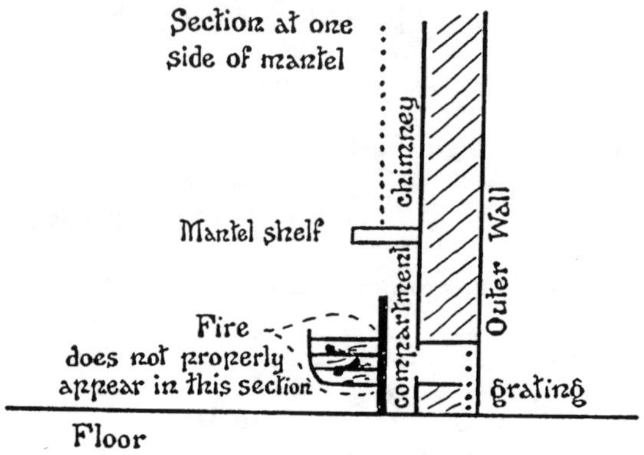

own use, but the patent laws may prevent your *selling* similar things to friends. I have not heard very lately from the Butlers.... I expect a long letter from Frank Butler in a few days.... Miss Elderton, of the Eugenics Laboratory, is staying with us for this week-end. She is a bright capable girl, and does her work excellently. I have not seen Evelyn Cunliffe* since her mother's death and doubt if she has yet returned home.

Ever affectionately, FRANCIS GALTON.

My section of the fire-place is vile, but I think you and Amy will understand it and could make a cardboard model for consultation with your builder.

MEADOW COTTAGE, BROCKHAM GREEN, BETCHWORTH, SURREY. *January* 31, 1909.

MY DEAR LUCY, Thanks for letter and enclosure of Gussie's which I return. Poor Erasmus! How stoical and how characteristic! The effects of the accident will take much pleasure out of his life of the sort he is accustomed to, but will I daresay lead to some compensations such as invalids learn to enjoy, as being taken care of.

I have intended to write to ask whether you would care to read, what to me is very interesting, the Journal kept by Sir Francis Darwin of his travels in Spain, Greece and Asia Minor, the first part in company with *Theodore* Galton†. It has been copied clearly in a limp quarto MS. book by Mrs Fellowes, a daughter of Mrs Oldenshaw, who has lent it to me. We are writing to her for permission to send it you. I was pleased to find confirmation of the fact that Dean Burgess of anti-Revised-Version notoriety *did* meet them abroad. There is not a word about eastern travel in his published life, but my recollection seemed certain that it was he, who spoke to me most appreciatively of Uncle Theodore at an Oxford dinner where I sat next to him. He thought him a man of rare promise, as so many others seem to have done. The pluck of Sir F. D. and of Th. G. was amazing. They travelled during most troublous times, viz. about 1809—brigands, pirates, and murderers everywhere.

Keir Moilliet's widow‡, her son Hubert and a daughter come to us to tea to-day. They are come to stay for a few days with a neighbouring relative of theirs, Miss Townshend. Is it not a pleasure that one more winter month has passed by? You both will get out soon I hope. I too have been much kept in by the weather. It seems that your foxes are not.

Ever affectionately, FRANCIS GALTON.

* Elder daughter of Sir Douglas and Lady Galton.

† See Vol. I, pp. 22–23. This diary has recently been published by the Cambridge University Press under the title: *Travels in Spain and the East* 1808–1810, *by Sir Francis Sacheverell Darwin*, Cambridge, 1927.

‡ James Keir Moilliet, son of Galton's sister, Lucy Harriot, married Sophia Harriet Finlay.

MEADOW COTTAGE, BROCKHAM GREEN, BETCHWORTH, SURREY. *February* 7, 1909.

DEAREST MILLY, Erasmus is, I believe, as comfortable as the case permits, and not in pain at all when still. I have not the least notion how soon he will be allowed to sit up and to attempt getting about in the smallest degree. It is difficult to see how his future life can be best arranged. Anyhow he has many comforts at the Regent Hotel which he could not easily meet with elsewhere, so at the worst he could remain where he is. Edward and M. L. were to have been with us to-day, but of course their plans had to be changed. The end of our stay approaches. The donkey and cart go to their home to-morrow and I part with them without much regret. Too much of a donkey like that would do permanent injury to one's temper, and make one revel in imaginary thwackings, pokes and imprecations. I have renewed a long-lapsed acquaintance here with the widow of a college chum, Mrs Ray, and find it very pleasant. There are many very nice people about here, as everywhere in England. My lease is out on March 1, but we shall leave a little earlier and go to Lyndhurst (or thereabouts) for March. In my walking tours of long ago, I came to the conclusion that the New Forest was the best place to find shelter from the cold March winds and to get sunshine. Besides, Lyndhurst is close to Eva's brother, the clergyman, who has a singularly pretty vicarage. You have not of late mentioned Hugh, *your* clergyman son. Doubtless no news is good news. The Eugenics Education Society is about (in a month or two) to publish a quarterly shilling publication, the *Eugenics Review*. There is a good Editorial Committee and it may be a success. I have just written a page or two as a "send off." Also I am again busy about the Feeble-Minded, being asked to help in a publication with Sir E. Fry and one or two others. This may possibly not come off. I have got the immense *Report*, eight folio volumes, of as yet unmeasured weight but certainly equal to that of a good-sized, well-packed portmanteau, out of which a few facts have to be gleaned. Best loves to you all. I heard of Guy at Claverdon, and hope that Amy has by now shaken off her cold. Ever affectionately, FRANCIS GALTON.

MEADOW COTTAGE, BROCKHAM GREEN, BETCHWORTH, SURREY. *February* 15, 1909.

DEAREST MILLY, Excuse this paper. I am writing before breakfast, and all of my proper letter-paper is finished. This squared paper is very convenient for my usual work and I always use it for that. Erasmus seems very comfortable, all things considered, and will I am sure find many compensations in the life of an invalid with attentive valet and nurses and plenty of friends. I hear of him from many quarters and all is to the same purport. The end of our stay is in sight, next Saturday will be our last Saturday here. I am in treaty for rooms in the Crown Hotel, Lyndhurst, which Eva went down to see. Though we are one hour S.W. of London, the only practicable way of reaching Lyndhurst by train is to go back to London and change there. It may prove more convenient by far to motor direct, and good motors are to be hired in Dorking. We shall see. A merit of the New Forest in March, which I realised in old walking days, is that the bare trees check much of the March winds and practically nothing of the March sun. Besides, Eva's brother, Walter, has his living two miles off. How the days roll on! I shall be 87 to-morrow and find on consulting *Whitaker's Almanack* that my "Expectation of Life" is now reduced to three years. In other words, that I am as likely to die before as after the age of 90. Also, that only one male out of every 50 reaches that age. Females are longer lived.

I have got off two little bits of work this week. One is the contribution I spoke of to the Cambridge hand-book, as it were (I don't know its title), to the *Report on the Feeble-Minded*. If they send me off-prints when it is printed, I will send you one. Also, I will send you the first number of the *Eugenics Review*, when it comes out in mid-April, that you (and I) may see whether it promises well. Heron (the "Research Fellow") has been week-ending here, and brings good reports of the progress of heavy work at the Eugenics Laboratory. Once again, about Erasmus' broken *neck* of the thigh bone. It is broken just below the cartilage; a common accident in old people. The fracture never *mends* but the broken end forms itself into a false joint, that is not wholly useless. I suppose that all inflammation has now subsided, so he can be moved, with much care, without pain.

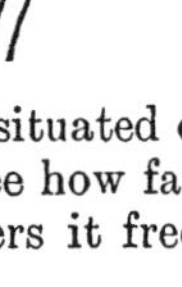

Such a capable man tea-ed here yesterday! Col. Melville, a doctor, the head of the Central Army Institution for tropical disease (chiefly). It is in fact what Netley used to be under Dr (? Sir) E. A. Parkes and is situated close by the Tate Gallery at Millbank. Heron is to go to his laboratory this week, to see how far the information collected there will be useful to the Eugenics Laboratory. He offers it freely to us. Guy knew Netley well, but

I think it was then somewhat in a transition stage, so far as its avowed object was concerned. Parkes was a first-rate, original man, and ranks easily as the founder of army and other medical hygiene in this country. Evelyn Cunliffe has gone with her husband to Switzerland to be set up after all her anxiety and cares. Himbleton is to be let. It now belongs to the Gascoignes. Best loves. Ever affectionately, FRANCIS GALTON.

Address: CROWN HOTEL, LYNDHURST. *February* 25, 1909.

DEAREST MILLY, Thanks for letter. Erasmus' death* is another great break. He died very peacefully. To-morrow, Friday, he is to be cremated at Birmingham, and, at his express wish, without any mourners or signs of mourning. His instructions were minute and unusual. Edward is doing his best to carry out his wishes.

Ever affectionately, with loves to you all, FRANCIS GALTON.

CROWN HOTEL, LYNDHURST. *March* 21, 1909.

DEAREST MILLY, I am up for an hour in my dressing-gown, fit to write a short letter, though weak and with the sense of lumbago being just round the corner ready for a spring. It would not have done for you to come here now, as you so very kindly proposed. When turned out from this hotel, it proves quite practicable to return to London, for the Cameron Galtons will have by then left our house and their servants behind, only too glad if we keep them on for our use. I dare say that I shall be fit to move then, without risk of sharp pain. We must arrange to meet before long. I lie in bed doing nearly nothing and fancy that illness exudes slowly all the time. Have you ever had the opportunity or patience to read the booklet that Jaeger printed and issued with his clothing? It is original and curious. He himself was the executive head of the Zoological Garden in Vienna, and is an enthusiast. His view is that all illness is *one* in essence, with many aspects, and, so far as I recollect, argues his point with some force, enough to make the view not wholly absurd. I read very little. J. G. Frazer has just sent me his *Psyche*, a quaint name derived from the task somewhere assigned to her of picking out the good seeds from a mixture of good and bad. He shows the important help given by superstition, even of the absurdest kind, in building up society. It is an interesting subject, which I thought much about many years ago on the occasion of a memoir being submitted to the Anthropological Institution on the paradox "Why Nations who believed in auguries, etc., overcame those who did not." I felt then that *any* creed was of more importance to a nation than none, in that it *saved* them from anarchy and disruption. Frazer's book takes the same line, with a wealth of illustration. I think frequently of Erasmus and feel that somehow he had not a fair chance in life. Circumstance and his own temperament were often much against him; and all that was visible ending in a small shovelful of ashes, scattered over the flower-beds of a crematorium! Edward, on the whole, liked the simplicity and common-sense of the last function. It is gratifying to know that many unexpected, kind remembrances of him were sent. They included one from the Committee of his London Club, to which he had belonged between 60 and 70 years. I wrote on this picture paper, partly as a safeguard against too long a letter. In fact, I have nearly reached the limit of my strength. With many loves.

Ever affectionately, good-bye, FRANCIS GALTON.

FOREST PARK HOTEL, BROCKENHURST, HANTS. *April* 16 (I think), 1909.

DEAREST MILLY, Read this please as though written with the whining voice of a beggar. For it is to ask if you will very kindly tell me exactly what the enclosed German letter conveys. I am ashamed at troubling you and will write a proper week-end letter all the same. The Whelers come here from Loxton for two nights, to-morrow afternoon.

Ever affectionately, FRANCIS GALTON.

Is "Werter Herr" all right?

* The last of Galton's brothers and sisters. It was a long-lived family. Putting aside two who died in infancy (Agnes and Violetta), Lucy Harriot (Mrs Moilliet) died in 1848, aged 39, but Bessie (Mrs Wheler) died in 1906 at 98, Adèle (Mrs Bunbury) in 1883 at 73, Emma (unmarried) in 1904 at 93, Darwin in 1903 at 89, Erasmus in 1909 at 94, and Sir Francis himself in 1911 at 89. His mother Violetta Darwin (Mrs Tertius Galton) died in 1874 at 91, and her grandmother Elizabeth Hill (Mrs Robert Darwin of Elston), mother of Erasmus Darwin, in 1797 at 95.

FOREST PARK HOTEL, BROCKENHURST, HANTS. *April* 19, 1909.

DEAREST MILLY, Thank you ever so much for translating that German imposter's letter for me, asking for pecuniary assistance in return for communicating early news of a big secret. I have sent him a card regretting that I cannot do what he asks. Yes, "Brennpunkt" is certainly "focus." The Edward Whelers are still with us, but leave to-morrow for Claverdon direct. He is busy planning small alterations to the house at Loxton, enough to make it liveable-in during occasional visits there of his wife and himself. It is a most difficult job, but he now has evolved something that is both simple and effective. We have had some beautiful weather here and the New Forest looks very attractive. But I get about with difficulty. Though quite at ease when still, the cramps and rheumatics are sharp, when I change position. I do hope your grandson will get over his present ear ailment. We go to Rutland Gate next *Wednesday*, the 21st. The landlord begged us to do so rather than stay a day longer as intended, on account of a customer in whose family there had been a death. So, as matters stand, we shall stay a full week in London, and I hope to see some old friends there, before going on to Cameron Galton's house at Eynsham. I occupy myself in muddling away at my hobbies. I am glad you like the look of the *Eugenics Review*. Eva and M. L. are just back from croquet at the Walter Biggs, five miles from here. Best loves. Ever affectionately, FRANCIS GALTON.

42, RUTLAND GATE, S.W. *April* 25, 1909.

DEAREST MILLY, We are back home and delighted that *your* home is on the point of becoming home-like. How beautiful this April is! We go next Thursday to Cameron Galton's house, Newlands, Eynsham, for a month; then back here. My rheumatics were driven away for one day by a wonderful medicine, some preparation of salicine (willow-bark), but it has returned, all the same. These sudden and transient effects are curious. I don't see much notice taken of the *Eugenics Review*, but it *is* noticed. It will, I trust, grow more solid. But how many people are ready to talk, and how few to work. There is no news for me to tell you. The parks here have already some beautiful bits of greenery and great patches of garden flowers. I was trundled* in Hyde Park for an hour this morning, and the same yesterday. Also I have been round Battersea Park in a taxi-cab. The old "growlers" and hansoms have almost disappeared from the cab stands. Spencer and May Butler and their classical son, Ralph, came to tea yesterday and Lady Isabel Richards (née Butler) came with her four little girls this morning. They don't look a bit as if they had ever been from England. She even looks younger, and fresher if possible, than when she started for India five years ago. Her husband returns next Sunday, "for good."

Ever affectionately, and I fear ever rheumatically, yours, FRANCIS GALTON.

42, RUTLAND GATE, S.W. *May* 3, 1909.

DEAREST MILLY, It would be very pleasant if you could come in June, when Eva would enjoy a sketching, etc., holiday. But she says that if I am not better then she would not like to leave me†. (But you will come all the same, I hope. I should anyhow stay the first week of June to enjoy your visit, but if Uncle Frank is really himself again I would like to go off sketching the three other weeks of June. At present he wants a great deal of care, and if left alone, quite forgets he isn't 25. Eva. This isn't *my* blot!) Let me then defer reply until next weekly letter. I am very helpless, and the swelling of one leg mends *very* slowly; still it mends, and the other is normal. So far as appetite, spirits and sleep are concerned, I am as well as I ever was, *now*; but how long this may remain so is a *quere*. I feel sure that I shall never be able to climb the stairs of this house again. I am carried up every night by Gifi and the man-nurse (Charman), who find me heavy, but I walk downstairs in the morning. I am thinking of parting with this house and of moving into a *flat*. Also of looking out for a well-built *bungalow* in some sunny place not far from London. Newlands, as I think I told you, has been quite given up. I did not feel fit to go, and the Doctor confirmed me. Poor Lucy Cameron, she is unfit to go to Aix, and I think may feel relieved that her house remains unoccupied while she is in Folkestone, in case she wants "home." Of novels, read *An Immortal Soul* by Mallock. It is

* This term now begins to appear in Galton's letters and signifies an airing in a bath-chair.

† The sentences in the brackets are in Evelyne Biggs' handwriting.

well written and healthy throughout, though the first chapters do not hold out that expectation, and his philosophy at the end of it is sound and interesting, and so far as it may clash with your own views, will at all events stimulate. I get through hardly anything. A sheet of scribblings comes upstairs every night and goes downstairs every morning with hardly any daily additions to it, and may come to nothing in the end. Best loves.

Ever affectionately, FRANCIS GALTON.

I had a blessed two hours' scientific chat with Karl Pearson, last Saturday, which was most cheering.

42, RUTLAND GATE, S.W. *May* 10, 1909.

DEAREST MILLY, It will indeed be a pleasure to us both to have you here on June 5. Eva has no definite plans yet for the end of the month, waiting, as she said, to see how I get on. But I get on *very* slowly, the rheumatics are as painful as ever. However I am engineered into my sister Bessy's wicker-work bath-chair, and am trundled in it to the beautiful parks. My doctor, McCaskie, told me to-day that he was about to leave off practice, and to hand his patients over to his sons. This is a grief to me, for he is a very capable man and I shall have difficulty in finding a substitute. My plans are gradually crystallising into a moderately sized London flat, and a *low* country house where I can live on one floor. In the meantime, I am going to live here on the drawing-room floor, turning the back drawing-room into my bedroom and having the man-nurse in the studio. Eva has been very busy planning this and the servants are now hard at work in carrying it out. Lucy Studdy comes here for a few nights next Saturday on her way to you. She will give you our latest news. I trust that Dim will soon get stronger. Ever affectionately, FRANCIS GALTON.

42, RUTLAND GATE, S.W. *May* 16, 1909.

MY DEAR MILLY, Lucy arrived at tea-time full of enthusiasm about the beauty, comfort and picturesqueness of your house and with the box of peppermints you sent me. Each of us three eat one after dinner with good stomachic comfort. I send with this a copy of the *Problem of the Feeble-Minded*, which contains a very useful analysis of the Royal Commission evidence. Please accept it. Lucy gives a good account of Dim*. I do hope the coming rest—you said she was going to Chester Square—will strengthen her as much as *you* could desire. She told me too a little about Guy, whom she just saw. To-day I have had two trundles in the bath-chair. The wind is cold but I go in a fur coat, which is a complete shield from it. I don't think that I am a bit better, though Eva and the doctor insist that I am. Anyhow, I do *not* gain in muscular strength, nor do the rheumatic cramps leave me. Perhaps they *are* a trifle better in the arms. Excuse bad writing, due to its being done on an arm-chair table to which I am not yet quite accustomed. The plan of turning the drawing-room floor into *my* floor answers perfectly. The man-nurse sleeps in the studio. The back drawing-room is my bedroom, and the front drawing-room is still the drawing-room. I get down to the dining-room for lunch and dinner, being carried upstairs after each. It all acts so well that I have given up the idea of a flat, but am looking out for a country house to rent on trial, with option of buying or of prolonging the lease. In this I should spend the winter. I have no news.

Ever affectionately, with loves to you all, FRANCIS GALTON.

42, RUTLAND GATE, S.W. *May* 22, 1909.

DEAREST MILLY, It will be nice when June 5 arrives. I *have* taken a house for August and September, near Cobham. Eva went down there yesterday to inspect and to report. Nothing is yet decided about next winter. I doubt if I am getting one bit better; they say I am, but I don't feel it. Of course, this warm weather is most grateful. This morning I have heard of the sudden death of my dear friend of very many years' standing—Mrs Hills, formerly of Corby Castle, now, or rather lately, of High Head Castle, near Carlisle. She was recently widowed, and spent a few days with us at Brockham Green, the last visit she ever made. She was the daughter of Justice Sir W. Grove, at whose house Louisa was taken so alarmingly ill more than 30 years ago, and who has been like a sister to me ever since. Very few friends are

* Pet-name of Amy Lethbridge, Galton's great-niece.

left to me like her. I can count *one*, but not more without some reservation. And in my own family there are none of my generation, and very few of yours besides yourself. *Sic transit.* I am just beginning Saleeby's new book, *Parenthood and Race Culture.* He dedicates it to me as "The August Master of all Eugenists." I read it in proof and, though there is much I would myself strike out, expect it will do good. He has eminently the art of popular writing with fluency. Mrs Horace Darwin spent a night with us, and her nice daughter* came to luncheon. She, the daughter, goes in for botany. Eva and I taxi-cabbed through Bushey and Richmond Parks a few days ago. It was town the whole way to one end of Bushey Avenue, but still countrified on the way back through Richmond Park. Love to you all three.

Ever affectionately, FRANCIS GALTON.

42, RUTLAND GATE, S.W. *May* 31, 1909.

DEAREST MILLY, You send me three grateful bits of news. 1. Frank's new appointment; 2. Guy's success in connection with the car trip; and 3. The new American grandson. Best congratulations thereon. I look forward to next Saturday with much pleasure. But you will find me a very helpless host. I can only walk a few paces and those totteringly and with pain, but get out sometimes twice in the day in my bath-chair. Kensington Gardens have been, and still are, lovely in parts, almost beyond expression. I am grateful to have lived to see this Elysian spring. Good-bye till Saturday.

Ever affectionately, with loves to you all, FRANCIS GALTON.

42, RUTLAND GATE, S.W. *June* 15, 1909.

MY DEAR EVA, The exact breadth of the arm-chair is 28 inches. If the garden gate be 30 inches wide, it will do well. 29 would *do*, but it would be rather a shave. I am so glad you have had an interesting day at Fox Holm. Curious that the Lawrence Jones's should have been there. Yesterday a letter came by post with "Prime Minister" printed on the cover and "Confidential" written inside. At first, I thought it must be some wine-merchant's circular, but its contents were "*Confidential.* My dear Sir, I have the pleasure with the King's approval of proposing that you should receive the honour of Knighthood on his Majesty's approaching birthday. Yours faithfully, (signed) H. H. Asquith." So I have to live till November 9 † and then shall blossom. Don't make any fuss about it. I told Gifi and Mrs Simmonds, as they would both like to know. I will keep this letter open till near post-time for news of the Cambridge Mathematical Tripos, which was issued this morning.

Tea-time, Charles G. Darwin is neither 1st, 2nd, 3rd nor 4th, but the 5th, 6th, 7th and 8th are bracketted equal, and he is one of these. Mrs Flack has been worse than useless. The result has been that, with my approval, Mrs Simmonds has got her 15-year-old sister to act as tweeny until we leave London. She will arrive to-night.

Affectionately yours, FRANCIS GALTON.

42, RUTLAND GATE. *June* 29, 1909.

MY DEAR EVA, I was so tired yesterday that I did not write; Milly helps greatly with my letters and I am at last nearly through with them, about 100. A letter came from the Home Office saying that I was to appear to receive Knighthood at a time to be fixed by H.M. I wrote back a piteous appeal against going to the ceremony, being unfit for anything of the sort, and to my joy a reply came this morning to the effect that I need not go, but that the patent of Knighthood would be sent me. Bad news for the tailors who sent circulars to fit me up for £32 odd. It was fair this morning and I went out in the bath-chair, but a sharp thunder shower suddenly came on, and Charman, I fear, got a little wet. Such nice letters come. I was wrong in rather ridiculing the Salvation Army missive, which, to tell the truth, I had not wholly read through. The last paragraph is very dignified and I respect their motives. So much so, that I am half inclined to frame and hang it up. Things go on here much as usual. McCaskie comes to-morrow. My legs are not one bit better, but I feel well in myself.

Bateson has *not* sent his book. I will wait a little longer and then buy it if he does not.

* Ruth Frances Darwin.

† The actual, but not necessarily the *official* birthday of King Edward.

Best love to Lucy and Cameron. I fear the weather has been too dull for your sketchings. It will be pleasant to have you back on Thursday. Leonard Darwin was here and we talked to him of the approaching dinner. Neither he nor I then remembered the day. Doubtless you and they are in complete accord, especially as to week! I said that the time was wholly in your hands. He will have a busy time on Monday next, with Lieut. Shackleton and Royalty at the Geographical. I saw something of the procession to South Kensington on Saturday, out of my bath-chair, from a path leading from the Park to opposite the Alexandra Hotel. It was very "spectacular" and well proportioned though not long and only three or four carriages.

Ever affectionately yours, FRANCIS GALTON.

42, RUTLAND GATE, S.W. *July* 8, 1909.

MY DEAR MELDOLA, Your kind and hearty appreciation is peculiarly welcome, for your judgment is especially valuable to me. The last time we met was at the final ceremony to Herbert Spencer. Nothing in that made so deep an impression on my imagination as the volumes of smoke rising from the chimney as we all went away. It meant "business." My time for leaving the world cannot be far off, for I am become very and increasingly infirm. It was a great grief being unable to join in any way in the Darwin celebration. How admirably it went off! How much there is to do in life and how very much has to be left undone! I follow your work, and from time to time I see accounts of it, always with keen interest.

Believe me, yours sincerely, FRANCIS GALTON.

That stupid error of address in the Royal Society *Year Book*! I took the house during the past winter and they printed it as my permanent address—I am still, as always, a cockney as to my home.

42, RUTLAND GATE, S.W. *July* 18, 1909.

DEAREST MILLY, It was but a scrubby message that I sent you through Eva, being at the time tired and sleepy. Pray, in your next, tell me the latest news of Jim*. His father kindly wrote me a very hopeful account. Has the visit to Folkestone confirmed it? And did you succeed in alleviating the "uncouthness" of Hugh's "solitary life"? You may recollect something of the sort about a great nonconformist Divine whose friends in consequence hunted out a wife for him, with great after happiness. I put the story into my *Hereditary Genius*, under the "Divines," but have not the book at hand. The cold weather has been against my mending, but has not made me worse. Are you a reader of *Peter Pan*? A group of small figures is to be set up in Kensington Gardens at the *Bay* where he landed. So Sir George Frampton told us, who is the sculptor. He came yesterday to tea, about doing a bust of me, which (but I shall hear more exactly to-morrow) he will begin at once. It will be amusing to witness the operation. It is to take place in my old dressing-room, above my present bedroom. I suppose a hodman will arrive with a bucket of clay. He will send the necessary furniture, stool and table, to work on. I miss you often. Do tell me how you found the garden, after the storm had treated it so rudely. We still get copious supplies of very big strawberries. It seems they send daily consignments of some sixteen tons of them from *Alnwick*, when the season in these parts is coming to an end. Loves to you all. Ever affectionately yours, FRANCIS GALTON.

42, RUTLAND GATE, S.W. *July* 24, 1909.

DEAREST MILLY, Good as your account is of Jim, I wish it could have been still better. There are so many dangers before abscesses in the bone of the ear can be wholly cured. Indeed the cessation of the wonted discharge is a danger in itself. It will be a pleasant experience for his brother to go to Claverdon. What an amusing but *terrible enfant* George must be! Poor Mary, pluck does much, but not everything. As regards home matters, one event last week was seeing a singularly beautiful black and white canary perched fast asleep on the frame of one of the pictures. He was carefully captured and put into the conservatory, pending inquiries. In the meantime I got him a cage, but in two days he died. I know not what from. It was a piteous sight. Sir G. Frampton gets on busily with the bust. He first gets the profile exactly, and does much of the side face, then goes on to the full face. He is wonderfully capable and painstaking. The time of our departure, August 9 or 10, approaches only too fast. Frank Butler comes next Saturday to clear out and arrange, as far as he can, my old instruments

* Second son of Mrs Lethbridge's son, Edward.

and papers. There is so much that I shrink from parting with, but which I can never use again and which would be useless to others. Mere rubbish in short to all but my past self, so it must go to the scrap heap.

All I saw of the fleet were the few destroyers and submarines moored at Westminster and the Tower Bridges. I ventured on that much in a taxi-motor. So sorry about your horse. He must be a great loss for the time, notwithstanding the occasional drives in Guy's motor. A *Eugenics Review*, under the title of "Rassen-Hygiene," has been started in Munich, by a very capable man, Dr Ploetz, who is the editor of a really solid anthropological periodical. The society that issues the review has five honorary members among whom are Haeckel and Weismann, and I am asked to be its honorary Vice-President, which honour I have gladly accepted. But I must work up my German! Both our loves to you all. You will be solitary when Amy leaves you for a week. Ever affectionately, FRANCIS GALTON.

42, RUTLAND GATE, S.W. *August* 7, 1909.

MY DEAR MILLY, Thanks many, *re* Miss Annie Butler. I fully believe that the National Institution for the Feeble-Minded is the Society in which Dr Alice Johnson is much interested, as the Eugenics Education Society also is. Its secretary, Miss Kirby, is a very nice and capable woman. The model of my bust *is* finished, except just the coat which is being modelled at the studio—I mean a little of it. Sir G. Frampton has had my coat and overcoat nearly two days already. I hope to get them back to-day. But it is so delightfully warm that I am putting on still cooler things. The bronze cast won't be ready till near Xmas. The operations take much time. I should like to see the white-hot metal poured in. All that part is done by specialists. We have quite fixed to go to *Torquay* for the winter. My two doctors severally and collectively recommended it. I don't think that in essentials I am any better, perhaps a little worse. No diminution of swelling of the legs, rather less strength, but happily much less rheumatics. Recollect my address after Monday will be Fox Holm, Cobham, Surrey, for two months. Eva will run down to Torquay in a few days to house-hunt, escorted by a lady friend who is staying there. Edward Wheler came up for a few hours to see and criticise the bust. Also, Cameron Galton slept a night here on his way to Geneva. They were of decided help. The likeness was brought out much better through their hints, and seems satisfactory to all now. Edward told me about Edward Lethbridge's visit to him. Ever affectionately yours, FRANCIS GALTON.

Loves to you all.

FOX HOLM, COBHAM, SURREY. *August* 29, 1909.

MY DEAR MILLY, What a house-full you have! Please tell Edward that my conscience pricks me at never having answered his letter about Jim, saying how much better he was. I was very glad to receive it. We get on very pleasantly here, especially when the sun shines. My man-nurse is an accomplished poacher and snares rabbits in our own grounds, which we eat. We see various members of the large Buxton family, all nice and interesting. The big house, "Fox Warren," now belonging to the Postmaster-General, Sydney Buxton, is occupied by a cousin. I get into a carriage most days, by means of an artful contrivance and the help of two men, and so for a $1\frac{1}{2}$ hours' drive. Also, my tent is an immense comfort. It keeps out the wind and lets in the sun. Yesterday I had a good 2 hours' steady work in it. Much love to you all. Ever affectionately, FRANCIS GALTON.

There is an appreciative article on Lucy Studdy as an embroidress, with an excellent portrait, in a monthly magazine called *Needlecraft*.

FOX HOLM, COBHAM, SURREY. *September* 6, 1909.

MY DEAR MILLY, Your account of Jim's health and hearing is unexpectedly grateful. Best congratulations to all concerned. You must indeed be a house-full. Here is a story sent me by Lionel Tollemache: Some French ladies were deploring the spread of scepticism (which word is, I think, pronounced with a softer *c* in French than with us), whereon one said "Heureusement on a inventé les antiseptics"!! Our plans are quite shaped now. We go from here straight to Haslemere and I lend 42, Rutland Gate during the winter to the Gibbons. It will be a great comfort to them. She was Gwen Chafy, as you know, and he will be a kind of man of business to me. Also my bedroom will be intact and I could come back at any time. The Buxtons' house close by is called "Fox Warren." It might appropriately be called "Buxton Warren." There are so many of them there and thereabouts, and all are attractive.

Eva is not as strong as is her wont. An old strain, at least nine years old, has been too much ignored of late, and she is ordered rest, and to get *fat*, if possible. I expect Lucy Cameron Galton to-day for the inside of the week. She will tell me about Cameron, who is somewhere in the Alps, and of Violet who was with a friend in Venice. Life goes on here much as usual; quietly and contentedly. My man-nurse continues skilfully to snare rabbits, which we continue to eat. Edward Wheler killed 131; they devour his corn. Loves to all under your hospitable roof-tree. Ever affectionately, FRANCIS GALTON.

FOX HOLM, COBHAM, SURREY. *September* 11, 1909.

MY DEAR MILLY, Yes, the *c* in the French "scepticisme" is "muet," which improves the anecdote. What a difference of sense one letter makes! I have no new ones to tell. To-day is glorious and I have had a trundle in my bath-chair, very successfully. Galtonias are sensitive to soil. Those hereabouts are very good. More than half of our stay here is over. I shall be glad to be safe in Haslemere before wintry weather arrives. I have been won over by a piteous appeal, in consequence of an offer from Karl Pearson, to accept a puppy. It is a pure albino of pure albino descent,—a Chinese pug with the name of Wee Ling. Albinism has been a recent study of Karl Pearson. This little creature may prove ancestor to a valuable breed of them; valuable, I mean, from a scientific point of view. Eva will rejoice in the young pet, of whom I have a photograph. Ever affectionately, FRANCIS GALTON.

THE RECTORY, HASLEMERE. *October* 4, 1909.

MY DEAR MILLY, We are just arrived after a very easy 1½ hours' motoring. "We" includes the puppy, who bewailed most of the way in an unknown dog-language, very like the self-made noises in a bad telephone. This place seems very suitable, but I have only seen this one room, the drawing-room, as yet. They are unpacking upstairs and in ½ hour after tea I shall be carried up to my bedroom. I am indeed grieved at poor Jimmy's renewed ear-trouble. It sounds so very serious. It feels very nice to be again near to many old friends and to have again seen familiar roads and scenery. The trees are just beginning to show autumn colouring, but some cottage creepers that we passed on the way were fully turned into gorgeous reds. Eva is distinctly better under her regimen of fattening food and rest. The latter will now, I hope, be taken in full doses, as Violet* will be here to-morrow afternoon for her long stay. She, Eva, has shown me your letter which was awaiting her. I learn now that the house is the "Rectory," not "Vicarage," as I had been told. The latter address has however sufficed. I am so glad you liked the Gibbon-Salomons. This is an ideal country parsonage, such as a cultured clergyman would enjoy. Excuse more. Ever affectionately, with many loves, FRANCIS GALTON.

THE RECTORY, HASLEMERE. *October* 19, 1909.

MY DEAR MILLY, Again I am unpunctual and blush (internally). All goes on steadily. Eva is happy in bed, and Violet seems to enjoy her double occupation of nurse and housekeeper and of companion to me when out of doors—twice yesterday, but often nil owing to bad weather. I have had calls from two scientific friends, full of information and pleasant talk. You will have received the *Eugenics Review*. Mügge's paper strikes me as very good. A substantial but comparatively thin book on Eugenics by the Whethams is just out. It is well written and well got up. He is a Fellow and Tutor of Trinity College, Cambridge, and his wife and coadjutor is zealous and able. He takes a broad and sensible view of the necessity for our Race Improvement. It is so well written that it may win its way. The albino puppy grows in body and mind. His tail curls tightly already, and he has had his first lesson in Sociology, through offending the cat and receiving a wipe of her claws upon his little pink nose. Lucy Studdy comes to Haslemere next Saturday. We can't take her in, but there is a fair hotel very near and she can meal with us. Three invalids in one house would tax domestic resources too much. There will be much "high faluting" in Birmingham this week. An extension of the University there will be opened and their power of giving degrees will be exercised for (I think) the first time. Oliver Lodge will be in his glory and will, I have no doubt, act his part exceedingly well. My horizon is now so narrowed that I have little to tell that would interest you. Eva would send her love if she knew I was writing. Ever affectionately, FRANCIS GALTON.

* Violet Galton, a sister of Cameron Galton; both were Francis Galton's first cousins once removed.

THE RECTORY, HASLEMERE. *November* 25, 1909.

DEAR MISS ELDERTON, Thanks for your book, now safely received and read through. I sincerely hope that it will have the success that it merits and which you and your brother and myself have all at heart. Enclosed I send in return a reprint, just out, of my own lectures. They might all have been improved much, but I thought it better to let them stand just as they were; besides, I am grown too infirm to do anything of value now, I fear.

The post has just arrived, and among press cuttings I found the enclosed by Dr Saleeby, which refers to your work. Also, I read lately a speech by the Poet Laureate, who gave the advice "Do not resent criticism and *never reply to it*." As a general rule this is excellent, but *if* you should care to reply, it would be well to confine what is said to justifying the *appropriateness of the data*, which Dr S. attacked in a former letter somewhere. As to the statistical treatment of them, it is needless to insist on its accuracy. For the life of me, I cannot yet see how a Mendelian objection can bear on the value of your work.

Very sincerely yours, FRANCIS GALTON.

THE RECTORY, HASLEMERE. *December* 12, 1909.

MY DEAR MILLY, January 3 would suit us particularly well. So let it be, unless as the time approaches you wish to change the date. You must be most anxious for a letter from South Africa. Pray let me know as soon as you receive one.

Matters go on here as usual. Janet Fisher is with us. She will probably be Eva's companion when Eva goes abroad. It appears that there is much to talk over first. How you *are* getting on with the Braille! A nice young fellow lunched here on Friday, who had been four years studying at the University of Freiburg (under Weismann). He told me about the vexed question of having boys and girls in the same school classes, and that doing so was considered a complete success in Baden, with *one solitary* drawback. You know what pretty blond tresses the Baden girls mostly have. It appears that the boys cannot resist the temptation of stealing up behind and dipping the ends of the girls' pigtails into an ink-pot. I suppose the pigtail of a girl in a fury would scatter its black charge right and left. What a picture! I have just procured a copy, 7*d*., of Harmsworth's *The World's Great Books* under which magnificent title four of mine are to be included. Certainly the No. 5, that I have, is uncommonly well done. It is not too snippety. The thing is published fortnightly and contains about 100 pages, good print. As I hear, 400 pages are to be assigned to me, 100 for each of (1) *Hereditary Genius*, (2) *Human Faculty*, etc., (3) *Eugenics* and (4) *My Memories*. They will take up four numbers or the greater part of them. We shall see. I am glad that Loxton* is within motor range of Minehead.

Ever affectionately, FRANCIS GALTON.

THE RECTORY, HASLEMERE. *December* 19, 1909.

MY DEAR MILLY, Only two days more, and the daylight will grow longer, and yet "as the day lengthens, so the cold strengthens." Thus far we have pulled through fairly well. Harmsworth has begun to publish epitomes of my books, on a much smaller scale than was proposed, and Dr Saleeby has put them together fairly well, certainly with much industry. You must be anxious about Frank, but a seldom-letter-writer accounts for much. You have, I suppose, no correspondents among his South African friends? So that astute, money-grubbing sovereign, Leopold II, is gone! If our Princess Charlotte had been his mother he might, according to the late Duke of Wellington, have been yet worse. I speak of Lord Stanhope's record of conversations with the Duke at Walmer Castle, in a passage that appeared in the 1st edition but was suppressed in the later ones. It stated that it was a mercy that Princess Charlotte died, for she had the vices of *both* her parents! However, she was idolised by our nation, which of course knew nothing of her real character. Thanks for Xmas greetings, which I heartily return to you all. My pigtail story must, alas, be discounted. I thought it had been a frequent occurrence, but it was apparently one solitary outrage which impressed itself deeply on the Baden mind. We have been chuckling over the *Caravanners*†. What an intolerable idiotic prig the German Major is made out to be, and yet not an impossible one. January 3 will soon arrive.

Ever affectionately yours, FRANCIS GALTON.

* Loxton had originally belonged to Samuel Tertius Galton, had passed to his son, Erasmus, and now belongs to Francis Galton's nephew, Mr Edward Wheler-Galton.

† By the Countess von Arnim, now Countess Russell.

THE RECTORY, HASLEMERE. *December* 26, 1909.

MY DEAR MILLY, So another Xmas has come and gone, and your peppermints have helped us to enjoy it. I am so glad that news had reached you from South Africa before the 25th, otherwise you must all have been worried. What you tell me of Fräulein Ronath's report of German opinion about my precious self simply amazes me. I feel sure there must have been imaginative exaggeration of merely civil answers to her leading questions. I doubted if twenty people in Germany knew of my existence. Anyhow, it is very kind of Fräulein R., though I cannot accept the over-flattering sentiments she conveys. Violet Galton is with us now. She was to have Xmas'ed with her sister Amy at Keston, but the domestic establishment got suddenly all wrong, I forget precisely what. Anyhow, another servant was wanted and could not be had. Eva is distinctly better and has lost her thinness; fat is a great help towards keeping truant kidneys at home. We look forward ever so much to January 3. With all loves.

Ever affectionately, FRANCIS GALTON.

THE RECTORY, HASLEMERE. *January* 6, 1910.

MY DEAR MILLY, Your welcome letter arrived this morning. I am glad that your long journey home ended without mishap. Thank you again, sincerely, for having come. My "one snipe" that has given me occupation every day for months past, namely the "Numeralised Profiles," is—to continue the metaphor—being *stuffed*. In plain words, it *is* printed and I send back the proofs to-day. You shall have a copy of it as soon as published, not improbably at the end of this week. In the meantime there is a placid interval, because I cannot write for material to work on until it, the article, is out.

Hesketh Pearson is here for the week-end. Pan (Josephine) Butler was to have come also, but is in bed with a cold. Last night we had jugged hare for dinner. I had insisted on its being jugged *before* otherwise cooked. It was excellent; twice, three-times, ten-times better than a jugged *réchauffé*. I believe this is the only event worth mentioning.

Ever affectionately, with loves to you all, FRANCIS GALTON.

THE RECTORY, HASLEMERE. *January* 6, 1910.

MY DEAR LEONARD DARWIN, I am very comfortably lodged here, and am pulling through the winter fairly well. It gave me much pleasure to read your proposal about honorary members to the Geographical Club. The rule *is* adopted, as you may know, at the Royal Society Club, but I think their limit of paying membership exceeds what you propose (? 20 years).

Oh dear! how people die. Life seems to me as occurring on an endless belt. Babies are

dropped on it through a hopper at *A*, they grow, frisk, and age, and drop off in senile imbecility at *B*. I don't yet feel my faculties to wane distinctly, but I tire very soon.

An article of mine, of which I return the proofs this day to *Nature*, may perhaps interest you. It is a literal fact that you can convey a very respectable profile likeness in four telegraphic "words"; that is, in four groups of figures, five figures in each group. I give illustrations. With kindest remembrances to your wife. Ever sincerely yours, FRANCIS GALTON.

THE RECTORY, HASLEMERE. *January* 9, 1910.

MY DEAR LEONARD DARWIN, It was a pleasure to hear some talk of you. I am settled here for the winter, very comfortably but increasingly feeble in body. The air of Haslemere suits me well.

I am very glad you continue well disposed towards Eugenics. The problems connected with it are difficult and statistically most laborious. I notice that in your lecture you do not take account of differential fertility, which to my mind is the most important of all factors in Eugenics. H. Spencer's law about the diminished fertility of the most differentiated animals seems to be an excellent guess founded on *à priori* data.

I read your excellent Geographical appeal for funds for a larger house*, and shall in due time send my quota. Just at this moment I am rather entangled with prospective obligations, or fancy that I am.

What an eventful Geographical Presidency you have had. I am very glad of it for your sake. With every appropriate New Year wish to you and your wife.

Ever sincerely yours, FRANCIS GALTON.

THE RECTORY, HASLEMERE. *January* 25, 1910.

MY DEAR EVA, By all means let Sir G. Frampton send the bust to the Royal Academy. My poor shrunken nose! I feel like Wee Ling looking into a glass. The little beast is as merry as can be, and we have a grand game of bob-*sugar* after dinner. He has grown disdainful of bob-indiarubber-ball. Yes, ask Pan and Hesketh for Feb. 5–7. Dr Barnardo must have been a wonderfully good organiser. I should be glad of particulars. Milly dragonises well. Sir A. Geikie tea-ed here yesterday and told me much in the scientific way. His book about Seneca's philosophy *is* printed, but held back until the election turmoil has subsided. Major Norris has got me a good account of Daddy Tin Whisker from Australia. It is aluminium, rubbed by an amalgam (= a metal combined with mercury). Its filamentous growth has been *noticed*, but no explanation is given. I can fancy a scab being produced, but don't understand the hairy growth.

I am quite in "my usual" again,—and Miss Jones is busy at Miss Baden-Powell's silhouettes (which you traced for me)—but I was trembling on the verge of being bad three days ago. Drives on two successive days, and an hour in the shed on the third, were too much for me. Milly seems quite happy, and I gather that you are also. Give much love from me to the Brees. I am wearing Adèle's muffetees with much sense of comfort. The partial discolouration has been washed out by Charman. Ever affectionately, FRANCIS GALTON.

Have you news of Bessie† yet?

THE RECTORY, HASLEMERE. *March* 13, 1910.

(42, Rutland Gate, S.W. on and after Monday 21st.)

MY DEAR MILLY, *Your* spring is a full week in advance of ours, I think. Here are lots of crocuses, but no green tips yet to the trees. William Darwin‡ is here with his motor for the week-end. Edward Wheler comes on Wednesday for a night or two. Then we pack up and send off most things by Gifi on Friday. Eva, man-nurse and I by motor on Monday, and the one remaining maid by train on that day. Amy and Guy will enjoy their Loxton picnic. Poor Frank! A man, Mr M. W., who was in office in the Cape, married a wealthy lady here and has now returned, did not speak in the same gloomy terms that Frank does. Probably he got his foot early in the stirrup and mounted a good horse, and so pushed forwards. Col. Melville spoke strongly in favour of Mr Haldane§, who he thinks has done and is doing wonders in the face of great difficulties of tradition, organisation and the like. Eva sends her best love to you all, so do I. Ever affectionately, FRANCIS GALTON.

42, RUTLAND GATE, S.W. *March* 27, 1910.

MY DEAR MILLY, Somewhat battered by coughing, mostly asthmatic, here I am, settled in home again. Everything looks homely and suggests old associations. Dim's portrait, in photo, stands conspicuously on the chimney-piece opposite. But the room has to be rearranged, owing to structural alterations in the form of a built partition between the front drawing-room, now my only drawing-room, and the back one, now my bedroom. It will take time to make it all comfortable, new bells, etc. I understood from a line in Lucy's letter that a picture of Ravenscourt is in this week's *Queen* newspaper. I will order it as soon as Bank Holiday is past. We are trying *Coalite*, said to burn more purely and with less heat than coal; a desideratum for my small bedroom. Do you ever use it? A friend comes to-day to show off his hearing apparatus, which, when in good

* Major Leonard Darwin was at this time President of the Royal Geographical Society.

† Now Mrs Simmons, Evelyne Biggs' former maid.

‡ Charles Darwin's eldest son.

§ Afterwards Lord Haldane, then at the War Office.

humour, acts very well indeed. It is a telephonic arrangement. That which I tried some months ago, made by ——, was always out of humour and made its own internal noises which overwhelmed what the speaker said. So my hopes are pessimistic, but I shall soon learn more.

Lucy will be with you now or now-abouts. Tell her that the Venetian window onto the balcony of my drawing-room, and the swing ventilator above, are a great success. They were put in after she left. With much love to you all. Ever affectionately, FRANCIS GALTON.

42, RUTLAND GATE, S.W. *April* 12, 1910.

MY DEAR MILLY, The time is in sight, though still a long way off, when I shall have the pleasure of having you here. Eva is looking forward much to Rome and to becoming a Roman Catholic. She is being "instructed" and I both hope and believe the change will suit her temperament. She is a very thoughtful and kind nurse to me. I don't get as quickly better as I hoped, but *am* stronger, a little. Yesterday I was able to sit half an hour on my balcony while the afternoon sun shone on it. Every day we shall get longer sunshine. I am doing as nearly nothing as can be, but began to revive yesterday on Molière. One advance is that I have at length got a really serviceable hearing apparatus, so that people can talk audibly to me without raising their voices, and Eva is reading out to me, each evening, a bit of Mrs Schimmelpenninck's biography. How vividly and well she tells her version of the tale. I heard from Edward of Guy at Loxton. What trouble the water supply gives. When you come, there will be a room available for Amy, if she likes to come too. She will of course be always most welcome. I am writing without your last letter at hand and mistrust my memory about many of your matters, your repairs, etc., and so do not write about them. With loves to you all.

Ever affectionately yours, FRANCIS GALTON.

42, RUTLAND GATE, S.W. *April* 16, 1910.

MY DEAR MILLY, It would give a very welcome addition to Eva's holiday, if you come on May 17 as you propose. She would welcome you and start the day after. Thank you very much. Miss Jones* will be here also during three weeks of your stay. She will do a good deal of pen and ink and pencil work, and she knows all my ways and the servants too. That will leave you your mornings free. Will Dim come on Friday, June 3, and stay up to the time of your return, Tuesday, June 14? Eva proposes to return on the 15th to get the house ready to receive her (and my) friend, Mrs Townsend, on the 17th. She will add a postscript. So you have really finished your long labour of "Brailling" my *Memories*! I trust that you will thereby give a pleasure to many by enlarging the choice of books readable by them. I am getting on a little, I think, and believe that, as you say, the coming summer will help. I got out in a bath-chair this morning, but the day, which had been brilliant, clouded over and a sharp hailstorm followed almost immediately my return. I had not time to get as far as the flower walk in Kensington Gardens and have as yet seen next to nothing of the glories of the coming spring. There has been much dusting and rearrangement of pictures and books in the drawing-room, which already looks quite pretty and harmonious.

Ever affectionately, FRANCIS GALTON.

Overjoyed you can come earlier for that gives me a nice jaunt. Also I hope Dim will now come for quite 10 days. Your loving Eva.

42, RUTLAND GATE, S.W. *May* 12, 1910.

MY DEAR MILLY, I owe you a letter, and, as the time is so near of your coming here, lose no more of it before writing this. All goes well, but I have been unable to face the cold winds and have been a prisoner in-doors for some weeks; but to-day seems milder. The King's death must bring forcibly back to you all his great kindness to your son Bob, when he was lying so ill. The act seems to have been a characteristic one on his part. What political storms, now temporarily lulled, are coming again soon! I hear that the new King and Queen will probably do much good by purging the Court of many undesirable persons and habits, and by

* Miss Augusta Jones, who in the last two years of Galton's life did occasional secretarial work for him.

introducing more simplicity where needed. Eva will receive you here on Tuesday and goes to Rome next day; in the first instance *through* to Baveno, where she will meet friends. I am rejoiced that she is now strong and very fit for travel. Miss Jones came here yesterday. She will take much trouble about me off your hands, being very vigorous and serviceable. I trust you will find all here as you want. The drawing-room is much improved, thanks to Lucy Studdy's idea of a solid partition instead of a curtain. Did I tell you that the bust, or rather a cast of it, is in the Royal Academy? Sir G. Frampton only sent one other exhibit, out of the six he is entitled to as R.A. Best loves to you all.

Ever affectionately, FRANCIS GALTON.

42, RUTLAND GATE, S.W. *May* 21, 1910.

MY DEAR EVA, You will be so immersed in Italian feelings and atmosphere, that news from here will seem petty and even profane, for a time at least, to you. The facts are: (1) I miss you. (2) Edward and M. L. returned last night and spent an hour here, looking very well and with much to say. (3) Grace Moilliet, by tact and enterprise, saw both the lying-in-state and the funeral procession. Both Milly and I, each in our several ways, went to Hyde Park to see the *crowd*. It was totally impossible to get near to the route. The crowd was singularly orderly and quiet and all in black. (4) Sir G. Frampton comes to-morrow to see where the bust is to stand and to fix for the pedestal accordingly. (5) The *Times* has a favourable leading article and a long analysis of Miss Elderton's paper about the children of drunkards, which will make Saleeby tear his hair.

All well and happy. Our best loves, F. GALTON.

(6) Miss Jones returned last night.

42, RUTLAND GATE, S.W. *May* 31, 1910.

MY DEAR EVA, I do most thoroughly enter into your happy feelings in this the crowning epoch of your life. Everything seems to combine to enhance its happiness—the air and climate of Italy and the sound of the language, the quiet affection in the convent, the ceremonial at the Vatican, all combining with the great function itself of your entering the church that your temperament most requires. I heartily congratulate you. It is pleasant to me to hear how helpful dear Louisa's relations have proved to you. I have no particular news. Sometimes more, *now* less of asthma—the usual round. A letter I wrote last night to the *Times* joining issue with one of Ray Lankester, is published this morning in biggish type. Miss Elderton and a sister of Miss Jones came yesterday to tea, and such-like events at present complete the round of my daily life. I have now no tearing wants or ambitions. My race has been run, and I have simply to await the close of life.

Milly seems very well and happy. Dim comes towards the end of this week. Miss Jones does all she can, and goes to-day for me to the South Kensington Art Gallery to add something to the tracings from Dance. Ever affectionately, FRANCIS GALTON.

42, RUTLAND GATE, S.W. *June* 26, 1910.

MY DEAR MILLY, It is well that your stay here has not, after all, interfered with the house-cleaning before your guests arrive. I miss the tapping of your tool for making the papers for the blind, and I often lift up my eyes and, not seeing you on the sofa, wonder for a moment if you are elsewhere in the room. Thank you again for coming. Eva has a "clergyman's" sore throat, brought on she thinks by talking too loud and long to me in eagerness to relate her story. She writes for me now what she wishes to say, but will probably be quite right again in two or three days. Mrs Townsend* is here. She has (while here) three big speeches to make, a masseuse to operate on her, and a weak heart. So she is in her room and takes her meals there most days, quaffing champagne (in moderation) which is her usual drink, and very grateful for the opportunity of being quiet. Beak's† wife was moved to St Mary's Hospital on Thursday. He will be allowed to see her to-day. It appears that a quantity of stale blood has to be drawn from the tumour on her wrist, and that she *may* get well afterwards quickly. Karl Pearson had a large reception last night at the Eugenics Laboratory. I am curious to hear about it. The Academic Registrar of the University of

* Well-known in relation to the "Girls' Friendly Society."

† At this time Galton's valet-nurse.

London (Hartog) called here to explain many matters. The authorities there are *most* friendly to the Laboratory, and, as funds permit, will increase its scope. Hartog is an excellent official, very able and of a very able Jewish family. One of his brothers was Senior Wrangler of his year at Cambridge and another is a distinguished professor in Ireland.

Ever affectionately, FRANCIS GALTON.

[In Evelyne Biggs' handwriting.]

Please thank Dim for her charming letter this morning and for copying out the Rosary so beautifully, it is sweet of her. Thank you also for your nice letter. Eva.

42, RUTLAND GATE, S.W. *July* 2, 1910.

MY DEAR MILLY, The artistic touch of Mrs St Maur will indeed be grateful to you, and leave abiding results in the garden. Beak's wife is steadily mending. The Doctor made a "culture" of microbes from the contents of the swelling, and injected it as an anti-toxin. She leaves hospital to-day and Beak is absent on the errand of escorting her home. I have had two rather bad days and the Doctor on each, but am now in a fair way of getting well. Thanks for the *Morning Post*. These journalists cannot write a column without blunder. It is so in this case, but I won't go into details. The *novel** has had a long set-back, having found the plot not to be as useful a one as was wanted. I have at last re-cast it in a better form, but written nothing yet. Some of what was written will still serve. I will keep you *au courant*. I expect Miss Elderton every moment for the week-end and have asked a few friends for a Eugenics tea to-morrow. Ploetz, the German author of the paper you kindly looked at for me, and who is in London now about a Race Congress, will come. It is long since I last ventured out of doors, but a convenient alteration has been made in my balcony fittings, which will make going in and out of it still easier. What an extraordinary cavern seems to have been discovered in Crooks' Peak, at Loxton. 300 feet long (as asserted) and wide in proportion. But measurement may greatly reduce the figures. Fancy in digging a deep well, suddenly breaking through the roof of a big cavern and tumbling through!! Eva is quite well again and sends her love.

Ever affectionately yours, FRANCIS GALTON.

42, RUTLAND GATE, S.W. *July* 11, 1910.

MY DEAR MILLY, Best congratulations on your motor. May it serve you well and safely to others! A dear old dog of Arthur Butler's has just been knocked down by one, much hurt, but they hope not mortally. You do not say who drives you in Guy's absence. Mrs Beak seems to go on well. I told *him* of your inquiries. She has to go weekly to the Hospital to be injected and seen to. Ploetz proved to be a pleasant acquaintance and full of "go." A fair-haired South German. My novel gets on but is quite re-written. I now get up an hour before breakfast and lie down for a bit after. The plan seems to suit well. A particularly good article in the *Westminster Review* on "The Scope of Eugenics," signed by H. J. Laski, was sent me among other Press cuttings. The name was unknown to me, so I wrote to him "Care of the Editor," and hear from Laski this morning in a very nice, modest letter that he is a school-boy at Manchester, aged 17!! It is long since I have been so much astonished. The lad has probably a great future before him and he will make a mark if he sticks to Eugenics, which he says has been his passion for two years. I as yet know nothing more about him, but hope to learn. Gertrude Butler has been staying for a few days with us. She and Eva get on very well indeed together.

You on the West Coast have sunnier weather than we have. I have not been able to get out for a full fortnight, more than once, and that only for half-an-hour, on the balcony. How good *strawberries* are this year! Best loves to you all.

Ever affectionately, FRANCIS GALTON.

* This is the first mention in a letter to Mrs Lethbridge of Galton's Utopia, *Kantsaywhere*. It had clearly been a topic of conversation between Sir Francis and his niece during her visit to Rutland Gate in May and June of this year. The first idea of the "Eugenic State" appears to have come to Galton in 1901, for I find in a note-book of that year, taken to the Riviera, the draft of the family characteristics and a description of the home of the "Donoghues of Dunno Weir" ("Don't-know-who's of Don't-know-where"). Galton had obviously been planning his Utopia for *nine* years. It was thus not a mere hasty product of his last days.

42, RUTLAND GATE, S.W. *July* 19, 1910.

MY DEAR MILLY, You have indeed a full programme. May the weather befriend the motor trip. I have practically taken a house near Haslemere for either a month or six weeks. It has a Roman Catholic chapel attached to it, for the use of which we pay in the rent! It is called "The Court, Grayshott, Haslemere, Surrey," and has excellent grounds and roads for my arm-chair on wheels. No news worth recording. My novel gets on and I live a new life with its characters. I have often read of this faculty but never experienced it before. An uncommonly neat and well got-up German translation of my *Hereditary Genius* has been published. It was translated by Dr Neurath of Vienna, aided by a Dr Lady of the same name*, presumably a near relation. Grace Moilliet comes to us early in August, and that is all that occurs to me to tell. I had a long morning in the Park to-day. Best loves to you all.

Ever affectionately, FRANCIS GALTON.

42, RUTLAND GATE, S.W. *July* 31, 1910.

MY DEAR MILLY, I do indeed sympathise with your having to forgo the long desired motor expedition to Tregeare. Perhaps it will come off after all when the bad colds are gone. An efficient ex-chief examiner of the Civil Service asked me whether I knew that the word "Whisky" appeared in the Bible? As I did not, he referred me to the 2nd chapter and 3rd verse of Hezekiah. Is it new to you, to Amy, to Guy, and to Hugh? When you write again, tell me the result; ask them *separately*. Our move approaches. In little more than a fortnight's time we ought to be settled at Haslemere, or rather Grayshott, which is in a different county, viz. in Hampshire. Haslemere is very near to the border of Surrey and Grayshott is over it. My cough has departed and now I am at "my usual" again, and have recommenced daily trundles in Kensington Gardens. Three Butler nephews have taken their several leaves, one to be a guest of a wealthy admirer in Canada (or ? the U.S.A.) for his holiday. Another for three weeks in the Auvergne country, and who will report on the state of dear Louisa's grave. Another, Ralph, to be assistant to the *Times* correspondent in Berlin, and I should add Harcourt, who will soon return to India, as a newly elected member of the Viceroy's Council. He has been made much of by people in high places and among others was struck by the keen interest of the King in Indian affairs. Howard Galton and his wife have just had tea here. He sticks to his story that the very oldest will in Somerset House is that of a Galton progenitor and that the wills of all his descendants are there in unbroken succession. The history of the first part of this is that the wills were registered at Old Sarum and that the earlier collection was transferred in recent years to Somerset House. There are hundreds of letters preserved at Hadzor that relate to the Galton family, who were not Quakers till after they settled in Birmingham†. Ever affectionately, FRANCIS GALTON.

42, RUTLAND GATE, S.W. *August* 7, 1910.

MY DEAR MILLY, Did any of you really look in the Bible for "Hezekiah"? I hope they did!! That is the fun of the thing. I should not have attempted such a "sell" without the prefix of the "Chief Examiner," which is literally true. It was G. G. Butler. *Gladstone* could not see the fun of the story about the Austrian Archduchess and her successful lottery ticket of No. 28. She had been inquiring everywhere for the holder of the then sold No. 28 and bought it back at much cost, and, lo and behold, No. 28 did really win the big prize. Her friends clustered round, begging to know her secret of divination. At last she said: "Well, I will tell you. I had dreamt it was 9, but said to myself 'a mere dream is nothing.' The next night, I again dreamt it was 9. Then I thought, this is becoming serious.—What will happen on the third night? It was 9 again. So I said 'as 3 *times* 9 *is* 28, that must be the lucky number; and so it was.'" Gladstone, who had no more fun in him than an average Scotchman, simply

* The title of the translation is *Genie und Vererbung* von Francis Galton, Autorisierte Übersetzung von Dr Otto Neurath und Dr Anna Schapire-Neurath. It was issued in 1910 by Werner Klinkhardt in his Vienna *Philosophisch-soziologische Bücherei* as Bd. XIX. The translators inscribed the copy sent to Francis Galton: "Dem Meister der Eugenik in Verehrung." O. Neurath u. A. Schapire-Neurath, Wien im Juli, 1910.

† I think this should at least read, "till after they settled in Bristol."

stared and said "but 3 times 9 is 27," which caused roars of laughter among the company, one of whom told me the story, which Gladstone himself was quite incapable of appreciating. Eva goes to Grayshott on Tuesday to lunch with my hostess and to learn about sundry details. My wonderful boy Jew, Laski by name, came here with his brother to tea. Eva was out, but Miss Savile fortunately called and did the necessary. The boy is simply beautiful. She is an artist and quite agreed. He is perfectly nice and quiet in his manners. Many prodigies fail, but this one seems to have stamina and purpose, and is not excitable, so he ought to make a mark. The two boys are grandsons of a famous Russian Rabbi, a mystic and a great Kabbâlist. They told me much about the Kabbâla; how only the initiated in it know the proper pronunciation of Jahveh. I told them about Professor Robertson Smith, who knew it, and on pronouncing it before a great Rabbi visitor at Cambridge was cursed by him from the crown of his head to the soles of his feet, and withered and died within three months!

Ever affectionately, FRANCIS GALTON.

42, RUTLAND GATE, S.W. Sunday, *August* 15?, 1910.

MY DEAR MILLY, I write full early, as to-morrow will be a busy time, though indeed I do next to nothing except saying "yes" or "no" when asked whether or no a particular thing is to be taken. Did you happen to read in one of Lord Morley's recent speeches that he looked upon having to say "yes" or "no" as the hardest part of his duties? In my case, I leave things very much to Eva, who works hard for me. I have been below par last week through the forceps of a dentist. The tooth had done good service for eighty-one years, so it was a moral as well as a physical shock to lose it. I am just going out for the first time for many days in my bath-chair into the park; among other things, to look at the four big beds of Galtonias by the Albert Memorial. They were beginning to flower when I saw them last a week ago. We start on Tuesday afternoon by motor for The Court, Grayshott, Haslemere, Surrey, which according to the Post Office Guide is the correct address. Please when you next write, send to that address.

The past week has been, as you have recollected, one of sad memories to me. One of the young Butlers is now in Brittany for a tour in Auvergne, and will visit the cemetery and report. The death was in 1897, thirteen years ago! I had a miserable week after, sorting out dear Louisa's trinkets, etc., but all her family were most helpful and affectionate. Harcourt Butler came two days ago to say good-bye. He starts in a day or two for India. He is given the control of education and sanitation, with his seat on the Legislative Council. It is a five-year appointment, that which he vacated was more important, but it was, I believe, terminable with the tenure of the Vice-Royalty. So he gains in one important way. He had 150 persons on his staff!! He is fairly satisfied with the interest and knowledge of influential persons about Indian matters. The King especially was keen and full of memories. Lord Morley seemed rather despotic, he thought. What an amusing story about your small American grandchild and the galloping pony! Best loves to you all. Ever affectionately, FRANCIS GALTON.

I have seen the Galtonias, which are good in a way, but warm rain and sun are wanted to plump them out.

(Post-card.) THE COURT, GRAYSHOTT, HASLEMERE. *August* 23, 1910.

MY DEAR MILLY, I am so asthmatic that you must excuse this card. We are in a beautiful and spacious house, high up, but for all that my asthma has been bad. I was in bed all day on Wednesday, and again all yesterday and some other half-days. But every now and then, it suddenly goes and I breathe freely. Some kind friends came on Friday and I was able to enjoy tea with them. One of the visitors was Captain Lyons, F.R.S., the retiring Surveyor-General of Egypt, who is full of interesting information. His work is reputed to be of the most thorough order. The plan of my novelette has been often altered, but is, I think, approaching its final form. I hope that Guy's fever is passing off. With best loves.

Ever affectionately, FRANCIS GALTON.

THE COURT, GRAYSHOTT, HASLEMERE. *August* 26, 1910.

DEAR LASKI, Were I to rewrite now the extract which you quote I should alter it. The exceptions I had then in mind were the large families of many conspicuous personages. Thus Maria Theresa had sixteen children. Of those of very modern times the Kaiser has a large family, and so on. The question might be usefully discussed by comparing the size of the families of newly created peers with those of other persons of the same date. The data for this are easy to get. A great error, for which I am partly to blame, has been in laying too much stress on breeding from the *very highest.* If the matter were so simple as to be reducible to this form :—such and such a sum is available to induce persons to marry, is it best policy to spend it on the few very best or to distribute it more widely? In that simple case the former of the two alternatives would be best. But the case to be dealt with is different. It is largely a question of social approbation or the reverse. I am thinking of writing on this subject and am getting the plan of what I want to say in a clear shape, before beginning. Anyhow, it would be excellent eugenic policy to favour the marriage of those who are somewhat but not necessarily much more likely than the rest to produce capable citizens. The average level might thus be raised a grade or two, with little difficulty, and sports from that level, two or three grades higher than it, would be common and would produce very able men. Whereas an equally high deviation from the lower level would be very rare. (By "grade" I have the Probable Error in mind.) Excuse more; I have had rather bad asthmatic troubles since arriving here, but am "at my usual" to-day, or nearly so. Very faithfully yours, FRANCIS GALTON.

THE COURT, GRAYSHOTT, HASLEMERE. *August* 31, 1910.

MY DEAR MILLY, Again I am late in writing and cannot excuse myself, especially as I am materially better (for the while), having thrown off I know not what, but anyhow a sense of illness and much asthma. We are house-hunting for the winter and know of likely ones, but have not yet seen all. I go to one in a quarter-of-an-hour with my nurse and a light carrying-chair on the box of a victoria, Eva and myself inside. I trundled in my bath-chair to one this morning, which had in its garden "a grove" of Galtonias. We are 800 feet high hereabouts. Edward Wheler was with us last Saturday–Sunday, looking very well. I am simply without news or anything of interest to tell you. I expect that Bob Butler will be at Clermont-Ferrand to-day. He will report to me about the grave*. He is a very nice young fellow, working at architecture, a son of Professor Stanley Butler of St Andrews, who is a nephew of Louisa, being a son of her brother George and of Josephine. He has been *via* Brittany and Angoulême, looking at wonderful architecture. His first stage from England was to *Chartres*, the finest, some say, of all cathedrals. We saw two old Bordighera friends yesterday and heard much of it from them. But this is all dull to you and, besides, it is written very badly, so I will stop short. With best loves to you all. Ever affectionately, FRANCIS GALTON.

THE COURT, GRAYSHOTT, HASLEMERE. *September* 6, 1910.

MY DEAR MILLY, I am late again in writing. The Frank Butlers and their last baby were here for the week-end, and there are *two* tempting houses to be had. I was to have gone to decide this very morning, but Eva is shut up in bed with a chill and I am rather afraid of the sunless cold of the day. I wish you had told me more of your impressions of Loxton, which I shall never see again. I often think of Erasmus, whose sterling qualities came out strongest towards the last. I have heard very favourably about the grave at Clermont-Ferrand, from Bob Butler, whose real Xtian name I doubt, but he "answers" to Bob. He will be here next week-end and will tell me more, but anyhow the grave is well attended to by the gardener there whom I pay for doing it, and the rose bushes by it are described as very pretty. A lady resident at Clermont-Ferrand, who taught French to Eva and who came frequently and with whom we have corresponded, looks after it from time to time. I have been occasionally not over well and done nothing, but as soon as this scribble is finished I shall begin again upon *Kantsaywhere.* How do your birds thrive? I was touched by the confidence of a *wren* here, who hopped about my feet while I was seated in the garden in front of a dense hedge. She popped in and out but brought none of her belongings with her. Oh—this horrid coming winter! I am about 750 feet above sea-level and one of the two houses is about 50 feet higher and the other as much lower, but they are well sheltered and look due South. Best loves.

Ever very affectionately, FRANCIS GALTON.

* That of his wife, Louisa Galton (née Butler).

THE COURT, GRAYSHOTT, HASLEMERE. *September* 19, 1910.

MY DEAR MILLY, Your sick list is sad. I hope Guy got comfortably to Loxton. Tell me the latest news about the well. Did you happen to see in the Obituary, or hear from Claverdon, of their sad adventure—of their friend, Mr Aylmer, *dying* at their house? I know no particulars yet. So Hugh is to be a curate in Exeter, good fortune to him! Devonshire air is certainly relaxing to most strangers, but what tough and hardy men have come from there! And Hugh is quasi-Devonshire in origin, having been reared just over the border. Beak's* wife is a slow case, but she is working towards the good, though it appears that her arm is permanently crippled. She is tortured by having it twisted and stretched weekly to prevent, I suppose, adhesions, and goes to the Hospital for the purpose. Bob Butler has been with us for the week-end and told us much about his tour besides Clermont-Ferrand. He is very observant and is already an advanced student in architecture. The country West and South-West of Auvergne has many architectural interests and some imposing situations, of which Rocamadour is one. It is built *against* a steep cliff, near its top. He and Eva went to London early this morning. She to inspect doings at Rutland Gate and to bring some things back with her. A rather pretty girl called here with her people, she wore a big hat at which I exclaimed saying it was as long round as she was tall. She wouldn't believe it, but it appears that on going home the measurements were made and I was right as she confessed yesterday, when she called with a still more absurd hat than before on her head. It was not so large but shaped like a cask. What absurdities abound just now. The *Times* will begin a weekly supplement of ladies' fashions in November, I believe.

Front

My old friend, General Sir Richard Strachey, was a famous Engineer Officer, and once upon a time when in a very out-of-the-way part of India his wife suddenly required a smart dress for some important function. He rose to the occasion and drew a pattern for it with the same care that he would have taken over the plan of a fortress and the gown proved a success. At least *he* said so. With loves to you all. Ever affectionately, FRANCIS GALTON.

It will be nice to have you both here, first Dim and subsequently you. Has the motor yet arrived? Of course you can do little with it while the gardener is ill.

THE COURT, GRAYSHOTT, HASLEMERE. *September* 28, 1910.

MY DEAR MILLY, Again I am late and without valid excuse! Beak is just off to fetch his wife here for a week or fortnight's fresh air and good food. We are going on as usual. What a tragedy it was at Claverdon. I hope M. L. does not suffer really much from her rheumatism, but the Harrogate treatment seems drastic. At the temperature of her air baths not only would a cup of water boil in her hands but a joint would roast and turn brown. I grieve at Lucy Cameron Galton's bad carbuncle. They ought to go to Bordighera in less than a month, but an actual or possible carbuncle is a bad travelling companion, and one never knows when the risk of a new one is over. I had a pleasant letter this morning from Mary Spencer Butler. Her son (the lame one), Geoffrey, has been having a lovely time in America. The equivalent there to our "Leader of the Parliamentary Bar" wanted a young Englishman as a guest, and hearing of Geoffrey and that he was looking out for a vacation pupil, invited him to his house and has taken care of him in every way, introducing him right and left and having him at a camp, where they live the so-called "simple life." The only ill-luck he had was through a misdirected letter from Roosevelt, asking him to stay a day or two with him. It was directed to Geoffrey *Baker* instead of Butler. This from M. L. to finish up with: *Scene*, Breakfast table, a small boy and his nurse who is reading the *Christian Herald*. *Boy*: "Nanny, I don't like this egg." *Nurse*, without looking up: "Be a good child and eat it." *Boy*, after a while: "May I leave half of it?" *Nurse*: "No, be good, eat it all up." *Boy*, after another pause: "Nanny, must I eat the *beak*?" Best loves, affectionately, FRANCIS GALTON.

THE COURT, GRAYSHOTT, HASLEMERE. *October* 3, 1910.

MY DEAR MILLY, Your account of Olympe† is pityful. I am vegetating on pretty happily, only vegetables don't cough in spasms and require cigarettes either of haschisch or of stramo-

* Galton's valet-nurse.

† Olympe Chapuis, a little Swiss girl who came as companion to Millicent Bunbury. She eventually married the Rev. T. K. Lethbridge.

nium to allay them, as I do. Our few peaches have just come to an end, but hope remains that six nectarines, still on their tree, will consent to ripen. We had an orange-coloured *turnip* for lunch to-day. I had never seen one before. It tasted just like a white one.

How is Bob getting on in Wall Street? Does he dream of millions? Geoffrey Butler is expected back this week from America with plenty to tell. His host took every care of him and he saw many people......I weary myself with devising a workable constitution for *Kantsawa*. After writing ever so much I find over and over again that some arrangement won't work rightly and everything has to be altered. I live quite as much in *Kantsawa* as I do in Haslemere. I go there continually, as on a villeggiatura by a suburban train. But I find it changed at every successive visit, and demolished quarters have to be replanned and rebuilt. It is very cheap to build castles in the air. With all loves. Ever affectionately, FRANCIS GALTON.

THE COURT, GRAYSHOTT, HASLEMERE. *October* 5, 1910.

MY DEAR LEONARD DARWIN, I can't help in solving your question. The answer must greatly depend on *where* the people live and *how*. In many villages, notably Scotch sea-shore ones, the fisherfolk never marry outside their immediate neighbourhood. In such an extreme case the number of their forefathers, any number of generations back, would hardly exceed that of the present villagers. On the other hand, a migratory population might have greatly intermarried with outsiders.

PROBLEM. Noah and his wife have an increasing number of descendants during n generations; find the rth generation in which the number of ancestors is largest. (Assume the problem in its utmost simplicity of every 100 persons becoming $100 + a$ in the next generation)—the figure is something of this sort. I worked it out once, but forget the result, except that r was *not* $n/2$.

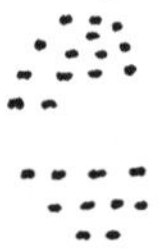

We are settled in the Hindhead district for the winter, in the above house till Nov. 15, and then in another close by. I pull on—sometimes rather badly, often rather well, but very infirm always, and am wheeled about and carried up and down stairs. But I have nothing to complain of. I sleep like Morpheus and enjoy a chastened dietary, and have had my day.

I hear, from time to time, personal and scientific news from men like Sir A. Geikie, who lives within a distant reach and there are many nice people about. My niece takes excellent care of me. The village is not far off where the following occurred—told me by the Vicar's son:

Vicar. "Why, Mary, is the old woman dead at last? she seemed to me fairly well yesterday."

Mary. "Yes, sir. Her cough had been bad and noisy at nights for long, and Jim said to me last night, 'I can't abide that cough; get up, Mary, and put the pillow on your Mother!' So I got up and put the pillow on her, and she was that weak, her spirit flew away like a bird."

Bhang—(Haschisch)—in cigarettes is, I find, a great solace in fits of bad asthma and cough. How good the photographs are in the Royal Geographical *Journal*. You will be hard at work soon with those stirring people. Remember me please most kindly to your wife and two brothers now with you. Ever sincerely yours, FRANCIS GALTON.

THE COURT, GRAYSHOTT, HASLEMERE. *October* 8, 1910.

MY DEAR MILLY, Dim left us this morning perfectly well. I wish she could have stayed longer. I had a "private" note last week from the President of the Royal Society to say that the Council had awarded me the Copley Medal, but that it was not to be publicly announced yet; it is, however, in the newspapers to-day! It is the "blue ribbon" of the scientific world, and I am of course deeply gratified. One is awarded annually, without distinction of *nationality* or of *time* when the scientific work was done, whether lately or some years back. As a fact, an Englishman gets the Medal not quite so often as once in two years. About five other living Englishmen have it. People are always very kind to me, but I wish my Father and Emma were alive. It would have given them real pleasure. We move to the new house in a month's time. Eva is off for the day to her old friend the blind Mrs A'Court.

Ever affectionately, FRANCIS GALTON.

What a nuisance about your motor! What a shameful blot!!

THE COURT, GRAYSHOTT, HASLEMERE. *October* 16, 1910.

MY DEAR MILLY, I am glad that you are safe home, though it be to a wind-wrecked garden, after so long and adventurous a journey. Did the motor go well? You had, I think, not a little beautiful weather. Lynton I know well, having spent a summer there with Louisa. That wonderful river teems with salmon. They pointed out a pool in it not much bigger than my drawing-room, so far as I recollect, out of which sixteen had been caught in the preceding year. They called it the "Slaughter Pool." It is wonderfully beautiful thereabouts, as you say. I am glad you liked the improvements at Tregeare. When such things *really* improve a well-known place how charming they are. This is the case with Claverdon, which I shall never see again! nor any of my old haunts, being so tied up by infirmity. Asthma comes and goes and I have frequent long respites, but it is always *en cache*, ready to spring. My *Kantsaywhere* gets on slowly, but I think surely at last. I want the Abbé Sièyes (spelling?)* to put its constitution into the best shape. I find it difficult to evolve a stable one out of nothing. Are your birds beginning to migrate? Ours have gone through various pranks. They seemed to be pairing again, mistaking the fine autumn for next spring. As I write, four of the plumpest of blackbirds are hopping in front of my window. I think it would be nice if they had been 24 and if the cook could make them up in a pie, at the same time putting a complete stop to their singing! Poor Portugal! The making of a respectable nation out there seems as difficult as that of a silk purse out of a sow's ear. But there *may* be good stuff left in their most ignorant peasants, though very little in their politicians. I have a Press cutting from New Zealand from which it appears that the Chief Justice, Sir Robert Stout, in his charge to the Grand Jury at the Autumn Sessions in Auckland, advocated the formation of Eugenics Societies, and that much is being done in that way there. I have forwarded the cutting to the Eugenics Education Society for them to keep the ball rolling.

Ever affectionately, with loves to all, FRANCIS GALTON.

THE COURT, GRAYSHOTT, HASLEMERE. *October* 19, 1910.

MY DEAR EVA, Your news of the acting is most gratifying. I saw in yesterday's paper that there were to be four Rosalinds this week and two of each of the male actors. The purpose is of course to test their relative merits.

Here is another gratifying incident. A long letter from Lionel Robinson describing how he and his wife had just attended service at St Margaret's, Westminster, where he heard Professor Inge preach on Eugenics. You know that L. Robinson is far from a gushing man but he fairly gushed over the sermon. He says it was "not only bold and eloquent but carried reflection if not conviction to every listener." Though he "had heard him on a previous occasion he never seemed so clear and attractive as then. He made no shifty evasions and spoke with startling clearness on many points which preachers as a rule evade or dilute beyond taste. My wife was delighted with the sermon," etc., etc. You shall see his letter. I will write to him.

Dakyns came yesterday and asked after you. He will gladly revise the MS. of *Kantsawah* which in a provisional sense has been finished. But I must interpolate some pages before sending it to the typist. Thus far, it would fill 17 pages of a magazine like the *Nineteenth Century*.

It is now sunny but uncertain. Whether it be fit for me to do more than trundle will be doubtful until after lunch. Yes, if it can easily be managed by the servants, ask Mrs Stanley Butler and Bob for the night as you propose. I am so glad you approve of Louisa and Mrs Phillips. Your bottle of Chianti was very characteristic!

Ever affectionately, FRANCIS GALTON.

Miss Jones is excellent, gives no trouble and writes from my dictation with almost short-hand swiftness.

* Galton is clearly referring to Comte E. J. Sieyès, the man trained as a Cleric, but he was one who had never preached nor confessed. He was the great constitution framer at the time of the French Revolution and after.

THE COURT, GRAYSHOTT, HASLEMERE. *October* 23, 1910.

MY DEAR MILLY, Eva returned yesterday and sends her love. We shall be so glad when the time arrives to see Dim. I have at last finished *Kantsaywhere*, if that be the best orthography. Miss Jones has been staying here during Eva's absence and copied it in fair writing. But I must keep it by for a little while and add and alter before sending it to be typed. Then I must ask you and other friends to kindly read and criticise. It would now fill about 20 to 30 pages of *Nineteenth Century* size and type. I have no news except of an invalid sort, so will not bother you with that. I quite agree with you that the re-visiting places one has known well is usually disappointing. The personal element has changed, and that counts for much more than one had anticipated. The feeling that a once familiar place "knows one no more" is disheartening. I am glad your motor went so well and falsified the grim remarks of the natives about the character of the roadway in front. I wonder whether Switzerland is all round such a good place as many say it is. The owner of the house we have rented for the winter is Swiss, named Le Pury, a very nice man. He married a daughter of Mr Whitaker, the big vine-grower of Marsala, in Sicily. He too has a big house whose grounds adjoin Le Pury's. Good-bye, loves all round. Ever affectionately, FRANCIS GALTON.

AMERICAN BREEDERS' ASSOCIATION—EUGENICS SECTION. EUGENICS RECORD OFFICE, COLD SPRING HARBOR, LONG ISLAND, N.Y. *October* 26, 1910.

MY DEAR GALTON, Your post-card of Oct. 14 just received. I thank you for taking the trouble to reply. You must think me a nuisance to add thus even a letter to your correspondence. But I must tell you of recent events here. As the enclosed printed matter will show in some detail, there has been started here a Record Office in *Eugenics*; so you see the seed sown by you is still sprouting in distant countries. And there is great interest in Eugenics in America, I can assure you.

We have a plot of ground of 80 acres, near New York City, and a house with a fireproof addition for our records. We have a Superintendent, a stenographer and two helpers, besides six trained field-workers. These are all associated with the Station for Experimental Evolution, which supplies experimental evidence of the methods of heredity. We have a satisfactory income for a beginning and have established very cordial relations with institutions for imbeciles, epileptics, insane and criminals. We are studying communities with high consanguinity also. Altogether the work is developing in a satisfactory and interesting manner. We have thought that, though our work is mostly in "negative eugenics," we should put ourselves in a position to give positive advice. We cannot urge all persons with a defect not to marry, for that would imply *most* people, I imagine, but we hope to be able to say, "despite your defect you can have sound offspring if you will marry thus-and-so."

I want to tell you how much I have enjoyed reading your autobiography. You have quite put yourself into it, and that makes it much more valuable than any "Life" by another hand. It would please you to realise how universal is the recognition in this country of your position as the founder of the Science of Eugenics. And, I think, as the years go by, humanity will more and more appreciate its debt to you. In this country we have run "charity" mad. Now, a revulsion of feeling is coming about, and people are turning to your teaching. With best wishes for continued strength and health, and with the expression of my profound esteem.

Yours faithfully, CHAS. B. DAVENPORT.

THE COURT, GRAYSHOTT, HASLEMERE. *November* 1, 1910.

MY DEAR MILLY, This will, I suppose, reach you about the time when Dim starts. We shall welcome her with all pleasure. Winter is now at the door. Gifi is for the moment in London looking out the warm winter clothing and we migrate to "Grayshott House" on the 15th. That beastly *Kantsawhere* (or whatever its name is to be) has been delayed by a $2\frac{1}{2}$ days illness of mine in bed. I am quite "at my usual," however, to-day. Some *talkers* knocked me over. It is odd how invariably one of these asthmas follows *any* form of fatigue, mental, vocal or otherwise bodily. The Doctor here, Lyndon, is as capable a man as could be found anywhere, which is a comfort. Do you care for the present Poet Laureate, A. Austin? Few do. This story against him was told me yesterday by the President of the Royal Society (Sir A. Geikie). It appears that the Scotch Judge, Lord Young, was noted for his sharp sayings. He met

Austin and said, "Well, have you been writing any more poems?" Austin replied, "Yes, a little; you know one must keep the wolves from the door." "How did you do it?" said Lord Young, "Was it by reading your poems to 'em?"!! What a contrast between Austin and his predecessor Tennyson. Ever affectionately, FRANCIS GALTON.

42, RUTLAND GATE, S.W. *November* 5, 1910.

MY DEAR MELDOLA, Excuse my not writing with my own hand. I am very glad that you are going to do justice to Herbert Spencer as an investigator, but I cannot help you with facts about it. I know of course about his experiments on the effect of wind on the upthrow of sap, but do not know where the account of the experiments is published. I feel myself unable to help you, as I wish I could. As regards his influence on contemporary science I feel it is small; on my own work it has been nil, but Romanes ascribed the idea of his beautiful experiments on the formation of nerves on medusae wholly to Spencer's published views. What a sad scene it was at Golder's Green*.

I am very infirm and have taken a house for the winter near Hindhead.

Ever sincerely yours, FRANCIS GALTON.

GRAYSHOTT HOUSE, HASLEMERE. *December* 8, 1910.

MY DEAR MELDOLA, Best congratulations on D.Sc., or Sc.D., as the case may be. It has been long deferred.

The enclosed letter is one you may like to read, and will I think sympathise with. Poor Collins; he gave, I know, much assistance to H. Spencer in revising MS., but my knowledge of this is not accurate enough to warrant my writing an obituary paragraph about him. Possibly you might be able and inclined to do so. His death *was* noticed in *Nature*, Dec. 1st, p. 146, 1st column.

I wrote to Miss Killick and mentioned that I had forwarded her letter to you. Poor Collins. His life was tragical. Extraordinary physical powers, shown in his first attempt at Alpine climbing; then, arm-chair-ridden by pleurisy. Next, an unhappy event in which the *She* was not to blame. Then, frequent failures to do good intellectual work, all combined with the most unselfish and eager wish to help others, by revising and criticising, which *within limits* he could do well.

Requiescat in pace. If you see your way to writing a brief memorial paragraph to the *Times*, poor Miss Killick would rejoice.

Please send me back her letter. I am wintering here, being now far too infirm for London fogs, etc. Very faithfully yours, FRANCIS GALTON.

GRAYSHOTT HOUSE, HASLEMERE. *November* 13, 1910.
(This will henceforth be my address.)

MY DEAR MILLY, Thanks for congratulations. Sir George Darwin will receive the medal for me on the 30th. People are very kind about it. There are only five other Englishmen alive who have received it: (1) Sir Joseph Hooker (Botany), (2) Lord Lister (Antiseptics), (3) Lord Rayleigh (Mathematical Physics), (4) Sir William Crookes (Molecular Physics and Radiometry), (5) Alfred R. Wallace (Zoology and Darwinism). I have pretty nearly finished *Kantsaywhere* in typewriting; but shall lay it by when quite finished for yet further revision, and then have two typed clean copies for friends to criticise, you to be one of them of course. I have no news. The weather has turned chilly and, according to advice, I spent most of yesterday in bed. They say it will be good economy if I lie up one day per week, selecting a nasty day, as yesterday was, for the purpose. We move on Tuesday. Love to those of your party who are at home. Guy wrote me a letter of congratulations from which I gather he is now at Claverdon, whence two fat pheasants have reached me. One is just eaten—so good!

Ever affectionately, FRANCIS GALTON.

* Galton's account of Herbert Spencer's cremation, which would have had historical interest, seems to have perished. I have added at the end of this Chapter his reminiscences of Spencer, found in rough draft among his papers: see p. 626.

GRAYSHOTT HOUSE, HASLEMERE, SURREY. *November* 29, 1910.

MY DEAR MILLY, There is nothing to say, the ups and downs of invalidism interest none except the persons concerned. I have been down and am up again, "le vieux (mieux) persiste." To-morrow the Copley Medal is given. The papers are too full of politics for anything about the Royal Society to be inserted beyond bare facts. It is very nice of Sir George Darwin to receive it for me. This morning's post brought the neatly typed *Kantsaywhere* revised and done up in book form. Methuen comes here if he can on Sunday or Monday afternoon, so I must keep it for him. This is, I expect, just the most awkward time for new publications, politics and Xmas both in front. I shall soon know more about all this. The Edward Whelers come here for two or three days about Xmas, and I shall hear much I hope then about Claverdon and Loxton. This house proves quite a success, but I have been very little out of doors, not at all of late.

Excuse if you can this extra dull letter and believe me all the same,

Ever affectionately yours, FRANCIS GALTON.

HIGHER COMBE, HASLEMERE, SURREY. *December* 2, 1910.

MY DEAR SIR FRANCIS, I very much enjoyed Professor Donoghue's account of *Kantsaywhere* yesterday. I like the additions, particularly about the resemblance of the young women to Guido's Hours. Wouldn't a reproduction of the engraving or a photogravure of the picture make a pretty frontispiece to the book? I hope you didn't bother over my minute criticisms. I can't recall the particular sentence verbatim where I boggled over the grammatical form, the sense being plain enough. "Its absence etc...." I wonder if "Its freedom from..." or "immunity from..." would please me better. I am looking forward to my next visit, but I find on Tuesday, Wednesday and Thursday I have to be in London. So it must be Monday?? or Friday? Perhaps Miss Biggs would kindly let me know if Friday suits.

Yours affectionately, H. GRAHAM DAKYNS.

Or that failing—Monday. You will have seen Methuen by that time.

GRAYSHOTT HOUSE, HASLEMERE, SURREY. *December* 6, 1910.

MY DEAR MILLY, Thanks many for all you say. The President of the Royal Society had tea here yesterday. I don't think I told you his last story, viz. that at the recent University celebrations in Liverpool, he stayed at Knowsley (Lord Derby) where Lord Morley and Lord Rosebery also were. He overheard this bit of conversation between them. *Rosebery*: Do you play at cards? *Morley*: No, it has never been my taste. *Rosebery*: But your Cabinet is keen upon a game, namely "Beggar your neighbour." Lord Morley tried to reply, but could not find a rejoinder. What a storm in politics. There must arrive a time for compromise. If so, I hope they will combine to diminish the Irish vote. The late Liberal candidate for this place, Methuen the publisher, had tea here two days ago. He asserts that Lloyd George has an extraordinary charm of manner in conversation, and that Lord ——, at the Conference, who is a stubborn Tory and hates him politically like Satan, was quite won over by him socially after three meetings. I doubt whether Methuen will take *Kantsaywhere*. I showed it to him and asked him to submit it to his reader, which he said he would do and ultimately marched off with it, but at first sight he was very dubious. He takes no interest in Eugenics. I have not ventured out of doors for a whole fortnight and crave somewhat for fresh air. Much love to you all.

Ever affectionately, FRANCIS GALTON.

Eva is in bed, recovering fast from a sudden chill (of no real consequence).

GRAYSHOTT HOUSE, HASLEMERE, SURREY. *December* 11, 1910.

MY DEAR MILLY, Methuen came to tea on Monday and took *Kantsaywhere* with him to submit it to his reader. He was not at all taken with the idea at first sight and may more likely than not decline it. We shall shortly see. What an inconclusive pother this election has stirred up. I wish both parties would agree to dock off the disproportionate number of Irish electors. They are a nuisance to both sides in turn. Nothing has occurred this week worth telling. I have not ventured out of doors for nearly a month. It *will* be a month next Tuesday. The Doctor inspects me and gossips once a week. I am grievously distressed at the tragic death of Sir Archibald Geikie's only son. I am glad the Coroner returned "accidental death." I never saw him nor any of his three gifted sisters nor his mother (who is mostly unwell), who live in Haslemere, though his father kindly comes pretty often and tells me scientific news. I write supposing you have seen what I am writing about in the newspapers. What a strange commercial world it now is. The defeated candidate here on the Liberal side is Davy, a son of the late great lawyer, Lord Davy, whom I used to know well. His income is derived from an agency to an American firm for making bandoliers for soldiers, on a principle that is partly patented, partly secret. The cylinders that hold the cartridges are woven together with the belt and so cannot come off. The firm has acquired the monopoly of supply, not only to the American and to our Armies, but also to most of the big continental nations. Each belt costs 2*s.* to make and is sold for £1. This was Methuen's story to me. He is a keen business man and stood as the Liberal candidate here last year, just as Davy is doing now; so probably his story is quite correct. Good bye, it is too early yet to send Xmas wishes; they shall go later. Eva, who has been shut up during the last week, is off to her Catholic chapel. She looks quite well, but the K is a little painful at times.

Ever affectionately, with loves to you all, FRANCIS GALTON.

GRAYSHOTT HOUSE, HASLEMERE, SURREY. *December* 20, 1910.

DEAR MR PERRY COSTE, Excuse writing by dictation. I return your very interesting account of your boy's colour-faculty. A very great deal has been written on the subject, especially in America, so it is not surprising that *Nature* is not able to find room for it, but I hope it will ultimately be utilised in some other way. You do not mention *how* the boy sees the colours, are they in coloured figures? or are they black or white figures on a coloured ground? or is there no figure at all, but a sense of colour conveyed by the sound alone? I think you ought to get this clear, even at the cost of making the boy somewhat introspective. There *must* be some explanation of the reason why particular colours adhere so readily to the several figures, but the whole thing is very mysterious as yet.

I hope you are all well at Polperro. I am grown infirm and have to winter out of town.

I hear from Karl Pearson frequently, and hope to have him here shortly for one night; he has done a truly great work. Very sincerely yours, FRANCIS GALTON.

GRAYSHOTT HOUSE, HASLEMERE, SURREY. *December* 28, 1910.

MY DEAR MILLY, (What a blot!) You and Guy more especially must have had a wretched time of floods and tempests. We on the high ground feel like Noah on Ararat. Edward Wheler left us yesterday for a night at Loxton and M. L. leaves us to-day. The glorious frosty sunshine of this morning picks me up. I have been "throaty" and obliged to rest a good deal. Karl Pearson comes this afternoon for one night. I am saving my voice for him. *Kantsaywhere* must be smothered or be superseded. It has been an amusement and has cleared my thoughts to write it. So now let it go to "Wont-say-where." My very best New Year wishes to all of you and best love. Ever affectionately yours, FRANCIS GALTON.

GRAYSHOTT HOUSE, HASLEMERE, SURREY. *January* 2, 1911.

DEAR MISS ELDERTON, First—my best new year wishes to you, to Dr Heron and to Miss Barrington, with many thanks for your joint Christmas Greeting. Enclosed I return both *The Child* and the New Zealand papers. It is gratifying that Eugenics has taken so strong a hold there. Professor Pearson will probably have mentioned that he has been with me. It gave me great pleasure to see him apparently not at all jaded by his hard work. I have read and shall re-read the recent Eugenics publications, full of hard and conscientious work.

Very faithfully yours, FRANCIS GALTON.

GROSS-LICHTERFELDE-WEST (WANNSEEBAHN), DRAKESTR. 37. den 6 *Januar*, 1911.

HOCHVEREHRTER SIR FRANCIS! Nach längeren Bemühungen ist es mir gelungen, eine Photographie des Otjikoto Sees im Damaraland zu erhalten. Ich sende Ihnen anbei zwei Photos der Stelle ein, auf denen Ihr Name eingemeisselt ist. Ich verdanke die Photos der Freundlichkeit des Herrn Tönnesen, des Direktors der South West Africa Co. Auf dem einen Bilde steht Herr Tönnesen und zeigt nach der Stelle hin, auf dem anderen Bilde ist der

PLATE LIX

The Rock on the Lake at Otchikoto, Ovampoland, from photographs of Herr Tönnesen. The upper picture shows the rock where Galton carved his name and below other names added since. The lower picture shows Herr Tönnesen pointing to the rock to which Galton swam out in 1851.

Name links deutlich zu sehen "F. Galton." Auffallend ist, dass der Wasserspiege um etwa 5 oder 7 Meter gestiegen ist. Vor 10 Jahren war er erheblich niedriger. Ich weiss nicht, wie Sie ihn Ihrer Zeit gefunden haben. Ich hoffe, dass Ihnen diese Erinnerung am Damaraland eine kleine Freude machen wird. Ich erinnere mich mit grosser Freude an den Mittag, den ich im Sommer 1909 in Ihrem Hause verleben durfte. Indem ich bitte mich Ihrem Fräulein Nichte empfehlen zu wollen, bin ich, mit ausgezeichneter Hochachtung, Ihr sehr ergebener, PH. KUHN.

GRAYSHOTT HOUSE, HASLEMERE, SURREY. *January* 9, 1911.

MY DEAR MILLY, I am not sure when I last wrote—possibly quite lately, for I was then writing many letters. If so, excuse repetition of *nothing*. You certainly have the art of attracting and taming birds. I can't induce them to come, when I try. But we have now put up coconuts for the tits, and we drink the fluid in them ourselves. Violet Galton is with us for the week. I hope Guy was none the worse for his long trudge in the flooded way, on returning from Loxton. I wish I had something interesting to tell you, but have nothing to say more, beyond affectionate wishes to you all, individually. FRANCIS GALTON.

[This is the last letter which I know of in Francis Galton's handwriting. ED.]

GRAYSHOTT, HASLEMERE. *January* 16, 1911.

DEAR MILLY, I am *so* sorry to hear of your illness now, and do hope you will pick up soon. I am thankful Dim is better.

I am sorry to have no good news. Dr Lyndon considered Uncle Frank worse this afternoon—his breath is so difficult to get, he is in great discomfort and very weak, but *so* sweet and cheerful, always saying something witty if he can speak a few words. Will write again to-morrow. Edward is such a comfort to me and to him also. Your loving EVA.

THE ATHENAEUM, PALL MALL, S.W. *January* 19, 1911.

DEAR MISS BIGGS, I am grieved to see the announcement in this morning's papers and send you truest sympathy. Sir Francis has been for so many years your charge and filled so large a part in your life that the loss of his presence, always so bright and kindly, will be a sore bereavement. I trust you may be enabled to bear up under so heavy a sorrow. I sincerely regret not to have been able to get to Grayshott for weeks past and have missed seeing your Uncle. But my own tragic bereavement, the illness of my Wife and the urgent business connected with my Son's death have kept me busy and much in London. I am still tied down here by business which I cannot shirk, otherwise I would come up to Grayshott to see if I could be of any service to you. With my sincere sympathy.

Yours very truly, ARCH. GEIKIE.

THE ATHENAEUM, PALL MALL, S.W. *January* 19, 1911.

DEAR MISS BIGGS, The telegram which you so kindly sent me yesterday reached me after I had written to you this morning. At the meeting of the Royal Society to-day the news of the death of Sir Francis Galton was received with the deepest regret. He was I think our oldest and certainly one of our most distinguished fellows, and the feeling was expressed on all sides that it was well that the Society even at the last had recognised his genius by awarding to him its highest honour, the Copley Medal. I shall never cease to regret that I was unable to pay him a visit during these last few weeks. But from the kindly note I had from him I knew that he understood how I stood. The Royal Society desires to pay the last tribute of respect to its venerated colleague by being represented at his funeral, and I made the arrangements this afternoon. I sincerely regret that I shall be prevented from attending myself. With renewed sympathy. Yours very truly, ARCH. GEIKIE.

ST RADEGUND'S, CAMBRIDGE. *January* 21, 1911.

DEAR DARWIN, I feel I must write to some one to express my sincere regret at the loss of our dear and venerated old friend Francis Galton. I don't know his own people. Ripe as his years were—and I am sure he would have hated to live in any crippled state—yet so sturdy and keen was he that his death seems a surprise and a shock. I had not seen him for some months and he may have been failing. What a splendid life it has been; personal courage and adventure, admirable mental and bodily endowments, and a powerful intellectual grip upon the problems and work of his time. And with all this *no freaks*—sane, humane and sociable.

Ever yours, CLIFFORD ALLBUTT.

To SIR GEORGE DARWIN, K.C.B., F.R.S.

42, RUTLAND GATE, S.W. *January* 25, 1911.

DEAR PROFESSOR PEARSON, If you were thinking of giving little amusing incidents in Uncle Frank's life, I wonder if you would like to mention a neat dodge he had for seeing comfortably in a London crowd. He got a wooden brick with a hole in it through which he passed thick string, with a big knot at the bottom. This he carried under his arm*, and if a tiresome tall person stood before him, he would gently and slowly drop his brick and stand on it with one foot, and when it was time to go, draw it up again by its string, and no one noticed anything. Also you know the "Hyperscope," I suppose, which he used for the same purpose. You put your eyes to the two holes and the matinée hat drops a few inches, and you see the lecturer quite clearly; the opposite side being arranged with a sloping looking-glass let in. He used this last, I think, when Queen Victoria came to open the Albert and Victoria Museum close by, and the whole of Brompton Road was crowded to see her pass by.

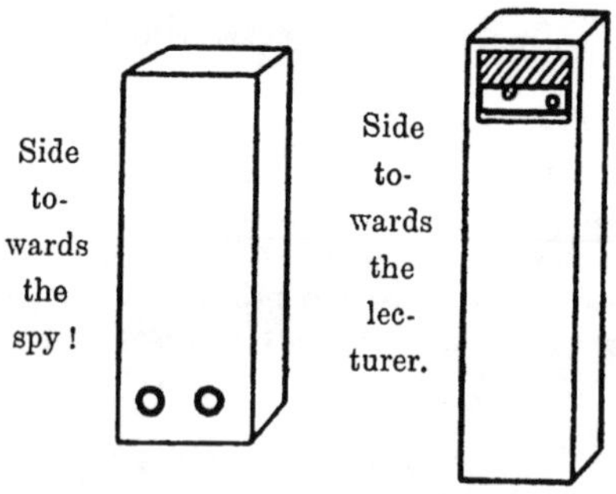

I wonder if you would mention his extraordinary good temper—it was quite a joke when he was a child, the boys at the school he went to used to stand round him in a ring trying to irritate him, but always failed. This was such an advantage in a household, as it made the servants love him; the Scotts at Bibury used to say they would like to work for him for love, because he was so delighted with every single thing they did for him, and yet they all had a reverence for him and no servant was ever impertinent. He was just like a child in his jokes and always said he was a tiny bit jealous of Wee Ling in the house! Another thing you might like to say is how extraordinarily keen he was about things, everything was so intensely interesting to him, any workman in a foreign country he would have a long talk with and ask how he did this, that and the other, and then tell the man how clever he was; he would then take a lesson himself from the man, or *child* as the case might be. Just before he died, when almost too feeble to speak, he was given a prick of strychnine in the wrist; this interested him intensely and though we didn't want him to exhaust himself talking, he wouldn't let the doctor alone without having it clearly explained what the strychnine would do for him. He was most excited about the oxygen they gave him and wanted Edward Wheler to tell Dr Lyndon all about *his* experiments with it—this an hour or so before death.

By the bye, it is a mistake to think, as some of the papers reported, that my uncle died in his sleep; he became unconscious about $\frac{3}{4}$ of an hour before death; Gifi looked in and Uncle Frank opened his eyes and smiled at him, and then never opened them again; he seemed in a sort of torpor. He looked so sweet and of such a good healthy colour after death, that I could not believe the doctor when he said the heart was not beating. I kept candles burning by him till the coffin was taken from the house and visited him continually in the nights to pray for his soul, and he was buried with my crucifix on his breast; he looked so sweet in his coffin with his own dear smile on his face, it was sad to leave him in that box, but he looked just like himself to the last......E. B.

42, RUTLAND GATE, S.W. *February* 26, 1911.

DEAR MR PERRY COSTE, Your kind letter has lain unanswered all this time simply because I have been so occupied, not because it was unappreciated. The sympathy of my Uncle's friends and admirers has been my great comfort.

How very curious that you should have been writing to him—you are indeed quite correct about his intellect, it was keen up to the day of his death, and when the doctor pricked

[* Done up as a brown paper parcel. ED.]

strychnine into his wrist, hoping to strengthen his heart which was rapidly failing, he would know all about it and exactly how it would act—this an hour before death. He died very peacefully and looked so natural and sweet after death, I could not believe he had gone. It is a terrible blow to Professor Pearson, who was his greatest friend. Thanking you for so kindly writing to me, and hoping all of you are well. Yours sincerely, L. E. BIGGS.

Extract from the Claverdon Parish Magazine.

The following has been written specially for this Magazine by the Rev. Dr H. M. Butler, Master of Trinity College, Cambridge:

Sir Francis Galton is best known to the public as an African traveller and a very eminent man of science. With certain branches of science his name is likely to be linked for all time to come as that of a leader and discoverer. In this short paper he will be sketched from another point of view by one who has known and loved him since the close of 1852. He was a man of singular sweetness of temper, courteous, considerate, prompt to sympathise in little things as well as great. He was a charming companion in any travelling excursion, at home or abroad, skilful in planning the various localities to be visited and the various stages, bright and resourceful in dealing with any incidents, imperturbable and amusing if any of these were of a troublesome or perplexing kind. In conversation he was keen, vigilant, always on the look out for something new or beautiful or wonderful. His interests were by no means only scientific. He had an intimate knowledge and an ever-fresh enjoyment of not a few of our greatest authors, among whom Shakespeare, Keats, Tennyson might be specially singled out, poets who had been among the favourites of the exceptionally able young men with whom he had lived during his happy days at Cambridge. He enjoyed greatly any novels that naturally stirred and encouraged thought, especially if these were read out of doors by two or three friends during a walking tour in beautiful countries. Among such novels may be named Kingsley's *Alton Locke, Yeast, Westward Ho!* He was a very faithful friend, and drew his many friends from many various quarters and very different lines of thought and creed. He was very happy in his long married life with the daughter of Dr George Butler, formerly Head Master of Harrow School and Dean of Peterborough. After her death abroad, as old age came gradually on, he retained all the freshness of his intellect and the warmth of his heart, but his bodily activities became less and less. As far back as July, 1908, he wrote to a near kinsman, "The sunset of life is accompanied with pains and penalties, and is a cause of occasional inconvenience to friends. But for myself I find it to be on the whole a happy and peaceful time, on the condition of a frank submission to its many restrictions." Two years and a half of life were still to be granted to him, but the words just quoted might have been written even to the close. He died in his 89th year, leaving behind him not only an abiding fame, but a beautiful memory, for he was in truth "a man greatly beloved." H. M. B.

CLAVERDON LEYS, WARWICK. *July* 11, 1911.

DEAR PEARSON, Sir George and Lady Darwin have been here, and he has taken much trouble re-drafting the proposed epitaph to my Uncle. I must say I like it much better than ——'s, but do not tell the latter! Sir George specially asked me to consult you about it, and if possible to suggest any amendment. I like it as being short and simple. What do you think?

I have to go to the Meeting to-morrow, but do not expect to see you as I hope you are still in the North and taking a rest, so I will write no more. There is no immediate hurry for an answer. Yours very sincerely, E. G. WHELER.

[The epitaph just as it now reads in Claverdon Church was enclosed: see Vol. III^{A}, p. 434.]

Extracts from Francis Galton's Rough Note-books.

It seems to me worth while illustrating in a single instance Francis Galton's method of work. He would take a note-book and write suggestions in pencil in it. In these his handwriting is very minute, very indistinct, and the text full of gaps to be filled in later when the required data might be accessible. He always carried such a note-book about with him, when on his travels or visits, and made rapid jottings in it. Often he would start the whole suggestion or idea afresh, sometimes it led up to the subject of a lecture or a paper; occasionally the matter is merely referred to in a paragraph of one of his publications, although it is far more developed, if disconnectedly, in the note-book.

To illustrate the whole process I publish here a few extracts from note-books on Eugenics from 1900 onward. Galton is concerning himself with the topic he often talked about—the average value to the State of the child of picked parents. The factors he has in view are: (i) The degree of superiority of the child of superior parents. (ii) The higher wage or income of such child over the average of its class. (iii) These differences which form the additional profit to the State of the child of superior parents. (iv) This additional profit capitalised is what the State ought to be willing to pay in order to obtain such children.

I have made no attempt to fill in Francis Galton's gaps, although it would in several instances be feasible. I have given these extracts from the note-books simply to illustrate Galton's method of research, but at the same time I have chosen a suggestive topic, of which the general purport is obvious notwithstanding the fragmentary state of the notes.

The Money Worth to the State of an Infant Male Child of Selected Parents. From the Papers of Sir Francis Galton, F.R.S.

Dr Farr has discussed this question with high actuarial skill in respect to the child of an ordinary Essex labourer, supposed to work and get on like the average of his class. He compares the present value of the expenditure incurred in his maintenance and that of the wages he will gain wherewith to maintain his own children, and striking a balance finds it £ to the good. This was in 18 The figures would now require revision.

My problem is of the same kind but depends on different data; it deals with the offspring of parents who have been selected for their civic worth at the rate of 1 in 50 or 1 in 20 of their class. In other words 2 per cent. or 5 per cent., as the case may be, have been picked out as the best by the judgment of the selectors, much as the 2 per cent. best pears might be picked out of a basket and charged for as being of a higher grade. It will be convenient to use the term "per cent. selects" with an appropriate figure prefixed, as 2 per cent. selects or 5 per cent. selects, to express both the fact and the rigour of selection.

In considering the money value of a select we may be guided by the wages he is likely to earn. If say 2 per cent. of the men of his class earn shillings a week, but that the remaining 98 per cent. of them earn shillings, then the excess of the former sum, duly capitalised to its value at the time when the calculation is made, represents fairly enough the superior worth of the children of the selects to the average worth of the children of the class*. The actuarial calculation must be difficult and take many things into account on which we need not now dwell, but the general principle will be intelligible from this outline sketch of it. The point immediately in view is that if A be the money worth of a child to the State, it would be good economy to spend any sum less than A in procuring and maintaining such a child, and bad economy to spend more than A. It is clearly important to ascertain the value of A in each particular case.

* The regression of the children of the selected parents referred to below has here been overlooked.

(*a*) How does a better workman obtain better wages and to what extent?
(1) By becoming a foreman or a higher grade workman;
(2) By change of occupation;
(3) By off work;

(*b*) Distribution of wages among the most successful 5 per cent.

On the reasonable supposition that the distribution of civic worth follows the Gaussian law, the value of the deviate corresponding to an *L*-select can be found from my small table on p. * or the larger one by Sheppard in *Biometrika*, IV. Then by the formula of Heredity as thus far determined, the mean deviates of the offspring of deviates, which we will call , can be found and its class place *n* from the above tables. This does not give the mean value of the offspring of *L* and all above *L* parents, but considering that fertility decreases as the severity of selection increases, and that we are as yet ignorant of the rate of decrease; also bearing in mind that an inferior limit of possible values of *L* is almost as serviceable in the argument as its exact value, it will suffice to say that the mean deviate of the offspring of *L*-selects and of all higher than *L*-selects exceeds *n*, and that their average value *v* is determinable. In short, it would be good economy to purchase infants whose cost of maintenance etc. capitalised to present value did not exceed *v*. Some purchases would turn out ill, others good, but taking them all round as in any large business, the rule would be founded on a statistical certainty. This general idea requires elaboration and a criticism by experts of the results reached.

Worth estimated either by Class Place or by Scale Value and their mutual Convertibility.

Ministers of State, Heads of Departments, Bishops, Judges, Commanders and Admirals in Chief, Governors of Colonies and other appointments. Foreign Ambassadors, Ministers and other diplomats.

Choice out of many applicants as Secretary, Clerk, superior servant.

Choice of candidates for M.P., Guardians and other municipal officers.

Choice of a Doctor, a Lawyer, an Agent, a School, a Governess, a Shop, a tour, a means of conveyance.

Selection of a Profession, a House, Investments, a Dress shape or colour, a book or any other purchase, an Hotel, food, wine, a dog, a pianoforte, a cigar, a horse.

Classification by marks at school or college examinations, and competitions for Government services.

Sorting fruit into classes differently priced.

Appraisement in money value of pictures, curios, horses, actors and actresses.

Pondering before choosing (Scotchmen).

Arrays—Class place and scale $\left\{\begin{matrix}\text{value}\\ \text{degree}\end{matrix}\right\}$, their convertibility into centile values, always feasible—Judgment by intercomparison.

Place in Scale of Distribution.
Position in scale 0° to 100° up to array.

Mid-Parental Deviations in units of q	Places of A in Class Scale 0°—100°	Mean Filial Deviations $=\frac{2}{3}A$ in units of q	Places of B in Class Scale 0°—100°
A	Separates from lower part of class the upper	B	Separates from lower part of class the upper
3°·0	2 = the upper 50th part	2°·0	9 = the upper 11th part
2°·5	5 = " 20 "	1°·7	13 = " 7 "
2°·0	9 = " 11 "	1°·3	18 = " 5½ "
1°·5	16 = " 6 "	1°·0	25 = " 4 "
1°·0	25 = " 4 "	0°·3	33 = " 3 "

* Galton is, I think, referring to Table 6 on p. 203 of his *Natural Inheritance*.

Worth defined by Class Place.

The phrase that so and so ranks among the upper half, quarter, tenth or other division of a class consisting of a hundred persons is a definite fact and of substantial importance.

I have often had occasion to comment on this, but propose now to elaborate the idea somewhat more fully.

The comparison of the merits of alternative objects is a familiar act and the classification of a large number of objects of like kind in order of merit, however defined, is merely a prolonged application of this power. Class lists are familiar in competitive examinations, when candidates are given marks, by which their order of merit is expressed according to the judgment of the examiner, but the faculty of accurate classifying is far more widely exercised when there are many competitors for a coveted place and only one or a few vacancies. No electorate doubts its capacity of so placing the men that the right ones shall be on the whole generally approved of.

The selective process is gone through in renting a house, or buying an article, a dress, wine, a horse, a pianoforte and, as a rule, whenever a purchase has to be made. It is gone through with care in selecting an agent, a governess, or other employees; Ministers of State, Heads of Departments, Bishops, Judges, Ambassadors and other diplomatic agents, recipients of honours, are all selected always with careful consideration, not seldom with anxious care. Appraisement in money value of curios and objects of art falls under the same head. If we please to take the trouble we may arrange a class in order of any specified description of merit.

I will now suppose this to be done for Civic Worth (a term that I need not now stop to define) and that examples have been recorded of the qualifications of those who stand at any two specified practical lengths of the array. It is convenient to take those at or about its middle and at or about its upper fourth division. Let us call them M and Q. The difference between M and Q we take as the unit of Civic Worth. This difference will be called q (describing briefly the Quartile difference).

All children of $\frac{1}{n}$ of all parental couples to be Wards of Government; $n =$ say, 10. No. of children to be provided for 4 per family (Average 4 children to parent, total children 40) $\frac{4}{10} \times$ population.

Expenditure on scale of upper artisan families say 5*s.* a week.

(p) at 5*s.* per week £13 a year.

Free from other expense „ 2*s.* 6*d.* „ £6. 10*s.* „ up to ... years.

Looked after without interfering with parental responsibility, unless grave faults of management. $\frac{200}{52} = 2$ ...... £10 a year.

Competitive insurance of male children when adults to partly repay at age......or death $\frac{1}{r}$ th of expenditure.

Marriages ...... per thousand of population.

4 times as many children, $\frac{1}{10}$ take or multiply marriages by 0·4.

5*s.* a week $= 5 \times 52 = 260$ shillings $=$ £13 a year,

to be continued for 15 years, $13 \times 15 =$ £195 say £200 total for each child,

in a population of 1000, 20 (say) marriages a year or a yearly capital to be put by of £4000, i.e. £4 per head. Army cost?

This is 4 times too much to be reasonable; make $n = 20$ to halve it,

$$p =$$

If 20 more per thousand and 1 in 20 taken, that is (4 children to 1 marriage) 4 per thousand of population to receive this at 4*s.* a head = £40 ann. $= 40 \times 15 = 600$ total per 1000.

Ignore compound interest, the are far greater than the allowance for it.

£10 annually $= \frac{200}{52}$ shillings $=$ 4*s.* about weekly.

It is deviation from M measured to units of q that we shall be solely concerned with here.

It has now become widely known, and is very familiar to modern statisticians, that the distribution of nearly all faculties among the members of an array, when referred to units of q, follows approximately, often very closely, the important theoretical law with which the name of the mathematician Gauss is associated. It is based upon the supposition that all variability is due to variety in the combination of a large number of small and independent elements. Be the cogency of the logic what it may, the Gaussian law is found to be an excellent approximation to observed fact in a multitude of cases, many of them analogous to the Civic Worth of which we are speaking. Assuming its application here and having determined q as above we can proceed with numerical precision.

I give a brief table by which the worth for any specified class place (or rather for the upper half of any specified class place) can be determined by adding to M (reduced to units of q) the value of the deviation stated, read in terms of q.

The partition values refer to a centennial scale (0° to 100°) placed alongside the array of class places which are described by the "ordinal" numbers of 1st, 2nd, 3rd, ... 100th and correspond to the partitions that separate the class places of the same name.

They are both reckoned from the lowest upwards.

Then the scale value divides the upper 25 per cent. or one quarter of the class places. That of 50 per cent. cuts off the upper half from the lower half.

Scale 0° to 1000°.

Evaluation which separates those beneath from the upper		Corresponding deviation in units of q
	very roughly	
21	1/47·6 or say 1/50th	3·0
46	1/21·7 „ 1/20th	2·5
89	1/11·2 „ 1/11th	2·0
156	1/ 6·4 „ 1/ 6th	1·5
250	1/ 4·0 „ 1/ 4th	1·0

Evaluation which separates the upper from those below it			
q	× ·674	points per mille 0°—1000°	say
3·0	2·02	22	2
2·5	1·68	46	5
2·0	1·35	88	9
1·5	1·01	155	16
1·0	0·67(45)	249	25
in σ	× ·666	rank per mille	
2·02	1345	88	9
1·68	1118	132	13
1·35	899	184	18
1·01	679	248	25
0·67	449	327	33

1 in 20 families of 4 each, or 1 in 5 [? 20] children
in population of 1000, 20 marriages per annum, children of one selected is 4 children at 4*s.* a week = 208*s.* a year = £10. 8*s.* say roughly £10 for 5 per cent. of child population for 10 years.
Total of £10 per child for 15 years = £150; per 4 children £600 max.,
3 £450 max.,
insurance might repay (in what part leaves max. £300)
or £300 annually to be extracted for each 1000 of population,
= £3 out of 10 = 66*s.* = £3. 3*s.* per head.
14 million for 45 say £1 to every 3 persons = about £3 per head.
15 ......

Selection.

1 in 20 families parent engineer family large,
total 6 in each family of whom 1 dies not counted owing to early death.
5 to be allowed for.
4*s.* a week for 52 weeks = £10. 8*s.* a year say £10
to be continued for 15 years = £150 each child
= £750 for each family of 5 children.

Marriages in a population of 1000 = say 20 or 1 in 50 of which one produces the 5 children.
Therefore the annual cost per head per thousand of population of 1000 would be £0·75 or 15*s.* per head of total population.
14 millions (and more could be had) so say 15 millions are expended annually in voluntary charities = £1 to each 3 persons = 7*s.* a head about or half the above.
What would be the money worth to the nation of each person selected at rate of 1 in 7?

Eugenic Administration. A Forecast.

It seems timely to put forward in no dogmatic way my own views on the possible future of Eugenics. They are submitted for extracting helpful criticisms and suggestions and for leisurely discussion, so that clearer ideas may be gained of the road in front before the time for marching arrives.

I will suppose (1) that Eugenics has taken firm hold upon the national conscience, (2) that large sums are in prospect for its support, of the same order of magnitude as those now devoted to charitable purposes or to old age pensions and to education, and that the point to be considered is how to administer these funds most wisely. The inquiry frequently bifurcates as it proceeds. I cannot follow all the roads but must pursue that which seems to be the main one.

The object briefly is to call into existence a large contingent of citizens, who are naturally endowed above the average of their contemporaries with health and vigour of mind and body (and of naturally good characters), and the question is how to spend money in the most economical way, in accordance with public sentiment, for doing this.

It seems to me that whatever is done should be tentative for some time, and yet be on a sufficiently large scale to give trustworthy results. How are we to begin?

I have already written somewhat to this effect in a paper read here on (...... Local Associations) which I need not recapitulate, but should be glad if it were referred to and taken into account together with what will now be said. There are two aims: (1) the most feasible plan seems to be in helping eugenic families to procure better house accommodation, food, general nurture, than they would otherwise be able to procure, and to make them feel that each additional child is a gain to them, (2) (which will here be treated briefly and incidentally) to promote early eugenic marriages. The best field of operation at first seems to lie in rural districts where the existing human stock is relatively good and to whom an extra few shillings a week is a potent motive. That points to the northern rather than to the southern counties.

Let us then confine our ideas for the present to those districts in which other conditions are also favourable, such as zealous residents of good social position, active and efficient professional and administrative workers, and so forth.

Money grants might I conceive be made, in a fair and judicious way, to eugenic families, as for example one shilling weekly for each child under 15 years of age. That would amount to £2. 12*s*. for each child annually, equal to a total sum of £39 for each child. This is a large sum, but not so very large considering that the value at birth of each male child of an Essex labourer was calculated by Dr Farr to be £20, and that the Old Age Pensions cost six shillings weekly. Looking at it from a national point of view the money would be well spent on the whole. There would of course be individual shortcomings, but an excess of individual merits above the present average.

The obvious remark is that if the money so spent be ultimately remunerative, the scheme could be made self-supporting. But the difficulties of doing this seem insuperable, to say nothing of the hardships of handicapping a youth who has his living to make with a serious debt. The difficulties of debiting [?] arise partly from the variability of the offspring, whom it would be not just to tax alike, and partly to the mobility of the population so that the whereabouts of men could not be followed without a large, costly and inquisitorial bureaucracy. So far as I can foresee, all attempts to recover the money spent for rearing must be abandoned and the charge be borne by the State, that is by the population at large.

A problem very desirable to solve is the average value to the State of each child, in any large group of them, who are born of parents exceptionally gifted in a specified degree with the qualities that make for civic worth. The hereditary element in the problem is already ascertained with adequate precision, the difficulty mainly lies in appraising the financial value of civic worth.

No one who is conversant with English history, can doubt that the immigration of the Huguenots—we need not stop to define the word—was of immense value to our country. If we were agreed as to the number of pounds it was worth on the whole and knew the numbers of immigrants, the average worth of each could be calculated. Thus it would be possible, though not easy, to divide that worth into its components of natural gifts and nurture with fairness. Dealing alone with the former and with its known intensity of hereditary transmission we could arrive at the *prairie value* of a Huguenot child. Call it x. Then it would be a fair financial transaction for the State to buy such children and to rear and educate them at a total cost of £x each. In default of other data we must try to get some idea of an x value in indirect ways, as by comparing the wages of picked men with those of the average. The crews of Arctic exploring ships are all picked men who are attracted to this work largely no doubt by a spirit of adventure, but to a considerable degree by increased pay. Whenever the attraction is greater, whether in pay or otherwise, there will be more applicants than places for them. So selection comes into play of corresponding degrees of rigour.

Picked Couples.

The offspring will be less exceptional on the average than the parents in a definite degree, and we can foretell the distribution of capacity in the children of any large number of parental couples who are all exceptionally gifted in any definite degree. Conversely we can tell what conditions must be fulfilled in order that an influx of persons may be called into existence whose average value is specified, while the distribution of capacity among them will be known.

It may be possible roughly to estimate the value to the State of such a group of persons proceeding on similar lines to those followed by the late Dr Farr in calculating the value at birth of a male child, son to an Essex labourer, but it is difficult. The problem which it is desirable to solve is: What would be the average money value to the State of each child of a large group of children, the average natural capacity of whose parents was superior to that of their contemporaries, and equal to that of a group picked out of them with a specified rigour of selection? Let x be the average value of each of these children, then it would be an advantage for the State to spend any sum not exceeding x in procuring and nurturing it. If x were known, it would be easy to consider how much the State might reasonably do.

Some clue towards the value of x is to be had by comparing the wages of picked workmen with those of workmen generally. In mental work of all kinds the difference is very great, whether we consider possible clerkships or the higher appointments.

Occupation picked	Sailor Arctic	Soldier Corp. Sergeant	Artisan Foreman	Policeman Sergeants, etc.

Occupation picked	Domestics [Housekeeper]	Railway [Guard]	Gardener [Head Gardener]

Huguenot deviation	Centennial graduations		Roughly approx. severity of selection	
0·5	63	37	1/3	
1·0	75	25	1/4	
1·5	85	15	1/7	1·5 × 2/3 = 1·0
2·0	91	9	1/10	
2·5	95	5	1/20	2·5 × 2/3 = 1·7
3·0	98	2	1/50	3·0 × 2/3 = 2·0

Galton's Characterisation of Herbert Spencer.

Among Galton's papers I find the following:

"Reminiscences of Herbert Spencer. Rough first draft of what I afterwards sent to Mr Duncan*."

Mr H. Spencer's magnificent intellect was associated with no small degree of oddity, obstinacy and even perversity, difficult to rate in their due proportions. My knowledge of him was chiefly due to a habit of spending an hour or two of the afternoon, during many years, in the then smoking room of the Athenaeum Club, when quiet conversation was easy. He was always interested in my various hobbies and though I did not always accept his criticisms, I received great benefit from them. Let me say parenthetically that to me one of the chief disadvantages of age lies in the diminishing number of friends who care for one's work and fearlessly speak their views. In those long bygone times I could go into the Club and talk with one man on this subject in which he was expert, and with another man on that; now it is all changed. Moreover, the relatively young are too diffident in freely pulling to pieces the arguments of a much more elderly friend, so that much wholesome correction is lost to him. Herbert Spencer had assuredly no diffidence in criticising others, though he was very thin-skinned under the converse process. He hated fair argument, and wicked friends asserted, not without grounds, that whenever he felt worsted he fingered his pulse and said abruptly, "I must talk no more." The fact was that excitement really harmed him. He was far too opinionated for candid argument. The following story is characteristic. Some years ago, when I was actively engaged in meteorology, he said to me that we were all wrong in forecasting weather through not taking preceding temperatures into sufficient account; that the earth became chilled by a long frost and its store of cold ought to be recognised, and conversely after a spell of hot weather. He said he would write me a letter on the subject, which he did and at length. My reply was to the effect that the influence in question was not wholly neglected, that it was a *vera causa*, but far less important than that of change of wind, as shown by the suddenness with which frosts and thaws often set in, and especially by the well-known effects of the south or "föhn" wind on melting Alpine snows. He was clearly imperfectly acquainted

* Mr David Duncan has published extracts from these Reminiscences in his *Life and Letters of Herbert Spencer*, 1908. Whether Galton or he modified the extracts used I do not know. As the Reminiscences stand they accord closely with Galton's opinions of Spencer, as expressed to me in conversation. He certainly would not have agreed with Dr Duncan's view that Spencer was "one of the greatest thinkers of this or any age" (*Life and Letters*, p. 477).

with the subject, but for all that he stuck obstinately to his conclusions and afterwards published the contents of his letter (at this moment I forget where) without any recognition of the facts that told against him. Another meteorological view to which he clung with some persistency, which in a narrow sense is right but in a broader sense is wrong, was that the fact of the weather having been, say, dry beyond the average for some months in some particular place, was no justification for the popular belief that the deficiency in rain would be made up later. He insisted on treating the past and future weather as independent variables, which they are not. Local deficiencies in one place testify to local excesses in others, and as the whole atmosphere travels on there is a tendency for the one to replace the other and for averages to be maintained.

He was a most impracticable administrator when put to the test. Thus, there were great complaints at the Athenaeum Club of the way in which the dining-room was managed. He, I and one of the chief of the malcontents happened to be members of its Committee at the time. Spencer argued that experience in dealing with such matters was of comparatively little importance, adducing examples in confirmation, and he finally carried his proposition that a sub-Committee of three should be appointed with large powers, and that it should consist of himself and the malcontent and myself as the third, being professionally unfettered and presumably having leisure. I did not much like the task, but accepted. We met, and a most comically inefficient group we proved to be. There was a continual perversity in Spencer's views, and yet it was always a defensible perversity. He gave what seemed to me a disproportionate weight to small questions, treating them as matters of deep principle to be set forth in ponderous words, with the result that we hardly got on at all. I recollect one amusing scene; our butcher was summoned to be admonished as to the quality of his beef. I forget the precise words used by Spencer, which the butcher rebutted in terms satisfactory to himself, to which Spencer replied with severity: "You seem not to appreciate the nature of our complaint; your beef has too large a proportion of cellular tissue." The butcher fairly collapsed under the weight of this accusation. He could not comprehend it but evidently believed that it might in some obscure way be justified.

As regards heredity—one day he spoke with surprised concern to me upon his learning that the weight of scientific belief was opposed to the inheritance of acquired faculties; for, if they were not inherited, much of his scheme of evolution would be invalidated. I spoke of many observations and arguments by which it seemed to be disproved, but he never I believe consented to go thoroughly and with open mind into this question. I am inclined to think that he unconsciously gave almost as much logical weight to one of his own deductions as he would to a well-observed fact. His over-tendency to *à priori* reasoning has been fully recognised. He came to me one day to have impressions taken of his fingers, I being at that time much occupied with finger-prints. I spoke of our ignorance of the object of the papillary ridges which form the peculiar patterns on the bulbs of the fingers and which are closely connected with the ducts of the sudorific glands, and said that more careful dissection was still wanted of the human embryo. He said: "You are studying the question in the wrong way, you ought to begin by considering the conditions that have to be fulfilled; the mouths of the ducts being delicate require the protection of the ridges"; and he then enlarged with ingenuity and elaboration on the consequences of this necessity. I wickedly allowed him to finish and then replied: "Your argument ought to be most convincing, but it unfortunately happens that the mouths do not open out in the valleys where they might be protected, but along the crests of the ridges in the most exposed position possible." He burst into a good-humoured fit of laughter and then repeated to me the now well-known story about himself, which curiously enough I have also heard from the other two persons present at the time. My version of it is more dramatic than that in the *Autobiography*. They formed a party of three, Huxley, Spencer and another, dining together at the Club. In course of conversational banter Spencer said: "You would little think when I was young I wrote a tragedy." Huxley instantly flashed out with "I know its plot." Spencer indignantly denied the possibility of his knowing it, he having never shown the tragedy nor even spoken of its existence to any one, before then. Huxley persisted, and being challenged to tell, said that the plot lay in a beautiful deduction being killed by an ugly little fact.

Spencer had never seen a race, so I succeeded in persuading him to go with me to see the Derby, and I got a clerical but large-hearted Don of a College to join us. Spencer proved rather a kill-joy. He summed up his impressions at the end, after careful thought, under three heads. First, that the general show was just what he had expected; secondly, that a crowd of men was a nasty object, like flies on a plate; thirdly, that he would never go again. However I was assured that he did, and that in the very next year.

I thought him a man of naturally a very strong constitution, ruined by over-work. When about to utter remarks he was apt to clear his throat by a deep "hem," that testified to a powerful chest. His natural strength is shown by the account in his autobiography of his extraordinary walk, when a boy of 13, while he was half-starved, of more than 40 miles the first day, 40 the second and 20 the third, to his destination.

The mental process I most admired in him was that by which he generalised. It is too common for persons to arrive at general conclusions through unconscious and unchecked steps, so that when asked for evidence they cannot give it Spencer had always a store of facts at hand whenever he wished to justify himself. His wealth of ready illustration was marvellous. Notwithstanding my admiration of his intellect and my sense of incompetence to treat subjects in the wide manner that he did so easily, I cannot say that I have profited much by his writings or taken pleasure in them. I rarely felt "forwarder" for reading them, least so in subjects with which I was familiar and where I felt somewhat entitled to criticise his results. I am far from being singular in saying this, as few of those with whom I have talked seem to admire his work whole-heartedly, and I have often expressed a wonder how far their non-appreciation would be justified by the judgment of posterity.

Note to p. 585, *Chapter XVII.*

Galton had sent Miss Elderton a ticket for a meeting at which sex-problems were discussed under the presidency of Dr Slaughter. The meeting was *not*, as Galton supposed, held under the auspices of the Eugenics Education Society. The exact origin of the latter Society is somewhat obscure. We have Galton's letter to Montague Crackanthorpe of December 16, 1906 (see Vol. III^A, p. 339), but we do not know what part the latter took in the matter until the spring of 1908. Meanwhile there existed in or before 1907 a body termed the "Moral Education League." At a meeting held on November 15, 1907, a section of this League reconstituted itself as a new Society—the "Eugenics Education Society." Members of the Committee of the League resigned their posts to become members of the Council of the new Society. Dr Slaughter was the first Chairman of this Council, and the guiding spirit of the infant Society during the early days of its existence. Galton did not join the Society until its practical control had passed into the hands of Montague Crackanthorpe, which was the state of affairs by June, 1908 (see Vol. III^A, p. 346).

Note to p. 618, *Chapter XVII.*

Plenty of illustrations can be given of Galton's good temper and sense of humour. He used to write in a minute diary $1''\cdot5 \times 1''\cdot7$ a brief record of events in the smallest of handwritings; some of these diaries have survived. Thus the entry for Easter Sunday in Seville, 1899, runs: "Cocks, Bulls, and Fire," which signifies a cockfight before breakfast, a bullfight in the afternoon and his niece Eva setting her room on fire while dressing for dinner in the evening. For the latter occurrence he had to pay eight guineas, and yet, Mrs Ellis tells me, he never said a word in reproof: see p. 508.

ADDENDA

Galton in the Appendix to his *Memories of my Life* gives a bibliography of 179 memoirs, books, articles and papers written by himself. I have been able to increase this by 59 titles, and am only too conscious that others may still have escaped me. It is indeed difficult in the case of a life as long as Galton's to discover all the side channels into which he poured the ideas of a fertile mind in the hope of reaching one or another section of the community, and so irrigating the arid wastes of prejudice. I can only trust that nothing of first-class importance may have escaped my notice. But neither Galton's own collections of memoirs and letters nor those preserved by his relatives cover by any means all that he wrote even in the years with which they deal. Just as I have closed my volumes with nothing I thought remaining but the indexing, I have come across two omitted papers of considerable interest.

In the case of the first paper—an important one—my excuse must be that while there are *two* papers by Galton in the XXVIth volume of *Nature* there is only *one* entry under his name in the Index, and having come across in opening the text one paper, I did not expect and look for a second. Of the other omitted paper I found the abstract given below among my notes, when checking certain entries in the index; it was marked for incorporation in Chapter XI of Vol. II, but a fitting place not having been found for it there, it had been overlooked and so omitted entirely.

I fear these two papers may not be the only omissions; if so, my sole excuse must be that working independently, I have been more comprehensive than earlier bibliographers, including even Galton himself.

Addendum I.

"A Rapid-View Instrument for Momentary Attitudes." This paper appeared in *Nature*, July 13, 1882*. In it Galton suggested a very simple mechanism for obtaining with direct vision an almost instantaneous picture of a moving object. His purpose was twofold, (i) to transmit a brief glimpse of a moving body—thus by aid of it he was able to see the wheel of a bicycle at full speed as a well-defined and apparently stationary object—and (ii) to transmit two or more such glimpses separated by short intervals, and to cause the successive images to appear as simultaneous pictures in separate compartments in the same field of view.

The power of the eye to be impressed by a glimpse of very brief duration has not, I think, been duly recognised. Its sensitivity is vastly superior to that of a so-called "instantaneous" photographic plate when exposed in a camera, but it is of a different quality, because the impression induced at each instant of time upon the eye lasts barely for the tenth of a second, whereas that upon a photographic plate is cumulative. There is a continual and rapid

* Vol. XXVI, pp. 249–251.

leakage of the effect of light upon the eye that wastes* the continual supply of stimulus, so that the brightness of the sensorial image at any moment is no more than the sum of a series of infinitesimally short impressions received during the past (say) tenth of a second, of which the most recent is the brightest, the earliest is the faintest, and the intermediate ones have intermediate degrees of strength according to some law, which an apparatus I shall describe gives us the means of investigating. After the lapse of one-tenth of a second the capacity of the eye to receive a stronger impression has become saturated and though the gaze may be indefinitely prolonged the image will become no brighter unless the illumination is increased. (p. 249.)

Galton next compares the sensitivity of the eye with that of an instantaneous plate as sold in the shops (1882). He says that given a dull day and an ordinary sitting-room, the window of which does not occupy more than $\frac{1}{30}$th of the total area of wall, ceiling, floor, etc., space, which is the light one usually reads or writes under, the eye takes about $\frac{1}{10}$th of a second to form a clear impression, but the "instantaneous" plate will not give an image under about 30 seconds. Hence Galton concludes that the eye is fully 300 times as sensitive as the usual "instantaneous" plate.

Referring to the effect of illumination our author considers that an object in bright sunlight may require no longer than $\frac{1}{1000}$th of a tenth of a second to be visible. Thus a cannon-ball of 10-inch diameter moving in mid-course at 1000 feet per second would in $\frac{1}{10000}$th of a second shift its place through one inch, and would present to the eye if it could be viewed "the appearance of an almost circular disc elongated before and behind by only a slight blur." Galton then proceeds to estimate roughly the speed of a very small stone flipped upwards from his finger, and that of the chiefly effective part of a pigeon's wing; he finds these to be respectively 288 inches per second and 1232 inches per second.

Now the duration of an exposure depends on three data, namely the rapidity with which the screen moves past the eye, the width of the slit through which the momentary glimpse is obtained, and the diameter of the available portion of the pupil of the eye. I prefer not to limit the pupil by using a small eye-hole, which is a source of much trouble in actual work, but to have as large an eye-hole as is in any way desirable. I find the width of the pupil of my eye in an indoor light, as measured by holding a scale beside it and reading off in the looking-glass, to be about 0·1 inch and I use a slit of the same diameter. The exposure begins when the advancing edge of the slit is in front of the near edge of the pupil, and it ceases when these conditions are reversed, in other words it lasts during the time the screen is moving through one-fifth of an inch. In the cases just taken of velocities of 288 and 1232 inches per second the duration of the exposure would be the 1440th and the 6160th part of a second, respectively. There is therefore no difficulty either theoretical or practical about shortness of exposure and sufficiency of illumination. The power exists, and can be utilised, of seeing bodies in motion by a rapid-view instrument, showing them in apparent stillness, and leaving a sharply defined image on the eye, that can be drawn from visual memory, which in some persons is very accurate and tenacious. (p. 250.)

Galton goes on to remark that for a galloping horse or a flying crow a great rapidity of exposure is not essential. He then describes his own

* Is "waste" the right word to use? Is not the "continual supply of stimulus" needful to maintain the sensation of perception of the object in the subject? The ideal photographic plate would be one which did "waste" this continual supply, as the eye does, so that there would be no risk of over-exposure. Who will discover it? ED.

rough pocket instrument, the duration of the exposure being a 360th part of a second under the action of its spring as computed, but its practical duration about one 500th of a second or rather less according to the nature of the tap on the stud; this arises from the fact that very little light passes through the edges of the pupil at the beginning and end of the exposure.

Galton's description of his rough instrument is as follows:

The instrument is shown in the figure below without its sliding lid, which protects it from injury in the pocket. *A* is an arm which turns through a small angle round *C*, its motion being limited by two pins. Its free end carries a vertical screen, *RR*, which is a cylindrical (or better, a conical) sheet described round an axis passing through *C* perpendicular to the arm. As the arm travels to and fro, this screen passes closely in front of the end of the box, which is cut into a hollow cylinder (or cone) to correspond. There is a slit in the middle of the screen and an eye-hole in the centre of the end of the box.

When the slit passes in front of the eye-hole, and the instrument is held as in the figure, a view is obtained. A stud, *S*, projects upwards from the arm, and an india-rubber band, *B*, passing round a fixed pin and a descending spoke of the arm, acts as a spring in causing the stud *S* to rise through a hole in the side of the box, where the finger can press it like the stop of a *cornet à piston*. In using the instrument, it is held in the hand as in the figure, with the eye-hole in front of the eye. Nothing is then visible, but on pressing or tapping the stud the slit passes rapidly in front of the eye-hole, and the view is obtained. After this, the stud is released and the arm springs backwards, when a second view can be obtained, or the eye may be purposely closed for the moment. (p. 250.)

This second view leads Galton to remark that the first view was invariably fainter than the second, showing that its brightness had faded in the brief interval that elapses before comparison can begin. Thus he suggests that the law of the rate of fading might be determined by an apparatus of this kind, and several arrangements for doing so are described (p. 251).

Galton further gives an account of various modifications of his instrument which he had made with revolving discs and multiple lenses. Also he explains how "to present the images formed by two successive glimpses as simultaneous pictures seen side by side in the field of view."

Finally we may note how Galton measured the velocity of his instrument. He put in temporarily a peg which checked the velocity of the recoil when the slit was opposite the eye-hole. Then the stud being held down, and the box fastened tightly to a support, the recoil was used to project a light weight into the air; this it did when the lever came against the temporary peg, and the weight was projected three inches: The velocity of the stud was therefore $\sqrt{2g\,3/12}$ ft. per sec. = 4 ft. per sec. = 48 inches per second. From this the velocity of the slit could be easily found by the known distances of stud and slit from the centre of the pin C (p. 250). Assuming that measurements can be safely taken on Galton's figure, I make the distances of stud and slot from centre of pin to be in the ratio of about 35 to 56 or 5 to 8 or the velocity would be 384/5 inches per second, and length of exposure $\frac{1}{384}$th of a second. Galton gives it in his actual instrument as $\frac{1}{360}$th of a second.

I know no reference by Galton to this paper in his later publications, nor am I aware of any work on Galton's lines done since its appearance.

Addendum II.

"Note on Australian Marriage Systems." This paper appeared in the *Journal of the Anthropological Institute,* Vol. XVIII, pp. 70–72, 1888 (published 1889). Galton gives a simple explanation of the well-known Australian Kamilaroi marriage system, which we can condense as follows:

Phratries	Subphratry	A Male	Marries a Female	Their children are
Dilbi (P)	Muri =	P 1	Q 2	Q 1
	Kubi =	P 2	Q 1	Q 2
Kupatkin (Q)	Ipai =	Q 1	P 2	P 1
	Kumbo =	Q 2	P 1	P 2

Galton illustrates this as follows: Suppose there were only two Universities (Oxford and Cambridge) and two University clubs (the Oxford and Cambridge and the Universities clubs), and assume them all open to men and women alike. Then a man (or woman) may not marry a woman (or a man) of the same university and the same club as himself (or herself). The children of Kamilaroi will be entered at the Mother's university (P or Q) and the Father's club (1 or 2). In the case of the Kiabara the children on the other hand would be entered at the Father's university and the Mother's club. If this be not the theory of the arrangement, it is, as Galton remarks, an easy way of remembering the complexities of the Australian system.

INDEX

d'Abbadie, Antoine, number forms of, II 206
Abbot, Rev. J., coaches Galton in mathematics, I 101, 103
Aberfeldy, Galton joins reading party at, I 168, 169
Aberystwith, summer holidays at, I 82
Ability, correlation of, with physical and moral characters, III^{A} 248; definition of, III^{A} 121; in divines, II 101; in dukes, II 93; and eminence, II 91, 92, 104, III^{A} 116; and environment, III^{A} 112, 116; and fertility, II 94, 96; Galton's veneration for, II 94; inheritance of, II 104–106, 128, 141, 151, III^{A} 102, 103, 108–113, 114–121, 347, 348; inheritance of, and Upper House of Legislature, III^{A} 32–34; and insanity, III^{A} 32; legal, II 93; measurement of, by reputation, II 89, 92; and power of mental imagery, II 243; need of, in our civilisation, II 108, 109, 112, 113; and the normal curve, II 90, III^{B} 623; and physique, II 94, 128; of various races, II 106–109; rarity of, II 89, 90; scientific, in father and son, II 97; and size or shape of head, II 94, III^{A} 247–249; and success, III^{A} 111, 112, 116; tests of, by examination, III^{A} 232; transmission of, through male and female, III^{A} 108, 109, 118. See also *Able Men, Civic Worth, Genius, Intelligence*
Able Men, individuality of, III^{A} 32; parentage of, III^{A} 27; wives of, III^{A} 32
Abney, Capt., Sec. to Roy. Soc. Committee on Colour-blindness, II 227; his rotating cylinder for tint-testing, III^{A} 308
Aborigines, Galton on, II 264, 265
Abrahams, amongst Galton ancestry, I 38, 39
Abyssinia, model of, II 34
Accidents, to Galton, I 116, 117, III^{A} 316, III^{B} 582
Achard, number forms of, II 205
Acquired Characters, inheritance of, II 147, 148, 169, 170, 173, 174, 182–184, 186, III^{A} 57, 59, 129–132, 335, 374; Dr Erasmus Darwin and, II 202
Acton, Lord, Hon. Fellow of Trinity College, III^{A} 236, 238
Actuaries, Galton on, III^{B} 589. See also *Insurance Offices*
Adrian, Prof., coaches Galton in Giessen, I 128
'*Adult elements*,' in genetic scheme of offspring, II 172
Ady, Mrs (*Julia Cartwright*), books of, III^{B} 546, 547
Aesthesiometer, tests on men and women with, II 222
Aesthetic training, Miss Hertz on, III^{B} 469
Africa, discoveries in, II 25–27, 30, 31; Galton's explorations in, I 214–240; Galton's diaries, etc. in Galton Laboratory, I 216, 217; and the Chinaman, II 33; and the Negro, II 32; trade products of, II 31
African Discoverers, memorial to, suggestions regarding, II 25, III^{B} 588
Age, disadvantages in advanced, III^{A} 318; at death of various classes, II 116; of parents and fertility, II 408, 409, 410; of parents and vigour of offspring, II 348, 349; slowing down of functions with, III^{A} 301; and youth, III^{A} 318. See also *Longevity*
Agility, measurement of, II 358
Agnetti, Dr, activities of, in Italy, III^{B} 541
Agnosticism, and Galton, II 102, III^{A} 424, 435
Agrippina, wife of Claudius, portrait of, and composite, II 296 Plate XLIII
Ague, Galton has, at Tripoli, I 204
Ailments, minor, Galton's list of, II 365
Air, possibility of its flying off earth, III^{B} 470
Airy, Sir G. B., hereditary traits in, II 208
Albinism, in dogs, III^{A} 356, 357, 389; Pearson's lectures on, III^{A} 360; memoir on, III^{A} 345, 390, 425, 430; Mendelian rules and, III^{A} 388; and mulattoes, III^{A} 370; in orchids, III^{A} 370
Albumen, and digestion, III^{B} 461
Alcoholic parents, children of, II 148, III^{A} 398, 405–407
Alcoholism, experiments to show effect of, on offspring, in mice, II 139, 140; and insanity, II 148, III^{A} 398; and impregnation, II 139
Aldershot, Galton's teaching of art of camp-life at, II 14–18
Alexander, Sir J. E., travels of, in Africa, I 214, 215
Alexander the Great, composite portraits of, II 295, 296 Plates XXXVI, XXXVII
Alexis, the mesmeriser, Galton sees, I 190
Ali (*Galton's dragoman*), I 200, 203; death of, I 197, 203, III^{B} 454 Plate XLV
Alix, work of, on finger-prints, III^{A} 142, 143, 161
Allardyce, connected with Galton's Barclay ancestry, I 33
Allbutt, Clifford, appreciation of Galton, III^{B} 617, 618
Allowance, given to the young Galton, I 154, 155
Alma Mater, and Englishmen, III^{A} 237, 238
Altazimuth, designed by Galton, II 50
Alveolar Point, use of, in composite portraiture, II 294
Amari, Weldon presents Galton with works of, III^{B} 526
America, and Galton, III^{A} 235, 243, III^{B} 509; Eugenics in, III^{B} 613; number forms and colour associations from, II 240
American Indian, disappearance of the, III^{A} 219
Analytical Power, and imagination in men of science, II 98
Ancestors, of man, II 85; direct of Francis Galton, I 10; distinguished distant, II 364; contributions of individual, III^{A} 58; and individuals, III^{A} 60, 61
Ancestral Heredity, law of, II 84, III^{A} 40–44, 60, 240, 251, III^{B} 503. See also *Heredity*, Law of ancestral
Ancestry, of Francis Galton, I 5–61, Pedigree Plates, pockets at ends of Vols. I and III^{A}. Importance of knowledge of, II 302, 303
Andamanese Skull, composite of, II 288 Plate XXXIII
Andersson, with Galton on African journey, I 218–220, 222–224, 226, 231–239; in Africa after the journey, I 240; death of, I 240
Anemometer, devised by Galton, II 44
Angina pectoris, deaths in Darwin family from, III^{B} 552
Anglo-Saxon, ability of the, II 106–108, III^{A} 252–253
Animals, hearing of, II 216; photographic measurement of, II 317, 318, 320–323
'*Annals of Eugenics*,' start of, III^{A} 363
Ansell, his 'Statistics of Families,' II 128, 129
Antecedents, of scientific men, II 177
Anthropological Institute, and Galton, II 182, 184, 188, 206, 334, 337, 361, 383. Galton's presidential addresses to, II 396–398, III^{A} 30; papers sent to, III^{A} 11, 12, 27; Mr Faulds' letter on finger-prints sent to, II 195, III^{A} 143
Anthropological Theory, of Max Müller, II 274
Anthropologist, special task of the, II 387
Anthropology, and Galton, II 4, 7, 13, 31, 32, 35, 62; Galton's early researches in, II 70–130; statistics in the service of, II 334–348, III^{A} 57; and anthropometry, II 380; and Eugenics, III^{A} 226, 227; in Japan, II 397

Anthropometric Characters, and mentality, II 232, 233, 388
Anthropometric Laboratory, Galton's, II 357–362, 370–386, II 371 Plate L, II 378 Plate LI, II 397, 398; in the Galton Laboratory, II 365, 373; in Cambridge, II 226, 379, 387, 388; at Oxford, II 379, III^{A} 328; at Eton, II 379; in Dublin, II 379; in Tokio, II 226; in schools, factories, etc., II 336, 337, 343–346; medico-metric section of, II 359; functions of, II 213, 346, 382, 395; instruments for, II 212, 213; cost of running, II 393; Galton's pioneer work regarding, II 226
Anthropometric Measurements, II 336, 337; value of, II 381, 382; the normal curve applied to, II 386. See also *Normal Curve*
Anthropometric Registers, need for, II 252, 398
Anthropometry, and Galton, II 334; aims of, II 380; definition of, II 345, 346; methods and scope of, II 346; progress of, II 211; in schools, II 336, 337, 343–345; industrial, II 358, 382; scientific value of, II 382
Anticyclones, first noted and named by Galton, II 39–41
Antiochus, composite portraits of, II 295, 296 Plate XXXVIII
Anti-suffrage Society, Galton on Committee of, III^{A} 359
Ants, from Petrie's settlement, III^{A} 251
Apes, finger-prints of, III^{A} 216
Apparatus, anthropometric, at Cambridge, II 388; for measuring index of mistakability, II 329, 330; for use in taking finger-prints, III^{A} 155, 177, 178; for refrigerating in the tropics, II 397; for measuring reaction time, III^{B} 514; for measurement of resemblance, III^{B} 576. See also *Instruments*
Aquarium, at Plymouth, Galton's interest in, III^{B} 579
Arabs, Galton on, II 28; and negroes, II 32, 33; profiles of, II 324
Arches, in finger-prints, types of, III^{A} 209, 213 Plate XXIII. See also *Finger-prints*
Ardennes, Château in the, Poem (? by Galton), III^{B} 459–460
Argyll, Duke of, at Darwin's funeral, II 198
Aristides, quotation from, III^{A} 239
Arithmetic, irksome to Galton, II 400; and imaginary scents, II 275, 276
d'Arnaud Bey, Galton meets, I 197, 200; influence of, on Galton, III^{A} 158, 159, III^{B} 455; Galton's tribute to, III^{B} 455; his portrait, III^{B} 455 Plate XLVI
Arnim, Countess von, Galton reads novels of, III^{B} 448, 601
Arsinoe, wife of Ptolemy II, portrait and composite of, II 296 Plate XLII
Arsinoe, wife of Ptolemy IV, portrait and composite of, II 296 Plate XLII
Art, and Galton, III^{B} 441, 448
Artistic Faculty, inheritance of, III^{A} 65, 66, 68, 69; in man and woman, III^{A} 66; in husband and wife, III^{A} 66
Artists, mental imagery in, II 243; accuracy in, III^{B} 533
'*Arts of Travelling and Campaigning*,' II 16–18
Ascain, Galton at, III^{B} 561–565
Asquith, H., and Criminal Identification Committee, III^{A} 140, 148, 153
Association, of ideas, Galton on, II 233–236; for promoting a Professorial University in London, III^{A} 289, 290
Associations, formed at periods of life, II 235; and education, II 235
Assortative Mating, II 79, 105, 106, 149, III^{A} 66, 67, 70, 102, 231; in temper, II 271; in stature, III^{A} 17; table showing statistics of, III^{A} 17; the source of varieties and, II 272, 273
Asthma, in the Galtons, I 52, 53, 123, 127, 156; Francis Galton suffers from, III^{A} 238, 249, 250, 339, 360, 430, 431, III^{B} 481, 520, 608; and gout, I 185; carpets and, III^{B} 520, 532
Astronomy, Galton on, II 335; need for statistical knowledge in, III^{A} 333; correlation in, III^{A} 370
Athenaeum Club, Galton nominated for, I 114, 145; Galton and Herbert Spencer at, III^{A} 317, III^{B} 626, 627; Galton's asthma at, III^{A} 238
Athenians, ability of, II 107, 108
Athletics, and tests for physical efficiency, II 394
Atkinson, with Galton on reading party, I 155, 159
Atoms, variation in, III^{A} 314, 315
Atwood, Rev., schoolmaster to Galton, I 77; sees Galton at Cambridge, I 150
Audibility, at limits of, II 215, 216; in men and women, II 221, 222. See also *Whistles*, Galton's
Auditory Sensation, experiments on the, II 272, 308. See also *Hearing*
Austin, Mr, letter from, on composite portraits, II 192
Australia, recent discoveries in, II 24, 25
Australian Native, ability of, II 106
Austria, colour associations and number forms from, II 240
Autobiography of Galton, III^{A} 329, III^{B} 585–588; appreciation of, III^{B} 613. See also '*Memories*'
Auvergne, Galton's trip to, I 93, 94
Avebury, Lord, takes chair at Galton's Huxley lecture, III^{A} 226; congratulates Galton on Copley Medal, III^{A} 400; grows older, III^{B} 534; on gout, III^{B} 590
Average, misleading use of, II 400, 401
Averages, and individual variation, II 174
Aversion, Galton on, II 258
Avignon, Galton at, I 199
Avuncular and cousin resemblances, III^{A} 329, 333, 334, 356

Bach, Sebastian, cited by Galton as a sport, III^{A} 85; not a sport, III^{A} 120
Bacon, Francis, ability of, II 107
Baconian method, and work of Galton and Darwin, I 58
Baden-Powell, Galton calls on, III^{B} 568
Baden-Powell, Miss, silhouettes of, III^{B} 577, 603
Bain, Prof. A., and tests on idiots, II 272
Baker, Sir Samuel, and African exploration, II 25, 30
Balance, invented by Galton, I 149
Balbiani, views and statements of, II 181, 182
Balfour, Earl, elected to Trinity Fellowship, III^{A} 236, 238; President of the British Association, III^{A} 276; speech of, III^{B} 518; appreciation of, III^{B} 528
Balfour, Francis, experiments of, on pangenesis, II 176
Ballad, of Whittier on David Barclay, I 28, 29
Ballater, fine men from, III^{B} 554
Banking, enterprise of the Galtons in, I 32, 33, 50, 51
Barbarians, and the barbarous, III^{A} 336
Barber, Mr, the African hunter, III^{B} 516
Barclays, in Galton ancestry, I 10, 26–35, also Pedigrees in pockets at ends of Vol. I and Vol. III^{A}
Barclay, David, I 10, 30, 33; sufferings of, I 28; slave emancipator, I 32; Thomas Young brought up in house of, I 48; portrait of, I 28 Plate XXII
Barclay, Hedworth, Galton up the Nile with, I 200–203
Barclay, Lucy (wife of Samuel Galton the second), parents of, I 33, 46; cousin to Priscilla Farmer, I 33; marriage of, I 33, 46; character of, I 48, 49; portraits of, I 44 Plate XXVI, I 46 Plate XXVIII; silhouette of, I 44 Plate XXVI; sampler of, I 46 Plate XXVII
Barclay, Margaret (Mrs Hudson Gurney), parents of, I 33; shows Galton over Ury, I 104, 105; portrait of, I 91 Plate XLVII

Barclay, Robert (the Apologist), I 10, scholarship and character of, I 27, 28; in London, I 35

Barclay, Robert (father of Lucy Barclay), I 10; portrait of, I 30 Plate XXIV

Barclay, Capt. Robert (the pedestrian and athlete), I 10, 30; portrait of, I 30 Plate XXIV

Barings, the, a noteworthy family, IIIA 113

Barlow, logograph of, II 192

Barometer, limited use of, II 55

Barometric prediction of weather, II 54, 55

Barrington, Amy, work and assistance of, I viii, IIIA 305, 314, 322, 368, 372, 387, 426

Barron, Galton consults with, IIIB 529

Barrows, Miss, work of, IIIA 426

Barth, and an African memorial, II 25; geographical discoveries of, II 31, 71

Bartlett, and experiments on pangenesis, II 159

Basques, finger-prints of, IIIA 193; language and habits of, IIIA 279; Galton's appreciation of, IIIB 564

Basset Hounds, law of ancestral heredity applied to, IIIA 40–44, IIIB 503

Bateson, W., on evolution committee of Roy. Soc., IIIA 126, 127, 290, 291; as referee at Roy. Soc., IIIA 241; and Galton's views, IIIA 81, 82, 86; and the foundation of 'Biometrika,' IIIA 100; on actuarial and experimental methods, IIIA 260, 288; on scientific knowledge, IIIA 288; hostility of, IIIA 383, 388, IIIB 528; in the Report of the Hybrid Conference, IIIA 314; on sports, IIIA 120; at the Darwin-Wallace celebration of the Linnean Soc., IIIA 340; and the Darwin Commemoration, IIIA 369; on horse colour, IIIB 561; Galton's letter to, on Eugenics, IIIA 220, 221; letter of, to Evelyne Biggs, IIIA 288

Bateson, Mrs W., assistance of, IIIA 220, 288

Bath Chair, Galton enjoys, IIIB 587, 595, 596, 600, etc.

Batt, Jaspar, a strenuous Quaker, I 35, 37, 59

Baxter, on hair colour and liability to disease, II 354, 371

Bayonets, source of name, IIIB 555

Beale, Dr J., in letter to Boyle, II 229

Bearcroft, W., meets Galton in Egypt, IIIB 519

'*Beauty-map*,' Galton's project of, II 341

Beddoe, Dr, and anthropology, II 334; on medical life histories, II 359; paper of, for 'Biometrika,' IIIA 249, 256; on Galton's sense of humour, II 310

Bees, and fertility, IIIA 218; and sweet-peas, IIIA 325, 326

Beetles and wasps, Wallace on, IIIA 370

Beheading, and the 'long drop,' II 407, 408

Beliefs, necessity for testing, II 297. See also *Creeds*

Benedict, on identification of criminals in Vienna, IIIA 148

Berbers, and Egyptians, characteristics of, I 203

Berenice, wife of Ptolemy III, portrait of, and composite, II 296 Plate XLII

Bergeret, his 'Des fraudes dans l'accomplissement des fonctions génératrices...,' II 142, 143

Berkeley, on generic images, II 298

Bernoulli, Chr., uses word 'Biometrie,' IIIB 500

Bertie Terrace (home of Galton's mother and of his sisters in Leamington), visits to, II 88, and in most years; last visit, IIIB 570

Bertillon, his system of identification of criminals, II 304, 305, 380, 383, 397, 398, IIIA 5, 55, 140–142, 144, 150–152, 155, 187, 188, 199; letter of, to Galton, IIIA 144, 145; letter of, to Dr Faulds, IIIA 144; vanity of, IIIA 249

'*Bertillonage*,' IIIA 140, 144, 148–153, 159, 187, 188, 200

Besant, Sir Walter, and the University of London, IIIA 289

Bess, Galton procures the vase with the God upon it, I 202; presented to the British Museum, IIIB 473; casts of the vase, IIIB 473

Beverley, and use of finger-prints in India, IIIA 147

Bewick, T., his thumb mark, IIIA 175; vignette of churchyard, IIIB 580

Biarritz, Galton at, IIIB 553–559

Bicknell, Mr, stays with Galton, IIIB 587, 588

Bidder, G., experiments with, II 195–196; number form of, II 242, IIIB 469; inherited faculty of, II 276; at Plymouth Aquarium, IIIB 579

Biggs, Evelyne ('Eva,' great-niece to Galton), plans to be much with Galton, IIIB 512, 520–522; to live with Galton, IIIB 520; with Galton, IIIA 279, 280, 282, 285, 287, 310, 324, 360, 374, 375, 388, 389, 391, 429, 433, IIIB 447, 449, etc.; travels with Galton, IIIB 507–512, 515–519, 520–522; her care of Galton, IIIB 595, 617, etc.; Galton's appreciation of, IIIB 512, 513, etc.; first ride in a motor car, IIIB 547; on Eugenics Education Soc., IIIA 433; Galton's letter to, on family prayers, IIIA 271; Galton's letters to, IIIB 577, 597, 603, 605, 612; Bateson's letter to, IIIA 288; Sir Archibald Geikie's letters to, IIIB 617; letters of, to Karl Pearson, IIIA 433, IIIB 618; to Lady Pelly, IIIA 433; to Perry Coste, IIIB 618; sketches of, IIIA 278; becomes a Roman Catholic, IIIB 604, 605; health of, IIIB 583, 605, etc.; portrays the personality of Galton, IIIB 618; portrait of, IIIB 507 Plate LIII; with Galton at Ockham, IIIB Plate LVIII; sketches of Francis Galton, II viii, 425

Biggs, Rev. G. H., father of Eva Biggs, IIIB 512

Billiards, and thought without words, II 274

Billings, Dr, and composite crania, II 294

Binet, A., and 'Psychologie des Grands Calculateurs et Joueurs d'Échecs,' II 275

Binomial Polygon, and normal curve, II 338, 339; and statistics of stature, II 339

Biographer, difficulties and duties of, I v–ix, 1–4, II v, 88, IIIA vii, 413

Biographers mislead, IIIA 277

'*Biographia Evangelica*,' of Middleton, II 100, 101

Biographical Register, Galton's, II 355

Biological Farm, proposition for, IIIA 133, 134. See also *Experimental Farm, Biometric Farm*

Biologists, mechanical aptitude in, II 151; and Biometry, IIIA 254, 256, 283, 286, 287

Biology, and Galton, II 62; source of Galton's interest in, II 201; future of, IIIB 501

Biometric Farm, need for, IIIA 251

Biometric Laboratory, IIIA 224; and Eugenics Record Office, IIIA 258, 297–299; and Eugenics Laboratory, IIIA 296–299, 303–305, 315; motto of, II 233; apparatus used in, IIIA 308; post-graduates in, IIIA 345, 356

Biometric Soirée, IIIA 335, IIIB 605

Biometricians, future outlook for, IIIA 381; on holiday, IIIA 277, 280, 322, 342, 368, 369, 388–390, IIIB 441, 527, 528, 583

'*Biometrika*,' and Galton, II 69, IIIA 248–250, 325; foundation of, IIIA 100–102, 235, 241, 243–245, 248, 281, 285, 319; guaranteed fund for, IIIA 244, 245, 250; early struggles of, IIIA 239, 254, 256; first proofs of, IIIA 246, 247; vol. II of, IIIA 251; progress of, IIIA 368, 425, 431; reconstitution of, IIIA 281, 302; skull photographs in, IIIA 257; definition of technical terms in, IIIA 334; new year greeting to Galton in, IIIA 394; Royal Society and, IIIA 283; Biologists and, IIIA 254; 'Journ. of Anthropological Institute' and, IIIA 247; Weldon memorial and, IIIA 285, 301

Biometry, and the Royal Society, IIIA 100, 101; early difficulties of, IIIA 101, 282, 283, 286–288; Bateson's contempt for, IIIA 288; Mendelism and, IIIA 357, 358; Eugenics a branch of, IIIA 309; loss of Weldon to, IIIA 280

'*Biometry*' and 'phylometry,' III^B 500
Bioscopes, of all that grows, III^B 576
Bi-projection, photographic, II 298–300
Birds, Samuel Galton on, I 48; Galton's early interest in, I 68; artificial nests for, III^B 547; eggs, measurements of, III^A 243, 244
Birmingham, Galton houses in, I 46, 49–51; photographs of, I 48 Plate XXX, 49 Plate XXXI, 50 Plate XXXII; Samuel Galton in, I 46, 47; Galtons buried in, I 52, 50 Plate XXXII; Galton at school at, I 81–90; medical student at, I 90, 99–104, III^A 452; mediaeval education at, I 81
Birmingham and Midland Institute, Galton lectures at, III^A 12
Birmingham Philosophical Society and Samuel Galton, I 47
Birth, of Francis Galton, I 62, 63; order of, and health, III^A 404, 405
Birth-control, and Eugenics, II 80; dangers of, II 111
Blending, and inheritance, II 173, 397
Blind, sensitivity of the, II 218
Blomefield, with Galton on reading party, I 155, 156
Blomfield, Admiral and Mrs, Galton meets in Egypt, III^B 518
Blood, and hereditary characters, II 157; experiments in transfusion of, II 156–169, 174–177, see also *Pangenesis*; examination of, Galton's prediction concerning, III^B 564
Blood-pressure, and measurement of emotional shock, II 270
Blow, force of, and pain occasioned by, II 408; rapidity of, and instrument to measure, II 220, 221, 374, 376
Blunt, Lady Anne, Galton meets, III^B 550
Boar, wild, as a pet, III^B 562–564, 566
Board School, degeneracy among children of, III^A 266
Boccaccio, on Dante, II 99
Boers, interfere with Galton's plans in Africa, I 219, 220; war with, III^B 515–517, 519
Bogatzki, Galton uses family copy of a work by, III^B 585
Boils, Galton's treatment of, I 107, 108, 110, 111
Bond, Dr, offers Galton a clinical clerkship, I 184
Bonner, Mr David, on trotting horses, III^B 498
Bonnevie, K., and measurements applied to finger-prints, III^A 168
Books, by Francis Galton: 'Art of Travel,' 1855 (and many editions), II 2–6; 'Decipherment of Blurred Finger Prints,' 1893, III^A 194–197; 'English Men of Science, their Nature and Nurture,' 1874, II 87, 130, 142, 145, References to, I 5, and to de Candolle, II 134, 149, 207; 'Record of Family Faculties,' 1884, II 363 etc.; 'Life History Album,' 1884, 1903, II 366 etc.; 'Finger Print Directories,' 1895, III^A 199–215; 'Finger Prints,' 1892, III^A 174–194, References to, III^A 141, 142; 'Hereditary Genius,' 1859, 1892, 1912, II 87–115, References to, I 5, 7, II 70, Charles Darwin on, I 6; 'Inquiries into Human Faculty and its Development,' 1883, II 248–267, References to, I 5, II 87, 207 (de Candolle), 212, 238, 241, 361; 'The Knapsack Guide for Travellers in Switzerland,' 1864, 1867, II 11; 'Memories of my Life,' 1908, III^A 354–355, References to, III^A 329, 339, 342–346, 354–355, III^B 585–588, 613; 'Meteorographica, Methods of Mapping the Weather; illustrated by upwards of 600 printed and lithographed Diagrams...,' 1861, II 38–43, Plate VII; 'Natural Inheritance,' 1889, III^A 57–77, References to, I 5, II 84, 87, III^A 79–82; 'Noteworthy Families (Modern Science),' 1906, III^A 113–121, Reference, II 149; 'Tropical South Africa,' 1853, 1889, I 215 etc., References to, I 215, 240; 'Vacation Tourists and Notes of Travel,' 1860–1862, ed. by F. G., II 6–7
Booth, Dr, medical sponsor to Galton, I 90, 99; consulted by Galton, I 152
Booth, C., his survey of the population of London, III^A 228, 231
Booth, Johnny, death of, from scarlet fever, I 81
Booth, Mrs (Adèle Galton, aunt of Francis Galton), at birth of Galton, I 62; at death of her father, III^A 45; old letters of, III^B 575; silhouette of, I 54 Plate XXXV; portrait of, I 54 Plate XXXVI
Border-line Cases, and 'mechanical selector,' II 305
Boredom, measurements of, II 277
Bortkiewicz, von, and Galton's difference problem, II 413
Boscastle, visit to, II 130
Botany, Galton studies under Lindley, I 121, 123
Boulogne, Galton at school at, I 68, 70–73
Boulton, H. W., at the Rev. Atwood's school, I 77
Boulton, Mathew P. W., Galton's friendship with, I 77, 188; at Cambridge, I 141, 143, 150
Boulton, Montague, at Cambridge, I 171; Galton up the Nile with, I 200–203; Galton meets at Beyrout, I 204; death of, I 204
Bourdillon, J., and use of finger-prints in India, III^A 147
Boutmy, É., honour conferred on, III^B 494
Bowen, and 'The Reader,' II 68
Bowman, Sir W., Galton's early tour with, I 92–97, III^B 565; Galton works at anatomy under, I 105; Galton's acquaintance with, I 126
Bowring, a wrangler at Cambridge, I 167
Boyhood, of Galton, I 62–91, III^B 449–452, 618; associations of youth and, II 235
Boyle Lecture, given by Karl Pearson, III^A 309, 313–315
Brabrook, Sir E., on low literary standard of scientific memoirs, III^A 330, 332
Bradford, effect of Factory Acts in, III^A 368
Bradlaugh, and Neomalthusianism, III^A 243
Bradshaw, Henry, advice from, III^A 322
Brain, Galton on activity of, II 234–236; selective action of, II 256; weight of, in eminent men, III^A 248; pigment of, and insanity, III^A 372; in the sane and the insane, III^A 298, 356
Braine, Elizabeth, marriage of, with Robert Barclay, I 35; ancestry of, I 31, 34, 35
Brainwork, unconscious, III^A 115
Bramah, his lock and Galton's, I 148
Branford, V. V., on architects of science of sociology, III^A 261; Galton consults with, III^B 529
Breadsall Church, sketch of, I 74 Plate XLIV
Breadsall Priory, home of Dr Erasmus Darwin, III^B 525; visits to, I 74; sketches of, I 74 Plates XLIII, XLIV
Breathing, its submission to the will, II 247
Bree, Adèle (great-niece to Galton), III^B 586, 587
Bree, Archdeacon, translates honorary degree oration at Cambridge as to Galton's achievements, III^B 494, 495
Bree, Sophy (great-niece to Galton), III^B 511, 512; sees Galton receive the D.C.L. in Oxford, III^B 494
Breeding, art of, and stud-books, II 321; experimental, III^A 128–131, 272, 273
Breeds, establishment of, and regression, III^A 31; suitability of, for experiment, III^A 135
Brewin, Mrs (Sophia Galton, aunt of Francis Galton), character of, I 54; mementoes from, III^B 585; portrait of, I 54 Plate XXXV
Brewster, Sir David, parentage of, II 137
Brewster, E. T., on a measure of variability, etc., III^A 95
'*Bridgewater Treatises*,' authors of, II 151, 152
Bristed, C. A., his 'Five Years in an English University,' I 171, 182

British Association, and Galton, II 13, 18, 20, 22, 27–30, 34, 35, 49, 51, 53–55, 59, 61, 70, 77, 233, 334, 347, 362, 386, 388; Galton's first attendance at, I 104; Tertius Galton attends meeting of, I 90; Galton's papers and addresses at, II 228, 238, 288, III^A 11, 57, III^B 461; Galton asked to accept office of Gen. Sec., III^B 458; Galton asked to stand as President, III^A 275, 276, III^B 488, 543; attack on Biometry at meeting of, III^A 297; Cambridge meeting of, III^B 528; meeting of, in Africa, III^B 548; Lister's address at, III^B 579; maps for the, III^B 462; collection of records of pedigree stock and, II 321; marks for physical efficiency and, II 394; experimental zoology and, III^A 129; finger-prints and, III^A 140, 148; on Section of Economic Science and Statistics of, II 347, 348
British Museum, Galton presents 'Bess' to, III^B 473
British Race, deterioration of, III^A 251, 364–367; defects and qualities of, III^A 252, 253; improvement of, III^A 253. See also *Anglo-Saxon*
British Type, union of, with Scandinavian, II 371
Broadley, Anne, describes a visit from the youthful Galton, I 98; amanuensis to Galton, I 217
Brodrick, G., death of, III^B 525
Brooke, Sir A., takes Galton to see a clairvoyant, I 190
Brothers, Galton's values of correlation and regression for pairs of, III^A 25
Browne, Rev. G. F., on Galton's Rede lecture, III^B 473
Browne, Sir T., and use of word 'aberrance,' III^A 99
Browning, Oscar, and a circular walk in Heidelberg, III^A 404
Brown-Séquard, experiments of, on guinea-pigs, Galton's criticism of, II 182–184
Bruce, and the African memorial, II 25; writes first on Khartoum, III^B 548
Brussels, Galton's early visit to, I 94
Buckland, F., and Galton, II 87; 'The Land and the Water' of, II 87
Buckle, achievements of, II 420
Buda Pest, Galton at, I 135
Budding, Galton on, II 190
Buffalo, African, in Italy, II 31, 32
Bull, episode of mad, I 82
Bull-dog, effect of continued selection of size of head in, III^A 94
Bull-fight, Galton's description of, III^B 508, 512
Bulloch, W., material from, III^A 359
Bunbury, Mrs (Adèle Galton, sister to Francis Galton), on birth of Galton, I 62, 63; her early training of Galton, I 63, 65, 66, 68, 69, III^B 446; early letters of Galton to, I 66, 71, 80, 86, 87, 95, 102, 103, 119; Galton advises, I 126; Galton's early bequest to, I 68; letters of, to her sisters, III^B 449–451; marriage of, I 193; Galton visits, II 130; portrait of, I 213 Plate LV bis; silhouette of, I 52 Plate XXXIV
Burbury, and technical scientific terms, III^A 334; work of, III^B 487
Burgess, Dean, meets Sir Francis S. Darwin and Theodore Galton abroad, III^B 592
Burglar, visitations of a, I 79; finger-print of, on window frame, III^A 160
Burial Grounds, III^B 532
Burnand, Miss, and a cartoon in 'Punch,' III^A 375
Burns, Mr and Mrs J., and Lady A., III^B 574
Burton, the African explorer, II 30; and Galton, II 25, 27, 28, 68; and Speke, II 25–27
Burton-Speke expedition, II 25, 26
Bury, Mr, Galton's schoolmaster, I 70
Bushmen, visualising faculty in, II 239, 240, 252; method of drawing in, II 239, 240
Business, success of Samuel Galton in, I 46, 48; capacity of Sir Douglas Galton, I 53
Busk, Sir E., and the Fellowship in Eugenics, III^A 222, 223; and Galton bequest, III^A 301, 302
Busk, Dr G., craniology of, II 334; Galton's obituary notice of, II 396
Bust, of Galton, proposals regarding, III^A 374, 375; Sir G. Frampton models, III^A 388, 389, III^B 598, 599; in the Academy, III^B 605; all aspects of a, on a single negative, III^B 520 Plate LIV
Butler, on mimicry, III^A 370
Butler, Arthur (brother-in-law to Galton), to stay with Galton, III^B 540; Galton to visit, III^B 573; death of, III^A 370
Butler, Frank, to act for Galton, III^B 507; with Galton, III^B 548, 549; assists Galton, III^B 598, 599; Galton to visit, III^B 569, 570
Butler, George, to be consulted regarding plans for eugenic certificates, III^A 296; sees Vesuvius with Galton, III^B 475; visits Galton, III^B 573
Butler, Harcourt, on finger-print system in India, III^B 591; member of Viceroy's Council in India, III^B 607, 608
Butler, H. Montagu, appointed Master of Trinity, III^B 476; writes to Galton on his election to a Trinity Fellowship, III^A 236, 238; writes on receiving portrait of Galton for Trinity College, III^B 551; on the characteristics of Galton, III^B 619
Butler, James (son of H. Montagu Butler), wins scholarship, III^B 558
Butler, Mrs Josephine, II 130; meets Galton in Italy, III^B 475
Butler, Louisa. See *Galton, Mrs Francis*
Butler, Maud, meets Galton in Egypt, III^B 518, 519
Butler, Prof. Stanley, at St Andrews, III^A 361; son of, III^B 609
Butlers, pedigree of the, III^A 343
Butterworth, captain of the 'Dalhousie,' I 217
Button, Elizabeth, I 36
Button, Robert, a strenuous Quaker, I 35–38, 59; and George Fox, I 37
Button, Sarah, married John Galton, I 36
Button, Admiral Sir T., travels of, I 36
Buttons, Quaker strain of, in Galton ancestry, I 10, 11; family of, I 36, 37; heavy infant mortality of, I 36, 37
Buxton, Charles, at Cambridge, I 141, 164, 166, 167; with Galton on reading party, I 168; his widow's love of animals, III^B 547
Buxton, Sir Fowell, with Galton on reading party, I 168; takes a poll degree, I 171; in London, I 188, 189; meets Galton at Trinity College, III^B 574
Buys Ballot, meteorological assistance of, II 39

Cairns, and 'The Reader,' II 68
Cairo, Galton at, III^A 240. See also *Egypt*
Cake, on scientific cutting of, III^B 579, 580
Calculating Boys and inheritance, II 276
Calorifère, use of, III^B 545
Calvin, influence of, II 137, 139, 142
Cambridge Anthropometric Laboratory, II 226, 379, 387, 388
Cambridge Union Society, and Galton, I 173–175
Cambridge University, Galton's plans for going to, I 106, 107, 110; Darwin advises concerning education at, I 110; Galton's mathematical studies and career at, I 140–195; his breakdown at, I 194; influence of, on Galton and Darwin, I 12; on Galton, I 141, 142, 194, 195, III^A 238; affection for, I 140, 141, III^A 237, 238; confers honour on Galton, III^B 494, 495; Galton gives Rede lecture at, II 268, 270, 271, III^B 473; Karl Pearson lectures on statistics at, III^A 314, 315; mathematical lecturers at, III^A 315. See also *Trinity College*

Camden Antiquarian Society, Galton becomes a member of, I 163
Camden Medal, Galton's poem in competition for, I 176
Camel, and gregarious instinct, II 73; incident with a, III^B 575
Camera, Galton's, for enlarging finger-prints, III^A 214, 215; panoramic, III^B 520
Cameron, the African explorer, II 30
Cameron, Sir Ewen (great-great-great grandfather of Galton), strength of, I 30; portrait of, I 27 Plate XXI
Camp Life, models illustrative of arts of, II 18; in Egypt, at Flinders Petrie's settlement, III^B 516, 517
Campbell, on generic image of 'man,' II 298
Campbell, Lord, at Cambridge, I 141; letters to, from Galton in Africa, I 224–226, 233, 234; on morality of judges, II 94, 95
Campbell, J. F., Galton's review of his 'On Frost and Fire,' II 53
Canals, Samuel Galton's interest in, I 48. See also III^A vi
Cancer, proposals regarding, III^A 72–73, III^B 525
Candolle, A. de, and Galton, correspondence of, II 131–149, 204–210, III^B 474, 476–481, 483; his criticism of Galton, II 145, 146; deficiency of method in, II 146, 147, 149; on inheritance of acquired characters, II 147, 148; on composite portraits, II 204; on dreams, II, 205; on the Jews, II 209; on the effect of maternal impressions, II 209, 210; on inheritance of eye-colour, III^A 34, 37; at age of 84, III^B 483
Cape Town, and Table Mountain, model of, II 34
Capri, Galton at, III^A 256, 257
Cardiograph, and measure of emotional shock, II 270
Careers, and natural dispositions, II 258
Carlyle, heroes of, II 94
Carnac, old temples at, I 200
Carnegie Institution, III^A 290
Carpenter, on ocean currents, III^B 461
Carpenter, Dr, on idiocy in children of drunkards, II 148; on transmission of acquired characters, II 148
Carpets, and asthma, III^A 238
Carter, Brudenell, on Roy. Soc. Committee on Colour-blindness, II 227
Carter, Frank, to copy the Furse portrait of Galton, III^B 550, 551; visits Galton, III^B 585; sketches of Galton, aged 88, by, II iv, III^A 432 Plate XL
Caseo-Tostic Club, last meeting of, I 181 Plate LIV
Caste, among the gifted, II 120, 121
Castle, Prof., Karl Pearson criticises, III^A 261
Casualties, minor, of Galton due to 'Eilwagen,' I 96, 97
Catholic Church, Galton on the, III^A 271; evils of policy in the past, II 111, 112. See also *Roman Catholics*
Cats, hearing of, II 216; Galton on, III^B 515
Cattle, of Damaraland, gregarious instincts in, II 73, 74; hearing of, II 216
Causation, and correlation, III^A 315
Cave, Miss F. E., meteorological paper of, III^A 282
Cayley, Prof., Senior Wrangler, I 164; Galton joins reading party under, I 168; not the ideal teacher for Galton, I 194; his size of head and stature of, II 150
Celandines, Galton to collect, III^A 251
Celibacy, evils of Catholic regulation of, II 111, 112; at Oxford and Cambridge, II 266; customs regarding, III^A 269.
Centiles, table of, III^A 303, 304. See also *Percentiles, Ogive Curve*
Cephalic index, values of, III^A 54
Certificates, proposal for Eugenic, III^A 272, 292–296. See also '*Kantsaywhere*'
Chamberlain, Joseph, never takes exercise, III^B 569
Chameleon, brought from Egypt by Galton, III^B 516
Chance, Galton's schoolfellow, I 160
Chances, paradox in theory of, II 405
Character, of Galton, early, I 64, 67, III^B 618; development of, I 112; Galton's stock-taking of, III^B 581, 582; sources of physical and mental, I 55, 56, 59, 60; mental, compared with that of Darwin, I 57, 58; compared with that of George Fox, II 122; of Galton, later, II 60, 64, 65, 385, 396, III^A 121, 239, 240, 262, 278, 279, III^B 441–443, 446–449, 486, 618, 619; independence of, in scientists, II 151; Galton on estimation of, II 208; measurement of, II 209, 268–270; inheritance of, II 269, 272; multiple factors of, II 269; correlation of factors in, II 269; innate, of different races, II 352, 353
Characterisation, of Galton by letters, III^B 441–619
Characteristics, and talents of Darwin and Huxley, II 178, 179; of Herbert Spencer, III^A 317, III^B 626–628
Charities, and Eugenics, III^A 234; evils of, III^A 243, 274, 323, 348, 352; future reduction of, III^A 273
Charterhouse, statistics from boys of, II 237
Chau-tao-Chen, visits the Galton Laboratory, III^A 385, 386
Chemistry, Galton studies, I 105, 109, 121; under Liebig, I 126–130
Chemists, mechanical aptitude in, II 151; lack of imagination in terminology, III^A 337
Chepmell, Dr, advises Mrs Galton, III^B 464, 473
Chess-players, blindfolded, II 252
Chesterton, G. K., on proposals of eugenists, II 365, III^A 374
Childhood, and boyhood of Galton, I 62–91; teaching in, II 127
Childlessness, of the male in Galton's ancestry, I 21
Chimneys, their faults and remedies, III^B 591
China, lessons from revolution in, III^A 90; use of finger-prints in, III^A 146
Chinaman, Galton on the, II 33, 85
Chinese, finger-prints of, II 195; identification of, III^A 175; honours among the, III^A 265
'*Chips from the Workshop*,' Galton asks for, III^A 300
Chree, Dr, and Kew Observatory, II 60
Christian, the, and sense of sin, II 102
Christianity, Galton on, III^B 582; and missionary enterprise, II 28, 32
Chronograph, designed by Galton, II 226, 227
Chumley, maid at Rutland Gate, III^B 517, 519
Church, effect of the, on evolution, II 111, 112; sequestration of property of, II 410, 411, III^A 370, III^B 591
Church, on Roy. Soc. Committee on Colour-blindness, II 227
Churchill, Col., Galton meets in Syria, I 203
Churchill, Mr, teaching of, I 78
Churton, Dean of King's College, Cambridge, III^A 345
Cinematograph, foreshadowed by Galton, II 284
Circle, photographic reduction of, to ellipse, II 300
Circular, issued by Galton regarding 'sports,' III^A 87. See also *Schedules*
Circulation, Galton's definition of, II 164
Circumstances, and success, III^A 111, 112, 116
Civic Worth, propositions regarding, III^A 227–229, 241, 242, III^B 620–626; classification of, III^A 351; and differential fertility, III^A 264; and caste, III^A 352; encouragement of, II 120, 121, III^A 231–233, 242, 352, III^B 609. See also *Eugenics*, '*Kantsaywhere*'
Civil Service, and marks for physical qualifications, II 387, 388, 394; examinations for, III^B 479–481, 483; as tests of ability, III^A 232; failure to demonstrate the efficiency of the examinations, III^A 232
Civilisation, rise and fall of, II 108, 109; influence of Church on, II 111, 112; effect of differential fertility on, II 112; best form of, II 112, 113; evil of centralising tendency of, II 118; and fertility, III^A 264

Clapperton, and African memorial, II 25
Clark, W. G., at Cambridge, I 141, 164; at meeting of Caseo-Tostic Club, I 181 Plate LIV
Clarke, Dr A. (Sir Andrew), advises Galton, II 130, 180; praised, III^{B} 463
Clarke, Sir E., on Roy. Soc. Committee for Measurement of Plants and Animals, III^{A} 127
Class, intermarriage within the, III^{A} 231
Class-representation, Galton's definition of, II 171
Classes, social, physical inequalities in different, II 35, 125, 126; relative fertility of, III^{A} 218, 219
Classical Scholars, fertility of, II 96
Classics, Galton works at, I 107, 146, 154; in education, I 88, 89, II 155; senior, at Cambridge, and heredity, III^{A} 347
Classification, of criminals, II 230; in pedigree work, III^{A} 343, 344. See also *Finger-prints*, classification of
Claverdon, purchased by Samuel Tertius Galton, I 51; a family centre, I 52; Galton's visits to, I 208, II 11, III^{B} 569, 577, etc.; Darwin Galton farms at, I 125; the Wheler-Galtons at, III^{B} 528, 529, etc.; sketch of house, I 48 Plate XXIX
Claverdon Church, Galton's memorial tablet in, III^{A} 434; sketch of, I 48 Plate XXIX; Galton's burial-place, III^{A} 433 Plate XLI; obituary notice of Galton in 'Claverdon Parish Magazine,' III^{B} 619
Clay, Rev. Mr, instructs Galton, I 67
Cleopatra, Queen of Egypt, composite portrait of, II 295, 296 Plate XL
Cleopatra, Queen of Syria, portrait of, and composite, II 296 Plate XLII
Clergy, among men of science, II 151; colour of dress of, III^{B} 534, 535
Clifford, W. K., unconventionality of, I 94; and agnosticism, II 102; as a scientist, III^{A} 333
Climate, Galton on, II 7, 35–49, 53–62; and geography, II 29. See also *Meteorology, Weather*
Clock, for cumulative temperature, designed by Galton, II 361
Clouds, height of, II 61
Cobalt, Dr, seen by Galton in Heidelberg, I 95
Cobbe, Frances Power, 'Hereditary Piety' of, II 160
Cockerell, T. D. A., on experimental zoology, III^{A} 129
Co-composite portraits, II 291, 292; examples of, II 296 Plates XXXVII, XXXIX
Coddington Lens, Galton purchases, I 115
Co-education, and pigtails, III^{B} 601
Coinage, decimal, II 21
Co-kinsmen, definition of, III^{A} 19
Colbert, law of, touching oak forests, II 122
Coleridge and Byron, Galton recalls, III^{B} 589
Collaterals, of great men, III^{A} 111
Collier, Elizabeth (grandmother of Francis Galton), ancestry of, I 18–22; birth of, II 193; descendants of, I 22–26, 74; characteristics of, I 49, 74; marriage of, to Colonel Pole, I 19; marriage of, to Erasmus Darwin, I 18, 74; death of, I 74; poem to, I 18 Plate XI; portrait of, I 20 Plate XVI; sketch of, I 18 Plate X; silhouettes of, I 14 Plate IV bis, I 21 Plate XVII
Collins, H., and 'Art of Travel,' II 6; researches of, on finger-prints, III^{A} 140, 174, 190, 194; letters to, from Galton, on finger-prints, III^{B} 484–486, 488–492; definition of Eugenics and, III^{A} 269; proof-reading of, III^{B} 525, 527, 548, 550, 556; and compass points, III^{B} 575
Collins, Inspector, and finger-prints at Scotland Yard, III^{A} 145, 151
Cologne, Galton visits Cathedral at, I 95
Colour, associated with number, II 214, 253; effect of, on irritability, II 214; associations, II 240, 241, 243, 253; measure of resemblance in, II 303; of hair, skin and eyes, tests for and standard scales of, II 223–226; hereditary in horses, III^{A} 95–98
Colour-blindness, Roy. Soc. Committee to investigate, II 227
Colour Mixing, experiments in, and Samuel Galton, I 47, 48
Colour Sense, tests for, II 223; Galton's measure of sensitivity of, II 226; in men and women, II 376
Coloured Light, and earthworms, II 196, 197
Colyear, Lady Caroline, and Elizabeth Collier, I 19
Colyear, Charles (Lord Portmore), and parentage of Elizabeth Collier, I 19; portrait of, I 18 Plate XIII
Colyear, Gen. Sir David (Lord Portmore), distinction of, I 19; portrait of, I 18 Plate XII
Commanders, fertility of, II 96
Committee, for Measurement of Plants and Animals ('Evolution Committee'), III^{A} 126, 127, 131, 133, 134, 135, 286–291, III^{B} 501; change of name, III^{A} 291; history of, III^{A} 290, 291; papers regarding, III^{A} 311, 312; for matters concerned with the fellowship of national eugenics, III^{A} 222, 223; to inquire into methods of identifying criminals, III^{A} 140, 148–153
Committees, difficulties of working through, II 362, 366, 367, 418, III^{A} 223
Communities, prosperous and decadent, III^{A} 402
Comparates, definition of, II 332
Competition, evils of, I 171
Composite Photographs, Galton's first announcement as to, II 229, 230; use of, II 230–233, 288, 294; of criminals, II 230, 231, 286, 293, 295, II 286 Plates XXVIII, XXIX; of the Darwins, II 192; of men and officers of Royal Engineers, II 290, II 286 Plate XXIX; of horses, II 399, II 288 Plate XXX; of members of a family, II 295, II 288 Plates XXXI, XXXII, XXXIII; of Welsh ministers, II 288 Plate XXXIII; of phthisical subjects, II 290–293, II 291 Plate XXXIV; of skull, II 288, 290, II 288 Plate XXXIII; of Jews, II 290, 293, 294, II 294 Plate XXXV; of Alexander the Great, II 295, II 296 Plates XXXVI, XXXVII; of Antiochus, King of Syria, II 295, II 296 Plate XXXVIII; of Demetrius Poliorcetes, II 295, II 296 Plate XXXIX; of Cleopatra, II 295, II 296 Plate XL; of Nero, II 295, II 296 Plate XLI; of Greek Queens, II 295, II 296 Plate XLII; of Roman ladies, II 295, II 296 Plate XLIII; of Napoleon, II 295, II 296 Plate XLIV; Quetelet's 'mean man' and, II 297; psychology and, II 297; racial, II 290; de Candolle on, II 204; American examples of, II 290
Composite Photography, II 283–299
Comte, A., religion of, III^{A} 93
Conception, influence of parents' states at time of, II 136, 138–140, 146
Condorcet, and application of mathematics to social phenomena, III^{A} 1; on correct judgments, III^{A} 320
Confinement, effect of, on fecundity, III^{A} 129
Congenital Peculiarities, and organic units, II 184
Conjugate Foci, of lens, mechanical determination of, II 50
Conscience, in criminals, II 230; inherited and acquired, II 257
Conscientiousness, correlation of, in pairs of brothers, III^{A} 247
Consciousness, antechamber of, II 256; and sub-consciousness, II 236
Conscription, effect of, on racial characters, II 191
Conscripts, French, unfitness of, II 120
Constable, Mr, criticises Galton, III^{B} 550
Constantinople, Galton's journey to, I 128, 131–138
'Consumptivity,' III^{A} 74

Controversies, pending, IIIA 326
Controversy, Galton's attitude regarding, II 27, 131, IIIA 138, 397–400, 404, 407, IIIB 590, 601
Convicts, freed, IIIB 549
Conway, Moncure D., letter of, to Galton on death of Darwin, IIIB 471
Conway, W. M., in the Inner Temple, IIIA 302
Cookes, Denham, travelling companionship of, I 199
Cooper, with Galton on reading party, I 155, 159; at Cambridge, I 164
Co-operation, in scheme for work on heredity, etc., IIIA 135
Coordinates of point in space, Galton's photographic determination of, II 318, 319, Diagram vii, Figs. 1–5
Copley Medal, conferred on Galton, IIIA 400, 431, 432, IIIB 611, 614, 615; other Copley medallists, IIIA 431, IIIB 614
Copts, profiles of, II 324
Corporal Punishment, Galton on, II 408
Correlation, and the application of statistics to the problems of heredity, IIIA 1–137. Galton approaching the idea of, II 55, 383, 384; first use of word by Galton, II 150; Galton reaches conception of, II 392, 393, IIIA 1, 2, 50; definition of, IIIA 50; of grades and ranks, II 393, IIIA 3; early methods of measuring, II 301; Galton's measure of, IIIA 50–56; coefficient of, IIIA 5, 9, 48; regression and variability, II 384, IIIA 3–5; multiple, IIIA 47, 55; multiple, and prediction of character of individual from study of kinsfolk, IIIA 27; and regression for pairs of brothers, IIIA 25; of factors of character, II 269; between characters of Bertillon's system, II 383; of finger-prints in right and left hands, IIIA 255, 256; of finger-prints on different fingers, IIIA 140, 161; of physical characters, II 390, 398; of eugenic qualities, IIIA 273; between physical and mental characters, II 301; between moral and physical sensitivity, II 408; between pain felt and force of blow, II 408; parental and fraternal, IIIA 329; of stellar characters, IIIA 326; first examples of, in characters of organisms other than man, IIIB 483, 484; Mendelian hypothesis and, IIIA 378; table, elliptic contours of, IIIA 13, 14; earliest for inheritance, IIIA 64; for stature and cubit, IIIA 52
Correlational Calculus, foundations of, II 380; Galton and the, II 357, 377, 378, 383; first stages in development of, IIIA 5, 6; scope of, II 383; psychology and, II 213; factorial genetics and, IIIA 3, 5. See also *Correlation*
Correlations, and their measurement, IIIA 50–57
Corrie, Mr, death of, IIIB 452
Cotton, Sir H., and finger-prints, IIIB 590
Country Life and Sir Francis S. Darwin, I 23
Court, Mrs A', lends donkey chair to Galton, IIIB 589
Courtney, Mrs L., Galton meets, IIIB 531
Courvoisier, Galton sees him hanged, I 126
Cousins, omission of, from Galton's records, II 363; marriage of, II 188, IIIB 470; resemblance of, IIIA 310, 322, 328, 329, 333, 334; in a double degree, IIIA 241
Cox, Sergeant, and spiritualism, II 63, 64
Crackanthorpe, M., and Eugenics Education Soc., IIIA 339, 345, 346; attacks Galton Laboratory, IIIA 405–408, 427–429, IIIB 586; paper of, IIIA 303, 322; on the feeble-minded, IIIA 343; on the sight of hawks, IIIA 380; visits Galton, IIIA 429
Cranial Composites, II 288, 290, 294, II 288 Plate XXXIII
Craniometry, debt of, to Galton, II 334, IIIA 256, 257
Crawley, A. C., on Anthropology and Eugenics, IIIA 268
Crayfish, concerning, IIIB 549
Creative Mind, a peculiarity of the, II 412
Creeds, founders of, IIIA 217; and nations, IIIB 594
Creighton, Bishop Mandell, honour bestowed on, IIIB 494
Cremation, concerning, IIIA 375
Cremorne, Viscountess, ancestry of, I 32
Creskeld Hall, Darwin portraits at, I 243
Crewdson-Benington, Dr, work and death of, IIIA 425
Crichton-Browne, Sir J., and Eugenics Education Society, IIIA 335, 339; material from, IIIA 356; Galton on, IIIB 587
Crime, effect of education on, II 417
Crimean war, and Galton, II 13, 14, 18
Criminal Anthropology, II 286, 293; Macdonell on, IIIA 247
Criminal Class, perpetuated by heredity, II 231
Criminals, identification of, II 326, 383, IIIA 55, 140–142, 144, 146–155, 176, 177, 195, 199, 249; errors in identification of, IIIA 153, 154, 249; from composite photographs, II 230, 231, 286, 293, 295, II 286 Plates XXVIII, XXIX; characteristics of, II 230; classification of, II 230; sensitivity to pain in, II 408
Criticism, Galton's attitude towards, IIIA 397–400, 408, 409, IIIB 550, 601; of Galton's work, II 249, 250, 258, 261; value of, IIIA 318; Galton welcomes from his friends, IIIB 586
Crookes, Sir W., and spiritualism, II 62–66, 167; radiometer of, II 63; a Copley medallist, IIIA 431, IIIB 614; golden wedding of, IIIB 568
Croom-Robertson, G., and Galton, II 212; and measurement of sensation, II 362; appreciation of, IIIA 355, IIIB 493
Cubit, and stature, IIIA 51, 52; statistics of, IIIA 54
Cuckoo, Galton on the, II 191; and foster parents, II 127; nature and nurture and the, II 258; eggs of, IIIA 247
Cunliffe, Evelyn, daughter of Sir Douglas Galton, IIIB 506, 549
Cunningham, Prof., and anthropometry, II 379, 380
Currents, of North Sea, experiments on, IIIB 579
Curve of Errors, used by Galton, II 90. See also *Normal Curve*
Customs, acceptance of, IIIA 217
Customs Officer, Austrian, Galton describes, I 96
Cuvier, and correlation, IIIA 2
Cyclones, Dove's work on, II 39; and anticyclones, II 39–41

'*Daily Telegraph*,' Galton visits office of, IIIB 574
Dakyns, H. G., to revise MS. of 'Kantsaywhere,' IIIB 612; writes to Galton, IIIB 615
Dalby, Sir W., and Galton's whistle, II 216; meets Galton, IIIB 509
'*Dalhousie*,' Galton's ship to Africa, I 217, II 54
Dallmeyer, assistance of, II 313
Dalyell, at Cambridge, I 153, 164; at meeting of the Caseo-Tostic Club, I 181 Plate LIV; friend of Galton, I 214
Damaraland, Galton proposes to cross, I 215, 217. See also *Travels*, African
Damaras, Galton's protection of the, I 226–230
Dance, use of profiles of, II 323, 326, 328, IIIB 578
Daniell, Galton studies chemistry under, I 105, 109, 126, 127
Dante, Boccaccio's account of, II 99
Danube, Galton travels down the, I 133, 134, 136
Darbishire, on influence of ancestry, IIIA 378
Darmstadt, Galton visits museum at, I 95
Darwin family, history of, I viii; pedigree of, IIIA 343, 345. See also Pedigrees of distinguished ancestors of, in pockets to Vols. I and IIIA

Darwin, Charles, Howard ancestry of, I 244–246; work of, I 15, 58; mentality of Galton and, I 57, 58; talents and characteristics of, II 178, 207, 208; early interest of, in entomology, I 68; early meetings of Galton and, I 51, 91, 108; gives advice on Galton's education, I 110; on spiritualism, II 62–67, 167, 168; on discoverers, II 1; and the pangenesis experiments of Galton, II 113, 156–197; influence of, on Galton, II 170, 200–202, 206; disagreement of, with Galton, II 156, 157, 162–165, 184–190; affection of, for Galton, II 197, 199; letter of, to Galton, on publication of 'Tropical South Africa,' I 240, 241; letter to Galton on receipt of 'Hereditary Genius,' I 6, II 115; correspondence of, with Galton, II 157–201; letters of, to 'Nature' on Pangenesis, II 163; to his Aunt Violetta Galton, II 183, III^{B} 460, 461; and on Galton's rabbit breeding, II 65; on Galton's theory of heredity, II 187; on Galton's eugenic policy, II 176; and statistical methods, III^{A} 246; on birth control, II 111; on the inheritance of acquired characters, II 147, 148, 170, 173, 174; views of, on latency, II 170; on elements of reproduction, II 174; on domestic animals, II 71; on earthworms, II 196, 197; on fertilisation of sweet-peas, III^{A} 325; on popularisation of science, III^{A} 333; on admission of women to examinations at Cambridge, II 134; 'The Reader' and 'Nature' and, II 68, 69; letter giving religious views of, II 102 Plate XII; on his grandfather Erasmus, II 192–194, 196, 204; visualising faculty of, II 194, 195, 207; size of head and stature of, II 150; health of, II 166, 175, 179, 197, III^{B} 461; death of, II 197, 361; burial of, II 198, III^{B} 471; memorial to, II 199, 200, III^{B} 476; fortune of, II 206; portraits of, I 56 Plate XXXVII, I 68 Plate XLI, III^{A} 340 Plate XXXV; study of, at Down, II 200 Plate XIX; Mr Faulds writes to, on finger-prints, III^{A} 143; de Candolle and, III^{B} 477, 478

Darwin, Mrs Charles, letter of, referring to the pangenesis experiments, II 158

Darwin, Charles Galton, wins scholarship, III^{B} 558; mathematical tripos of, III^{A} 368, 385, 386, III^{B} 597

Darwin, Col. C. W., assistance of, I viii, 243

Darwin, Edward (uncle to Francis Galton), I 22

Darwin, Edward Levett (cousin to Francis Galton), early bequest to, I 69; death of, III^{B} 552

Darwin, Emma (aunt to Galton), I 22; sketch of, I 18 Plate X

Darwin, Dr Erasmus (grandfather to Galton), descendants of, I 15, 22; influence of, on Darwin and Galton, I 13, 14, II 201; mechanical ability and scientific tastes of, I 16, 49, 58, 60; marriage of, to Elizabeth Collier, Mrs Pole, I 18, 22; life of, II 192–194, 196, 202; natural daughters of, I 17, III^{B} 462; and Samuel Galton, I 46, 47; and Watt, I 16; the Lunar Society and, I 61; and Samuel Johnson, II 194; and mental imagery, II 196; visualising faculty of, II 194; on extinction of families, II 95; doctrine of evolution of, II 202, 203; poetry of, II 206; lines of, on air-ships, I 83; poem of, to Mrs Pole, I 18 Plate XI; contribution of, to posterity, II 204; monument to, II 202–204; reproduction of tablet to, II 204 Plate XX; portraits of, I 13 Plate III, I 76 Plate XLVI, II 192 Plate XV; sketch of, I 18 Plate X; silhouettes of, I 14 Plates IV and IV bis; medallion of, III^{B} 473 Plate L; visiting card of, I 195; armchair of, III^{B} 571; Dr Krause's life of, II 192

Darwin, Erasmus (son of Dr Erasmus Darwin), genealogical notes of, I 244

Darwin, Erasmus (brother of Charles Darwin), and spiritualism, II 66; and life of Dr Erasmus Darwin, II 192

Darwin, Sir Francis (son of Charles Darwin), assistance of, I viii; on Committee for Measurement of Plants and Animals, III^{A} 126, 290; on farm for experimental breeding, III^{A} 134; achievements of, II 208

Darwin, Mr Francis Rhodes, assistance from, I viii

Darwin, Sir Francis Sacheverell (son of Dr Erasmus Darwin), character and tastes of, I 22–24; interest of, in plague, I 23; travels of, I 22, 23; journal of travels of, III^{B} 592; resemblance of Galton to, I 24, 137; Galton sends to Darwin a portrait of, II 192; portraits of, I 22 Plate XVIII, II 192 Plates XV, XVI; sketches of, and by, I 18 Plate X, I 22 Plate XIX

Darwin, Sir George Howard (son of Charles Darwin), on Galton's wave machine, etc., II 52, 53, 55; on maps, II 22, III^{B} 461, 462; on spiritualism, II 62, 66; on marriage of first cousins, II 188; and the pangenesis experiments, II 189–191; on Galton's 'transformer,' II 315; on finger-prints, III^{A} 153, III^{B} 590; on technical scientific terms, III^{A} 334; on interpolation, III^{B} 467; correspondence of, with Galton, II 179, 180, 188, 190, 191, 200, III^{A} 276, III^{B} 461–466, 469, 470, 474, 475, 505, 584, 590; receives Copley Medal for Galton, III^{B} 614, 615; gold medal of Astronomical Soc. conferred on, III^{B} 488; receives the K.C.B., Galton's pleasure, III^{B} 552; President of British Association in Africa, III^{B} 533, 536, 543, 546; at the Benjamin Franklin commemoration, III^{B} 568; Galton on work of, II 206, 208; assistance from, I viii, 244; family portraits in possession of, I 243

Darwin, Lady George, assistance from, I viii, 244. See also III^{A} 340 Plate XXXV

Darwin, Harriet (daughter of Dr Erasmus Darwin), I 22

Darwin, Henry (son of Dr Erasmus Darwin), I 22

Darwin, Sir Horace (son of Charles Darwin), anthropometric instruments of, II 226, 227, 388; on Galton's chronograph, II 227; interest of, in the feeble-minded, III^{A} 373; Galton on, II 208; assistance of, II 311

Darwin, Mrs Horace, visits Galton, III^{B} 597

Darwin, John (son of Dr Erasmus Darwin), I 22

Darwin, Major Leonard (son of Charles Darwin), interest of, in Eugenics, III^{A} 311, 312, 323, III^{B} 602; address of, III^{A} 426; the L.C.C. and, III^{B} 515; Speke's memorial and, III^{B} 588; Pres. of Roy. Geographical Soc., III^{B} 603; letters to, from Galton, III^{B} 588, 602, 611; Galton on, II 208; assistance of, II vii

Darwin, Reginald (cousin to Galton), death of, III^{B} 488

Darwin, Robert, of Elston (great-grandfather of Darwin and Galton), portraits of, I 16 Plate VI, I 243 Plate LXII

Darwin, Robert Waring (brother of Dr Erasmus Darwin), 'Principia botanica' of, I 15; portrait of, I 16 Plate VI ter

Darwin, Robert Waring (father of Charles Darwin), portrait of, I 16 Plate VIII; character of, I 51; Galton visits at Shrewsbury, I 186, 187; assists Darwin Galton, III^{B} 450; letter of, to his sister Violetta Galton, III^{B} 454

Darwin, Violetta (mother of Galton). See *Galton, Violetta*

Darwin, William Alvey (brother of Dr Erasmus Darwin), portraits of, I 16 Plate VI bis, I 243 Plate LXII

Darwin, William Erasmus (son of Charles Darwin), birth of, III^{B} 453; portrait of, III^{A} 340 Plate XXXV; assistance from, I viii, 243, 244

Darwin Commemoration, at Cambridge, III^{A} 369, III^{B} 590, 598

Darwin's House. See *Down*

Darwin Medal. See *Medals*

Darwinian Hypothesis, and man, II 109; a religious creed, II 263; statistical methods and, III^{A} 126; grave reaction against, III^{A} 432

Darwinian Institute, a future possibility, III^A 311
Data, storing of, III^A 101
Davenport, C., as American editor of 'Biometrika,' III^A 244; on Inheritance of Eye Colour, III^A 376; letter of, to Galton, III^B 613
Darcy, Sir Horace, honour conferred on, III^B 494
Davies, Llewelyn, meets Galton at Trinity College, III^B 574
Day, Mr, marries Miss Parker, III^B 462
Dead Sea, Galton proposes to navigate the, I 205
Deaf-mutes, speech of, II 192; pigmentation of membrane of perilymph chamber in, III^A 372
Deafness, of Galton, II 280, III^A 276; aided by imagination, II 308; apparatus for, III^B 603–604; Galton's sympathy for sufferers from, III^B 584
Death, age at, of various classes, II 116; statistics regarding, III^A 70, 71
Debenham, W. E., on Galton's composite portraits, II 293
Decaisne, J., names a South African hyacinth 'Galtonia H. candicans,' III^B 533, 534
Degeneracy, effect of differential fertility on, III^A 9; a theory of, III^A 372
Degenerate Stocks, social danger of, III^A 373
Degeneration, of race, effect of conscription on, II 191, 192
Deity, Galton's conception of, II 114
Demboa Lake, Galton determines to reach, I 220, 231
Demetrius Poliorcetes, composite portraits of, II 295, II 296 Plate XXXIX
Democracy, and evolution, II 385; and Eugenics, III^A 348, 349
Democratic Judgment, trustworthiness of, II 403–405
Demographers, Galton's address to, III^A 218
Démolins, 'Anglo-Saxons' of, III^B 553
Dendy, Miss Mary, data of, on the feeble-minded, III^A 373
Denison, E. B., and Galton, on ability and fertility of judges and peers, II 93
Derby, The, Galton at, with Herbert Spencer, III^A 123, 124, III^B 627; flush of excitement at, III^A 124
Descent, on value of, II 84, 93, 364; diagram illustrating scheme of, III^A 230, 231
Desert, Galton's experiences of the, I 201–203
Desirables, and undesirables, III^A 348
Deviation, from an average, and degrees of independence in cattle, II 73. See also *Variation, Standard Deviation*
Devil, Catholics and the, III^A 432, 433
Dew-Smith, anthropometric instruments of, II 226; photograph of Galton by, III^A 217 Plate XXXI
Diagram, of instrument for compounding six objects, II 285; of camera, for composite portraiture, II 289; illustrating photographic reduction of circle to ellipse, II 300; of 'absolute values at each rank,' II 390; illustrating standard scheme of descent, III^A 230; illustrating graphical process of finding slope of regression line, III^A 52
Diamandi, arithmetic of, II 275, 276
Diary, Galton's, of school days, I 83–86; tiny diaries of his late years, III^B 618
Dice, suitability of, for verification of laws of frequency, II 405
Dickinson, Lowes, portrait painter, III^B 513
Dickson, J. D. H., and frequency surface, III^A 12; and mathematical analysis of Galton's problems, III^A 12, 13
Diet, effect of, on stature of races, II 210
Difference, just perceptible, II 307, 308; 'Greek girl,' to illustrate, III^A viii Extra Plate, facing Table of Contents
Digital Defect, in twins, II 181
Dinners, at Trinity College, Galton as undergraduate, 1844, complains of, I 182; in 1899, III^B 513; in 1908, III^B 574
Diplodocus, Galton calls on the, III^B 543, 544
Diplomas (or register), for eugenically fit young people, II 120, 121, III^A 231, 232, 234, 241, 242, 388. See also *Certificates, 'Kantsaywhere'*
Discontinuity, in statistical frequencies, II 411–414; in evolution, III^A 31, 32, 79–82, 84–87
Discoverers, plans for, II 27; scientific, and statesmen, II 135
Discovery, geographical, II 1, 2
Diseases, and piety, II 101; liability to, and hair colour, II 354, 371; screening of liability to, II 360, 368, 369; heredity of, III^A 70–76; Galton's schedules for, III^A 71–73; antagonistic, examples of, III^A 73
Disraeli, not as other men, III^B 568
Dissolute Lives, effect of, on weight, III^A 136, 137
Distance, photographic measurement of, II 316–318
Distant Ancestry, appearance of traits from, II 84, 364
Divers, spectacles for, II 34
Divines, Galton on, II 99–103; of Middleton's Collection, II 100, 101; fertility of, II 96, 101; parentage of, II 101, 102; marriages of, II 101; sons of, II 101–103, 137; moral oscillations of, II 102; health of, II 101, 116; age at death of, II 101, 116
Dogs, hearing of, II 216; suggestion of, for breeding experiments, II 76; transmission of acquired habits in, II 147, 148; judge by smell, II 275; albino Pekingese, III^A 356, 357, 389; and a pair of new trousers, III^A 431
Dohrn, III^A 290
Domestication, historical, II 70–72; Galton on, II 70–73; 258; suitable animals for, II 71; in Damaraland, II 72–74
Dominance, and the Mendelian theory, III^B 535
Donkeys, Galton's use of, and interest in, III^B 515, 519, 589–591, 593
Donovan, the phrenologist, on Galton's character, I 157
Dots, and continuous lines, II 308, 309. See also *Difference*, just perceptible
Doubleday, work of, III^B 577
Double-image Prism, for optical superimposition, II 287
Dove, meteorological assistance of, II 39
Down, Darwin's home at, as station for experimental evolution, III^A 133, 134, 287; a national possession, III^A 135; photograph of, II 134 Plate XIV; Darwin's study at, II 200 Plate XIX
Dowsing, III^B 585
Drapers' Company, grants of, to Biometric Laboratory, III^A 368, 384
Drawing, in education, II 155
Dreams, genesis of, II 247; de Candolle on, II 205
Drowning, Galton's narrow escape from, I 116–118
Duddeston (or Dudson), home of Samuel Galton, I 49, 50; visits to, I 74, 76; sketches of, I 48 Plates XXIX, XXX
Duncan, on medical histories, II 359
Duncan, Dr J. M., data of, on fertility, II 265
Dusing, C., work of, on sex-incidence, etc., II 210
Dyer, Thiselton, on Committee for Measurement of Plants and Animals, III^A 127, 128, 290; resigns from, III^A 291; on experimental breeding, III^A 130, 131, 287; on Galton's contribution to theory of natural selection, III^A 369

Ear, convolutions of, and identification, II 306; syringing the, III^B 464
Earle, Ann, and Galton ancestry, I 17
Earle, Erasmus, I 17; portrait of, I 246 Plate LXV
Earle, Thomas, portrait of, I 246 Plate LXVI

Earp, 'tremendous rows with,' I 84, 86, 87
Ecclesiastics, English and Swiss, II 142
Eclipse, Galton sees total, II 6–10
Economics, and ideals, II 254, 255
Eddis, the Cambridge tutor, I 155, 156, 160, 163
Edgeworth, Prof., work of, III^B 486
Editors of 'Biometrika,' III^A 244, 245, 281
Edouard, silhouettes of, II 309
Education, of Galton, I 110, at Cambridge and London, I 141, 142; of Darwin and Galton, II 155, 156, 179; power of, II 91; and Catholic control, II 139; reform of, II 155, 156, 344; and schools, II 344; and primary schools, II 416, 417; science, classics and mathematics in, II 155; visualising faculty and, II 241, 253
Educational Systems, value of, III^A 233
Edward VII, kindness of, III^B 604; funeral procession of, III^B 605
Edwards, at Cambridge, I 167, 171
Eggs, measurements of, III^A 243, 244
Egypt, Galton's travels in, I 197–203, 205, III^B 515–519; Galton's sketchbook in, III^B 454 Plate XLIV; need for identification office in, III^A 157, 158
Egyptians, some characteristics of, III^A 157
Eichholz, Mr, on physical deterioration, III^B 542
Eilwagen, Galton's experiences of, I 96
Einstein, ability of, II 107; his theory foreshadowed by Galton, II 263
Elderton, Ethel M., work of, III^A 328, 329, 356–358, 360, 371, 376, 384, 385, 387, 392, III^B 605; on finger-prints, III^A 140, 258; on resemblance of cousins, III^A 310, 322; on employment of mothers, III^A 345; career of, III^A 258; at Eugenics Record Office, III^A 278, 279, 300; as Francis Galton Scholar, III^A 305, 307, 330, 332; 'Primer of Statistics' of, III^A 317, 320, 363, 364; appreciation of, III^A 358, 359, III^B 583, 584, 589; visits Galton, III^A 336, 361, 429, III^B 592, 606; letters of Galton to, III^B 585, 601, 616; assistance of, II vii, III^A vii, 53
Elderton, W. P., work of, III^A 383; on heredity and environment, III^A 73, 260; on data for measurement of heredity of disease, III^B 537–539; 'Primer of Statistics' of, III^A 320, 363, 364; correspondence of, with Galton, III^B 537, 538
Elephants, in Africa, I 237
Elliptic Contours, of correlation table, III^A 13, III^A 14 Fig. 4
Ellis, Mrs. See *Biggs, Evelyne*
Ellis, J. A., and Galton's whistle, II 216
Elston Hall, original home of the Darwins, sketch of, I 30 Plate XXIII; Darwin portraits at, I 243
Ely Cathedral, Galton's sketch of, I 167 Plate LII
Emerson, R. W., on evolution, III^B 471
Eminence, appreciation of, II 89, 104; and ability, II 91, 92, 104; and mathematics, II 97, 98; criteria of, II 135; inheritance of, II 104–106; assortative mating and, II 105
Emotion, measure of changes in, II 270
'*Encyclopaedia Britannica*,' advertisement for, III^A 251
Endurance, measurement of, II 358
Energy, Galton on, II 251; in scientists, II 151, 251; and size of head, II 149, 150; inheritance of, II 251; need of tests for measurement of, II 358, 395
Engine, Galton's rotatory steam, I 150
Engineers, mechanical aptitude in, II 151
English, language, in education, Galton on, I 88
English Epigram Society, Galton gets up, I 176, 178, 187, 189
Englishmen, colour associations and number forms of, II 240; type of skull of, III^A 253; diversity of type among, III^A 257
Enthusiasm, Galton on, II 260, 261
Entomological Society, Galton reads a paper to, on moth breeding, III^A 47
Entomology, Galton's early interest in, I 68
Environment, limited influence of, II 118, 127, 128, 146, III^A 348; and history of twins, II 126–130; and scientific achievement, II 96, 97, 148, 149; and ability, III^A 112, 116; and deterioration of the British race, III^A 251, 252; effect of, on eyesight, III^A 345; and heredity, III^A 260. See also *Nurture*
Epigram. See *English Epigram Society*
Epileptics, issue of, III^A 373
Epitaph, I 189, III^B 562, 567; Galton's in Claverdon Church, III^A 434, III^B 619
Error, Law of. See *Normal Curve*
Eschbach, Galton's courier on journey to Spain, III^B 507–512
Eskimo, drawings and mental imagery of, II 240, 252
Ethnology, Galton and, II 68; composite photography and, II 294
Eton, Galton lectures at, II 361
Ettington Church, stained window for, III^B 569, 571, 572
Euclid, and modern geometrician, II 107; Darwin on value of, II 179
Eugenics, and Francis Galton, II 74, 77–80, 86–88, 110, 113, 114, 117–122, 174, 176, 231, 249, 264–267; the founder of, III^A 217; definition of, II 249, 251, 252, III^A 221–225, 262, 263, 269, 305, 318, 321; positive and negative, III^A 350; problems of, III^A 274, 275; coining of word, III^A 318; naming of science of, II 249; needs of science of, III^A 265; origin of, as academic study, III^A 259; science of, III^A 372; official recognition of, III^B 531; a social code, III^A 321; conscience, growth of, III^A 268; as a creed, III^A 217–432, 348; religion of, II 249, 250, 261, III^A 267, 272–274; a social and religious programme for, III^A 87–93, 355; Galton's policy and plans for, II 110, 128, 139, 176, 266, III^A 220–226; means of promoting, III^A 264, 265; certificates, proposals regarding, II 386–396, III^A 272, 292–296, III^B 573, 574; grant for scientific study of, III^A 258; plans for chair of, III^A 381–383; fellowship in, III^A 221–223, 300, 301, 305, 306; grant for scientific study of, III^A 258; plans for memoirs in, III^A 305, 313; institutes for research in, III^A 217; popularisation of, III^A 339, 351; administration, a forecast, III^B 624–626; national and international, III^A 220; future of, III^A 219, 220; difficulties of, II 120, 121, 176, 249, 252; biometry and, III^A 309; and actuarial methods, III^A 221, 274, 320; qualities, correlation of, III^A 273; and civic worth, III^A 227–229, 231–234, 241–243; and birth control, II 80, 110–112; and charitable expenditure, III^A 234, 243, 273, 274, 323, 348, 352; and the modern woman, II 133, 134; and evolution, II 267; and public opinion, III^A 321; anthropology and, III^A 226, 227; Darwin on, II 176; passages in Plato, III^A 312; need for knowledge before action regarding, III^A 253; sociologists' views on, III^A 259–261; Galton lectures on, III^A 261–265, 318–321; German translation of Galton's memoirs on, III^B 562–564; in Norway, Switzerland and Roumania, II 267; in America, III^B 613
Eugenics Education Society, foundation of, III^A 332, 335, 346, 348, 349; activities of, III^A 371, 378; origin of, III^B 628; Galton's address to, III^A 345–350, III^B 586; Galton and, III^B 585, 586; Hon. President of, III^A 355; Journal of, III^A 362; Galton's 'Essays in Eugenics,' III^A 367; Biometry and, III^A 404, 405; Eugenics Laboratory and, III^A 362, 363, 372, 379, 397, 404, 405, 407, 408, 409, 427, 430, 431; unfriendly members of, III^A 398–400; in last year of Galton's life, III^A 404–407, 433

Eugenics Laboratory (Galton Laboratory of National Eugenics), work of, and plans for, III^{A} 345, 349, 356, 358–362, 368–373, 376, 377, 381–389, 425, 426, 432; early publications of, III^{A} 259; and biometry, III^{A} 315, 428; reception at, III^{B} 605; Darwin relics left to, III^{B} 571; plans concerning, III^{B} 581; origin of, III^{A} 223, 299, 332, 333; directed by K. Pearson, III^{A} 299–302, 304, 322; early plans of, 304–307, 330: Galton's gifts to, III^{A} 304; work of, III^{A} 327–329, 342

Eugenics Record Office, original, 1904, Galton's plans for, III^{A} 258, 259, 274, 276–279, 296–299, 303, 339; reconstitution of, III^{A} 296–299; Americans start a, 1910, III^{B} 613

'*Eugenics Review*,' Galton's foreword to, III^{A} 362, 371, 380, III^{B} 593; progress of, III^{A} 378, 379, III^{B} 595

Evans, Sir J., on Roy. Soc. Committee on Colour-blindness, II 227; size of head and stature of, II 150

Eve, on measurements taken on Marlborough boys, II 396

Events, the observed order of, II 262

Everest, Sir G., and triangulation, II 23

Evidence, from finger-prints, III^{A} 160, 182, 183, 195, 196

Evolution, and Galton, II 13, 33, 171, 263, 264, 267; by mutation, or continuous variation, III^{A} 31, 32, 79–82, 84–86, 99, 126; regression and, III^{A} 48, 58, 60–62; and natural selection, II 79, 80, III^{A} 170; progressive, III^{A} 94; and law of ancestral heredity, III^{A} 48; of man, II 86: of mankind, II 74, 75, III^{A} 219, 220; and the average man, II 385; and Eugenics, II 249, 267; of the germ-plasm, II 171; source of, III^{A} 431; duty of man regarding, II 263; purpose of universe in, II 261, 262; doctrine of, and religious belief, III^{A} 89–93; effect of the Church on, II 111, 112; palaeontological evidence of, III^{A} 82; Erasmus Darwin's doctrine of, II 202, 203; Max Müller's theory of, II 274, 275; need of experimental farm for research into, III^{A} 128

Evolutionary Study, tribute to Galton's contribution to, III^{A} 236

Examinations, Galton's place in, at Cambridge, I 154, 164, 165, 179; evils of competitive, I 170, 171; value of, III^{A} 232, 233; as test of ability, III^{A} 247; personal equation of examiner and, II 388, 389, 407; physical and mental proficiency and, II 382, 386–396; marks of, and number of candidates, II 89

Exceptional, and habitual occurrences, strength of impression in, II 296

Exceptionality, definition of, III^{A} 121

Excitement, flush of, at the Derby, III^{A} 124

Experimental Farm, need for, III^{A} 81, 128–135; disadvantages of, III^{A} 130; Galton's desire to establish, III^{A} 135

Experiments, Galton's, on pangenesis, II 156–177, 181–183, III^{A} 129; in moth breeding, III^{A} 40, 45–47, 49, III^{B} 484; with sweet-peas, II 180, 181, 187, 189; on association of ideas, II 233–236; concerning free-will, II 245–247; on interference with automatic breathing, II 247; with auditory sensations, II 272, 308; with scents and arithmetic, II 275; concerning insanity, II 247, 248; on fetishism, II 248; on tea-making, III^{B} 456–458; Galton's, on himself, II 270

Exploration, Galton's, in Africa, I 214–240

Explorer, career of, II 29–31

Exploring Expeditions, equipment of, II 34, 35

'*Expression of the Emotions*,' Galton on the, II 175

Extinction, of inferior races, II 264; of families of great men, II 341; of surnames, II 341–343; of families, II 141, 143; prediction of, II 360

Eye Colour, II 223, 224, 226; inheritance of, III^{A} 34–40, 60, 376; Mendelians on, III^{A} 324; percentage of types of, in successive generations, III^{A} 35, 36; prediction of, in offspring, III^{A} 39; hazel, III^{A} 37, 38; blue and green, III^{B} 474; in British, II 371, 397

Eyesight, defective, investigations regarding, III^{A} 322, 368; test for acuteness of, III^{A} 382

F., Miss, and spiritualism, II 53, 62–65

Factory Acts, effect of, on fertility, III^{A} 368

Faculties, human, variability of, II 274

Faith, of Galton, when at Cambridge, I 176, 177

Falbe, Mme de, with the Galtons at Royat, III^{B} 502, 503

Fallacies, in considering totals instead of rates, II 76

'*Fallow Years*,' of Galton, I v, 4, 196–210, III^{B} 454, 455

Familiarity, measurement of, III^{B} 493

Families, extinction of, II 141, 143; the fate of large, III^{B} 577

Family, size of, in urban and rural populations, II 123–125; and age of mother, II 123–125; variation in size of, III^{A} 391, 392; criteria for merit in a, III^{A} 121; composite photograph of members of a, II 290, 295, II 288 Plates XXXI–XXXIII; links, III^{B} 575

Family, Limitation of, II 110–111, 132–134, 265, III^{A} 243, 304, 322

Family Records, Galton's collection of, II 356, 357, 363–370; prizes for, II 359–362; photographic, II 302, 356; in Biographical register, II 355

Fantasies, in children, II 244

Farmer, James, marriage of, I 33

Farmer, Joseph, and the iron industry, I 38, 39; a gun-smith, I 49

Farmer, Priscilla, cousin to Lucy Barclay, I 33

Farmers, in the ancestry of Galton, I 11, 31–34, 38–40; banking enterprise of the, I 32: gun factory of the, I 32

Farming, of Erasmus and Darwin Galton, I 125

Farquharson, Dr. on Roy. Soc. Committee on Colour-blindness, II 227

Farr, Dr, and statistics of population, II 123: on statisticians and science, II 348; estimates the value of a labourer's baby, III^{A} 228

Farrar, Canon, at Darwin's funeral, II 198, 199

Fassie, his medallion of Erasmus Darwin, III^{B} 473; reproduction, III^{B} Plate L

Fathers, of Fellows of the Roy. Soc., III^{A} 118: sons of gifted, III^{A} 102, 103; aged, and male offspring, II 210. See also *Parents*

Fatigue, mental, II 276, 278, 351, 352; physical, II 278; measure of, II 277, 278: reaction time and, II 277

Faulds, Mr H., work of, on finger-prints, III^{A} 141, 143–148, 150, 175, 177: and Galton's work on finger-prints, III^{A} 176; letter from, to Darwin, on finger-prints, II 195

Faustina, wife of Antonius Pius, portraits of, and composite, II 296 Plate XLIII

Faustina, wife of Marcus Aurelius, portraits of, and composite, II 296 Plate XLIII

Fawcett, Cicely, work of, III^{B} 504

Fazakerley, Galton visits, III^{B} 456

Fear, in mother at conception, II 209, 210

Fechner, Galton on his 'Elemente der Psychophysik,' III^{B} 464, 468

Fecundity, effect of confinement on, III^{A} 129

Feeble-minded, work on the, III^{A} 371–373, III^{B} 593; fertility among the, III^{A} 373; voluntary homes for, III^{A} 366, 367

Fellowship. See *Trinity College, Galton Research Fellow*

Female, transmutation of measurements of, III^{A} 59

Females, excess of, in West Indian islands, II 337

Fencing, of Galton at Angelo's rooms, I 109

Fergus, W., on statistics from Marlborough College, II 343

Ferris, Major, and finger-print identification in India, III^A 187

Fertilisation, of sweet-peas, III^A 325, 326

Fertility, differential, II 79, 80, 110–113, 118, 123–125, III^A 9, III^B 577, 602, 608, 609; intelligence and, II 77–79, 94, 96; of able men, II 341; of judges, II 93–95; of statesmen, II 94; in families of Fellows of Roy. Soc., III^A 109, 110; of heiresses, etc., II 95, 96; of various professional classes, II 96, 99, 101; of urban and rural populations, II 123–125; of twins, II 128; of latent germs, II 186; of hybrids, III^A 130; of highly bred animals, II 264; and divergence from mediocrity, III^A 48; relative, of classes and of nations, III^A 218, 219; suggestions for selective, III^A 233, 241–243; in a stable community, III^A 117; civilisation and, III^A 264; effect of small causes on, III^A 368; observations regarding, III^A 135; feeble-minded and, III^A 366, 373; age of parents and, II 110, 408–410; hair colour and, II 354; inheritance of, II 95; Dr Duncan's data for, II 265

Fetishism, Galton's experiments on, II 248

Fidget, measure of, II 277

Fiducial System, for profiles, II 294

Finger, middle, measurements on, III^A 54

Finger Impressions, on ancient pottery, III^A 174, 175

Finger-Prints, historical, III^A 138–154, 215, III^B 590, 591; early use of, III^A 174, 175; Galton's investigations on, III^A 254–258, 369, III^B 485; summary of Galton's work on, III^A 140, 176, 177; classification of, III^A 139, 140, 152, 153, 158–165, 167, 171, 172, 179–181, 189, 190, 199–202, 204–207, 210–214, III^A 213 Plates XXIII–XXV, XXVII–XXX; outline of patterns to assist classification, III^A 180 Plate X, III^A 181 Plates XI, XII, XIII; defects of Galton's classification of, III^A 207, 208; analysis of, III^A 161, 162, 163, 167; indexing of, III^A 140, 149–153, 159, 164, 165, 167, 170–174, 186, 187, 197–214, III^B 488; permanence of, III^A 142, 143, 145–148, 152, 161, 166, 170, 176, 181, 182, 195–197, 257, 438; persistence of minutiae in, III^A 166 Plates VII, VIII, III^A 182 Plates XVI, XVII; patterns of, III^A 178, 179–181, 183–186; ambiguous patterns in, III^A 212, 213; standard patterns of Purkenje, III^A 179 Plate IX; transitional patterns, III^A 181 Plates XIV, XV; types of loops, arches and whorls, III^A 162, 167, 209, 210, 213; relative frequency of types of, III^A 161, 184, 185, 192; correlation between type and finger, III^A 140, 161, 203, 255–257; of different digits, III^A 171–173, 183–186, 203; on right and left hands, III^A 168, 184, 185, III^B 485; illustrative examples of, III^A 139, 142, 156, 160, 162–165; Galton's symbolic notation for types of, III^A 214; origin of ridges of, III^A 161; measurable characters in, III^A 254; Galton's method of counting ridges in loops, III^A 213 and Plate XXVI; blurred, III^A 194–197; Galton's treatment of, III^A 197 Plates XIX–XXII; method of taking, III^A 161, 177, 196; apparatus for taking, III^A 155, 177; method of comparing, III^A 155, 156, 163; Galton's collection of, II 378, III^A 139, 140, 161, 193, III^B 485; racial collection of, III^A 139, 193; racial differences in, III^A 139, 143, 193, 194; in like twins, III^A 191 and Plate XVIII; effect of injuries on, III^A 154, 155, 167, III^A 154 and Plate VI; effect of occupation on, III^A 155; in infants, III^B 496–499, 524; after death, III^A 181; of different classes, III^A 194; normality of distribution of, III^A 167, 168; in the Chinese, II 195; heredity in, III^A 140, 168, 189–193, 203, III^B 485, 488–491, 522–524; correlation of, with other characters, III^A 168, 178; touch and, III^A 168, 169; sexual selection and, III^A 168–170; natural selection and, III^A 169, 170; left-handedness and, III^A 169; of Bengal criminals, III^A 254; camera for enlarging, III^A 197, 214, 215; criminal identification by, III^A 140, 141, 144, 146–149, 176, 195, 249; identification by, II 307, 380, 398, III^A 156–159, 176, 195, 369, III^B 572; use of, in China and Japan, III^A 145, 146, 148; use of, in India, III^A 146, 147, 151, 153, 155, 157–159, 176, 187, 195; as evidence, III^A 160, 182, 183, 195, 196; of Galton's friends, III^A 216; letters of Galton to Collins regarding, III^B 484–486, 488–492; Mr Faulds' letter on, II 195; folding sheet of Galton's classified types, Pocket in binding, Vol. III^A

Fire, destruction from, III^B 566; in Eva Biggs' room at Seville, III^B 508, 511, 618

Fischer, E., hair scale of, II 226

Fisher, Dr, Galton dines with, I 165

Fishing, of Galton at Duddeston, I 76

Fitzroy, Admiral R., and English meteorological office, II 43, 53

Fleas, in Lebanon, I 203; the stimulus of, to dogs, II 251

Flower, Sir W., craniology of, II 334; proposes Galton for President of British Association, III^A 276

Fluency, Galton on, II 256

Flying Machine, Galton's 'Aerostatic Project,' I 83

Foley, Thomas (ascendant of Charles Darwin), portrait of, I 245 Plate LXIII

Folk-lore, in Italy, III^B 540

Ford, E. Onslow, medallion of Dr Erasmus Darwin, II 202

Forecaster of Stature, III^A 13, 15, III^A 16 and Fig. 5. See also *Genometer*

Forecasting, from promise of youth, III^A 232

Forecasts, anthropometric, II 346

Forefingers, finger-prints from, III^A 140, 171–173, 186

Forensic Medicine, Galton delights in, I 105, 121, 126, 127

Forgeot, work of, on finger-prints and prints of whole hand, III^A 145

Forster, L., Galton calls on, I 113

Forsyth, Mr, death of, III^B 516

Foster, Sir Gregory, on Committee of Galton Laboratory, III^A 385, 386

Foster, Michael, as to Committee for Measurement of Plants and Animals, III^A 128; as to Evolution Committee, III^A 287, 289, 290; on Committee on Colour-blindness, II 227; foundation of 'Biometrika' and, III^A 100; letter of Galton with memoir on smallpox sent to, III^B 482

Fox, George, compared with Galton, II 122

Fox, W. Darwin, cousin and friend of Charles Darwin, I 68

Frampton, Sir G., models bust of Galton, III^A 388, III^B 598, 599, 603, 605

France, use of finger-prints in, III^A 144; identification of criminals in, III^A 149. See also *Bertillon*

Frankfort, Galton at, I 95, 132

Franks, W., and identification by finger-prints, III^A 176

Fraternal Correlations, for mental characters, III^A 247

Fraternal Means and variability, III^A 221, III^B 503

Fraud, words to express, III^A 157

Frazer, J. G., congratulates Galton, III^A 239; 'Psyche' of, III^B 594

Frazer, P., composite photographs by, II 290

Freames, in Galton's ancestry, I 10, 31–34. See also Pedigree Chart C, pocket at end of Vol. I

Freedom, of man, Galton on, III^A 91

Freemasonry, Galton's initiation into, I 187, 189

Free-thinkers, destruction of, II 111, 112; and religious dogma, II 257

Free-thought, Galton on, II 97

Free-trade, III^B 561

Free-will, Galton's introspective inquiry concerning, II 245–248

French, Mrs, Galton at school of, I 67

French, the, visualising faculty of, II 239; colour associations and number forms of, II 240

Frequency Distributions, of sociological phenomena, II 227, 228; in nature, II 338–340; first U-shaped, III^A 74

Frequency Surface, asymmetrical, Galton's introduction to, II 344

Frere, Lady, death of, III^B 511

Frere, Temple, Galton meets at Malta, I 199

Friends' Meeting House, Birmingham, where Galtons were buried, I 50 Plate XXXII

Fritton, Mr and Mrs, Galton's neighbours, III^B 585

'*Frost and Fire*,' J. F. Campbell's, II 53

Fruit Farming, III^B 543

Fry, Sir E., pedigree of, III^A 343, 345; on the feeble-minded, III^A 365, 371, 373; on the age of the inhabited world, III^A 122; controversy with Galton, III^A 122–123; honour conferred on, III^B 494

Fry, Elizabeth, ancestry of, I 33; relationship of, to Hudson Gurney, I 64; and the infant, Francis Galton, I 65; visit of, to Samuel Galton, I 64, 65; portrait of, I 91 Plate XLVII

Funerals, Galton on ceremonies in the Abbey at, II 198

Furrows, on hands and feet, and identification, II 306

Furse, C., his portrait of Galton, III^A 125, 379, III^B 531, 551; death of, III^B 531

Galton, origin and ancestry of family, I 31, 34, 35, 39, 40, etc.; banking enterprise of, I 32, 50, 51; gun factory of, I 32, 45, 49; slaves of, I 32, 40

Galton, Adèle (aunt to Francis Galton). See *Booth, Mrs*

Galton, Adèle (sister to Francis Galton). See *Bunbury, Mrs*

Galton, Agnes (sister to Francis Galton), I 63

Galton, Arthur, III^B 566; delivers Galton's 'Herbert Spencer' lecture, III^A 316, III^B 583

Galton, Cameron, spends Christmas with Galton, III^B 589

Galton, Lucy Cameron, visits Galton, III^B 575, 576, 589, 600

Galton, Father Charles, Galton hears him preach, III^B 544

Galton, Darwin (brother to Francis Galton), I 63; early bequest to, from Galton, I 69; farms at Claverdon, I 125; desires to enter the army, III^B 450; marriage of, I 218, III^B 454; portrait of, I 76, Plate XLVI; letters to, from Galton, early, I 76, 77, from Africa, I 219–221, 231, 232; death of, III^B 521

Galton, Diana (daughter of Hubert Galton), death of, I 123

Galton, Sir Douglas (cousin of Francis Galton), ability and fame of, I 53; advises Galton, I 214–215; letter to, from Florence Nightingale, II 416; illness and death of, III^B 506, 509

Galton, Edward Wheler, nephew to Francis Galton, III^A 281, 433, III^B 449, 571, 617, etc.; work of, III^B 563–566, 580; at Claverdon, III^B 528, 545, 570; Galton's appreciation of, III^B 531; letters to, from Galton, III^B 558–560, 563–567, 579–581; assistance from, to biographer, I viii, II vii

Galton, Emma (sister to Francis Galton), early bequest to, from Galton, I 69; attends a meeting of the British Association, I 90; visits Galton at Cambridge, I 162; travels of, I 193; in Dresden with Galton, I 178, 179; visits the Gurneys, I 191; 'Guide to the Unprotected' of, I 162; letters of—to Galton, I 62, III^B 524; to her sisters, III^B 451; letters to—from Galton, I 95, 106, 151; on death of Charles Darwin, II 198, 199; on award of the Darwin Medal, III^A 237; on election to a Trinity College Fellowship, III^A 238, III^B 473, 476, 481, 494, 495, 502, 503, 506–513, 515–519, 527; facsimile of Galton's letter to, III^B 494 Plate LI; aged 91, III^A 238; death of, III^B 447, 527–530; tombstone of, III^B 533–535; portraits of, I 213 Plate LV bis, II 198 Plate XVII, III^B 531 Plate LV; silhouettes of, I 52 Plate XXXIV, I 96 Plate L; sketches of, by Galton, I 180 Plate LIII. See also II 70, 88, 130, 175

Galton, Erasmus (brother of Francis Galton), I 63; early bequest to, from Galton, I 69; farms at Loxton, I 125, III^B 528, 543; aged 87, III^A 238; breaks his leg, III^B 591–593; characteristics of, III^B 594, 609; death of, III^A 374, III^B 594; letters of, to Galton, III^B 543, 575, 579; silhouettes of, I 52 Plate XXXIV, I 69 Plate XLII

Galton, Francis, ancestry of, I 5–61; childhood and boyhood of, I 62–91; medical studies and flight to Constantinople, I 92–139; mathematical studies and Cambridge pleasures, I 140–195; 'Fallow Years,' I 196–210; scientific exploration, I 211–242; transition studies, art of travel, geography, climate, II 1–69; early anthropological researches, II 70–130; early study of heredity, II 131–210; psychological investigation, II 211–282; photographic researches, II 283–333; statistical investigation, anthropometry, II 334–425; correlation and the application of statistics to problems of heredity, III^A 1–137; personal identification, finger-prints, III^A 138–216; Eugenics as a creed, and last decade of Galton's life, III^A 217–436; characterisation, by letters, III^B 441–619; 'Wanderlust' and travel of, I 24, 55, 58, 92–97, 199–205, II 1, 6, 7, 11, 14, 21–25, III^A 279, III^B 507–512, 515–519, 566–568; mechanical ingenuity of, I 60, 83, II 3, 18–21, 35, 44–53, 59–61, 219–221, 226, III^A 177, 279, etc.; mentality of, I 56–59, II 1, 3, 4, 10, 12, 19, 27, 75, 88, 98, 106, 134, 153, 157, 165, 308, 317, III^A 50, etc.; generosity of, II 60, 65, III^A 244, 245, 250, 257, 258, 284, III^B 588; humour of, I 59, 60, II 5, 310, etc.; eugenic ideals of, III^A 87–93, etc.; religious views and philosophic outlook of, II 102, 117, 119, 261–263, III^A 271, 272; correspondence of, with his father, I 92, 94, 97–111, 113–115, 118, 122–137, 142–181, 184–190; with Charles Darwin, I 6, 240–241, II 156–197; with Karl Pearson, III^A 224, 225, 240–251, 254–258, 261, 266, 277–291, 297–317, 322–336, 339, 342–346, 349, 350, 355–361, 368–400, 408, 409, 425–432, III^B 501, 502, 504–506, 513–515; with de Candolle, II 131–149, 204–210, III^B 474, 476–483; with Mrs Lethbridge ('Milly'), III^A 412, III^B 471, 472, 520–522, 528–536, 540–617; early letters of, I 65, 66, 71–80, 82, 86, 88, III^B 449–452. (For other letters or correspondence see names of individuals.) Married life of, I 241, 242, II 281; as to honours conferred on—Huxley Medal, III^A 226, 235; Darwin Medal, III^A 236–238, 249, 250; Medal of Linnean Society, III^A 340, 341; Royal Society Medal, II 201, III^B 476; Copley Medal, III^A 400, 431, 432, III^B 611, 614, 615; Medal of Royal Geographical Soc., I 239; Medal of the French Geographical Soc., I 239; knighthood conferred on, III^A 386, III^B 597; Hon. D.C.L. at Oxford conferred, III^B 493, 494; Hon. D.Sc. at Cambridge conferred, III^B 494, 495; Hon. Fellowship of Trinity College, Cambridge, III^A 236–238, III^B 521; see also III^A 388, III^B 509, 551, 552, 599; health of, I 103, 145, 146, 166, 167, 170, 173, 238, II 280, III^A 249, III^B 454, etc.; death of, III^A 433, 434, III^B 617, 618; wills of, III^A 224, 225, 299–303, 382–384, 437–438. See also under *Books* of F. G., *Character* of F. G., *Memoirs, Papers and Letters to Journals* of F. G., *Portraits* of F. G., *Eugenics, Finger-prints, Composite Photographs*, etc.

Galton, Mrs Francis (Louisa Butler), marriage of, I 241, 242; extracts from 'Record' of, II 11, 51, 53, 59, 70, 88, 130, 161, 245, 268, 281, 361, 362, 393; illness of, II 130, 179, 180, 196, 245, etc.; death of, II 281, III^{B} 502, 503; grave of, at Royat, III^{B} 609; portraits of, I 241 Plate LX, I 242 Plate LXI, II 88 Plate X
Galton, John (married Sarah Button), I 36, 37, 49
Galton, John Howard (uncle to Francis Galton), I 53, 91; promises to propose Galton for the Athenaeum Club, I 114
Galton, John Hubert Barclay (uncle to Francis Galton), I 53
Galton, Lucy (sister to Francis Galton). See *Moilliet, Lucy*
Galton, Mary Anne (aunt to Francis Galton). See *Schimmelpenninck, Mary Anne*
Galton, Robert, of Bristol, I 31, 39, 40
Galton, Samuel (great-grandfather of Francis Galton), appreciation of, I 41–43; home of, I 49, 50
Galton, Samuel (grandfather of Francis Galton), some characteristics of, I 43, 46, 49; marriage of, to Lucy Barclay, I 33, 46; friendship of, with Dr Priestley, I 44–46; supporters of, for Fellowship of Roy. Soc., I 44; Dr Erasmus Darwin and, I 46, 47; associates of, I 47; gun trade of, I 32, 45, 50; disowned by Society of Friends, I 45; banking of, I 50; career of, I 46; Lunar Society and, I 46, 47; scientific tastes of, I 47, 48; statistical bent of, I 48; on colour mixing, I 47, 48; on canals, I 48; on birds, I 48; homes of, I 49, 50, I 48 Plates XXIX, XXX, XXXI, I 50 Plate XXXII; death of, I 74, III^{B} 450, 451; wealth of, I 74, 75; portrait of, I 43 Plate XXV; bookplate of, III^{A} iv (facing preface)
Galton, Samuel Tertius, father of Francis Galton, characteristics of, I 52, 53, 59; banking of, I 50, 51; relationship of, to Hudson Gurney, I 34; homes of, I 51; attends a meeting of the British Association, I 90; the education of Francis Galton and, I 87, 88; letter of, to his son, on his admission to hospital studies, I 91; wishes Galton to concentrate on medicine, I 191; requires accounts from his son, I 107, 109, 112, 114, 115, 121, 123–126, 143, 154; early letters of Galton to, I 65, 76–78, 82, 88, 92, 94, 97–106; letters of Galton to, whilst at King's College, London, I 105–111, 113, 114, 115, 118, 122–137; letters of Galton to, whilst at Cambridge, 142–181, 184–190; warm affection of, for his son, I 192; health of, I 161, 185, 191, 192; death of, I 193; portrait of, I 52 Plate XXXIII; silhouette of, I 52 Plate XXXIV; lines of, on his sister Sophia, III^{B} 450; book plates of, III^{A} xii, III^{B} iv
Galton, Sophia (aunt to Galton), lines on, by Samuel Tertius Galton, III^{B} 450; plans for, on death of her father, III^{B} 450, 451; portrait of, I 54 Plate XXXV. See also *Brewin, Mrs*
Galton, Theodore (uncle to Galton), travels with Sir Francis S. Darwin, III^{B} 592; portrait of, I 54 Plate XXXVI; silhouette of, I 54 Plate XXXV
Galton, Theodore Howard (cousin to Galton), at Cambridge, I 141–143, 145, 149, 162, 172, 173
Galton, Violet, visits Galton, III^{B} 600, 602, 617
Galton, Violetta (Violetta Darwin, mother of Galton), characteristics and tastes of, I 25, 26; her accounts of her son, I 63, 67, 68, 70, 71; sample page of 'Life History of F. G.' by, I 63 Plate XL; early letters of Galton to, I 72–74, 79, 80, 112, 113, 120, 121; letters to, from F. Galton, on his South African expedition, I 217–224, 231, 234–239; letters of Charles Darwin to, II 183, III^{B} 460, 461; letter of Dr R. W. Darwin to, III^{B} 454; of Galton to, III^{B} 456; illness of, II 160; death of, II 130, 179; portraits of, I 26 Plate XX, III^{B} 531 Plate LV; sketch of, I 18 Plate X; as child, I 246
Galton, Violetta, sister of Galton, I 63
Galton Bank, I 49–51, I 50 Plate XXXII
Galton Laboratory. See *Eugenics Laboratory*
Galton-MacAlister Curve, II 228
Galton Research Fellow, III^{A} 258, III^{B} 530–533, 535; first appointment of, III^{A} 232, III^{B} 535, 544; discussions regarding, III^{A} 296–301, 305, 306; duties of, III^{A} 222, 223, 225, 226
Galton Scholar, III^{A} 222, 306
'*Galtonage*,' III^{A} 149, 152
Galtonia H. candicans, III^{B} 533, 534, 608, III^{B} 534 Plate LVI
Galtoniana, relics in—concerned with finger-prints, II 378, III^{A} 139, 178, 191, 216; with composite portraits, II 288, 290, 356. See also II 3, 11, 36, 41, 93, 230, etc.
Galton's 'Toys,' II 49
Gametic Characters, correlation of, with somatic characters, II 171–173
Gametic Elements, II 170
Gardiner, Mrs, takes finger-prints of her baby from the sixth day, III^{B} 496–499
Gardiner, Prof. John, Galton writes to, on measurements of infants, III^{B} 496
Garson, Dr, on identification of criminals, III^{A} 5; scientific adviser to Convict Office, III^{A} 151, 153; at Scotland Yard, III^{A} 249
Gassiot, J. P., and Galton, II 11, 59
Gaussian Hypothesis, and Weber-Fechner Law, II 227
Gavr'inis, III^{B} 543; finger-print ornamentation on stones at, III^{B} 543 Plate LVII
Gedge, Mr, reports on Galton's progress at school, I 89; assists Galton, III^{B} 452
Geikie, Sir A., on low standard of scientific literature, III^{A} 329–332; friendship of, with Galton, III^{A} 393, 431, III^{B} 603, 615, 616; letters from, on death of Galton, III^{B} 617
Gemmules, and Pangenesis, Darwin and Galton on, II 113, 156–190; Galton's theory of heredity and, II 185, 186, 189, 190
Genera, and Species, II 171
Generalisation, an easy vice, II 270; in artists, II 296; regarding man, II 298
Generant, Galton's conception of the, III^{A} 20; K. Pearson's definition of, III^{A} 29
Generic Images, Galton's paper on, II 295–298, III^{B} 469
Genetic Characters, Galton on, II 146, 148
Genetics, and composite photography, II 294; human, material for study of, II 359
Genius, Galton on, II 91, 92, 94, 98, 99, III^{A} 115, 116; and imagination, II 98; inheritance of, II 106; and sterility, II 341; descendants of men of, III^{A} 278; hereditary, II 70, 87–115
Genometer, construction of, III^{A} 30; use of, III^{A} 243; proposals regarding, III^{A} 335, III^{B} 514; illustration of, III^{A} 30 Plate I
Geographer, aids to the, II 23–25; goal of the, II 29
Geography, and Galton, II 4, 21–35; in schools, II 27, 28; human side of, II 70
Geologists, mechanical aptitude in, II 151
Geology, Galton's early interest in, I 68
Germ-Plasm, continuity of, II 81, 82, 114, 147, 169–171, 174, 186, 187; priority of Galton's idea regarding, III^{A} 340, 341; selection and evolution of, II 171; of individuals, III^{A} 60
German Language, Galton studies, I 99–101, 129–131
Germans, Galton on the, I 95–97; colour associations and number forms of, II 240; interest of, in Eugenics, III^{B} 545, 546
Germany, Galton's visits to, I 178, 179, II 140; hints of unfriendliness in, III^{A} 379, 380

Germinal Selection, and Evolution, II 171; and development, II 185, 186
Germs, latent and developed, II 182, 186
Gestation, observations regarding, III^{A} 135
Gibbs, W. F., Galton takes rooms with, I 196
Giessen, Galton's vacation in, I 128–132
Gifi, devoted servant to Francis Galton, III^{B} 449, 503, 508, 519, 582, 583, 587, 595, 597, 618; tablet at Rutland Gate and, III^{A} 311; portraits of, II 11 Plate III, III^{A} 390 Plate XXXVIII
Gilbey, Mr, collects pedigrees of deaf-mutes, III^{A} 380
Gilson, Mrs, at Khartoum, III^{B} 548
Giraffe, shot by Galton, I 223
Girgenti Cathedral, acoustic properties of, III^{B} 563
Gissing, G., noteworthy book of, III^{A} 312
Gladstone, W. E., large head of, II 379, III^{A} 248; not as other men, III^{B} 568; lack of humour of, III^{B} 607, 608; at Galton's South Kensington laboratory, II 379; H. Montagu Butler and, III^{B} 476
Glaisher, and Galton's exponential ogive, II 191
Glanvill, J., and use of word 'aberrance,' III^{A} 99
Godliness, and material well-being, II 100, 101
Godman, F. D., on Committee for Measurement of Plants and Animals, III^{A} 127, 291
Goethe, as example of sane exercise of senses? II 99; pictures seen with closed eyes by, II 244
Gold, in the world, II 21
'*Golden Book*,' proposals regarding, III^{A} 264, 265
Goldie, G., to send finger-prints of Africans, III^{B} 485
Goring, C., criminal investigations of, II 232, III^{A} 377
Gorst, Sir J., Galton meets, III^{A} 266; his 'Children of the Nation,' III^{A} 310
Gotto, Mrs, and Eugenics Education Society, III^{A} 342, 350, 371, 372, 379, 380, 427, 428; visits Galton, III^{A} 370
Goulds, marriages of, with the Freames, I 32
Gout, III^{B} 590; and asthma, I 185; strawberries and, III^{A} 124
Governments, democratic, and the elimination of 'weeds,' III^{A} 349
Grade, of individual, and the statistical scale, II 337. See also *Percentiles* and *Ogive Curve*
Graduates, Cambridge, measurements on, III^{A} 247
Graef, portrait of Galton by, III^{A} 125, 126
Grahame, Dr, on cousin marriages, III^{B} 470
'*Grammar of Science*,' by Karl Pearson, received by Galton, III^{A} 240, 241
Grandsons, and nephews, nearness in kinship of, III^{A} 33
Grant, and Galton, II 25; and African exploration, II 30, 36
Grassi, a Darwin medallist, III^{A} 237
Graunt, Capt. John, on Bills of Mortality, II 123
Gray, anthropometry of, II 380
Great Barr, home of Samuel Galton the younger, I 49; sketch of, I 49 Plate XXXI
Greek Girl, of 'Just Perceptible Difference' lecture, II 309, III^{A} ix (reproduction facing)
Greek Queens, composite portrait of, II 295, 296 Plate XLII
Greeks, ability and culture of, II 107–109
Greg, 'Enigmas of Life' of, II 176
Grew, N., work of, on finger-prints, III^{A} 141, 142, 176, 177
Grey, Mrs, seen by Galton in Italy, III^{B} 474
Griffin, N. W., his 'Optics', and an epitaph, I 189
Griffith, G., Sec. of Brit. Assoc., III^{B} 458
Griffiths, A., and identification of criminals, III^{A} 148
Grosvenor, Lady Constance, eugenic proposal of, III^{A} 377
'*Groundwork of Eugenics*,' Lecture of K. Pearson, III^{A} 380
Grove, Sir W., Galton's visits to, II 130, 161, 180, III^{B} 465; appreciation of, III^{B} 531; size of head and stature of, II 150, III^{A} 248; death of, II 280
Grove, and Galton's 'Efficacy of Prayer,' II 131
Growth, statistics of, in plants, II 191, 192; laws of, II 380; bioscopic representation of, III^{B} 576; curves of, for vital capacity, II 377
Grundy, Mrs, power of, III^{A} 342
Gruppe, his definition of 'religion,' III^{A} 89, 90, 93
Guido, Reni, his picture of 'Aurora and the Hours,' III^{A} 422 Plate XXXIX
Gull, on medical life histories, II 359
Gumption-reviver, machine used by Galton at Cambridge, I 144
Gun Trade, of Farmers and Galtons, I 32, 45, 49, 50
Gurney, Emily, death of, II 280
Gurney, Hudson, on the future fate of the Barclays, I 30, 33, 34; relationship of, to Elizabeth Fry, I 64; Thomas Young and, I 48; portrait of, I 91 Plate XLVII
Gurneys, Russell, and Galton, II 11, 88, 161

Haddon, A. C., and anthropometry, II 379, 380; on social evolution and marriage customs among primitive peoples, III^{A} 267, 268; to procure finger-prints of native races, III^{B} 485; letter of, to Galton, as to Huxley Lecture of Roy. Anthrop. Inst., III^{A} 235, 236
Hadleys, genealogy of the, III^{B} 462; sketch portrait of Mr Hadley, surgeon, I 18 Plate X
Hadzor, Galton's visits to, II 11, 70; letters preserved at, III^{B} 607; sketch of; I 48 Plate XXIX
Haeckel, E., medallist of Linnean Soc., III^{A} 340; hon. member of German Eugenic Soc., III^{A} 388; death of, III^{A} 342
Hahn, disastrous expedition of, I 240
Hair, on male face, III^{A} 321; colour of, II 223–226; in husband and wife, II 149; and liability to disease, II 354, 371; and fertility, II 354; darkening of, in the English, II 353, 354; in the British, II 371; analysis of pigments in, III^{A} 97, 98; investigation into, III^{B} 474; of Galton in childhood, gold brown, I 71, cf. III^{B} 456 Plate XLVII; Sorby's paintings of trees from pigments in, III^{A} 97 Plates III, IV
Haldane, Lord, army achievements of, III^{B} 603
Hallam, Henry, at Cambridge, I 140, 141; sees Galton in Dresden, I 179, 180; Galton's friendship with, I 190, 191; letters of, to Galton—refusing invitation to travel, I 198, 199; after Galton's travels in Egypt and Syria, I 205–207; characteristics of, III^{A} 354; death of, I 238, III^{A} 354; Bristed's obituary notice of, I 171
Hallam, Julia, sketch of, I 180 Plate LIII
Hallams, the graves of, Galton's hour of grief alongside, III^{A} 354
Halsbury, Lord, on supposed deterioration of the British race, III^{A} 364
Hamilton, Dr, on louping-ill, III^{B} 566
Hamilton, Dr Lillias, III^{B} 541, 542
Hamilton, Sir W., on generic image of man, II 298
Hampstead, Galton considers moving his home to, III^{A} 310, 311, 316, 317
Handwriting, and identification, II 306
Happiness, of segregated feeble-minded? III^{A} 366, 367, 373, 374
Harcourt, Sir W., elected to a Trinity College, Cambridge, Hon. Fellowship, III^{A} 236, 238
Hardy, Thomas, and London University, III^{A} 289
Harmsworth's 'The World's Great Books,' Galton's work included in, III^{B} 601

Harrison, Frederick, gives first 'Herbert Spencer' lecture, IIIA 313
Hartog, and London University, IIIA 224, 225, 279, 300–302, 429, etc.; Galton's appreciation of, IIIB 605, 606
Haslam, Lewis, assistance of, II v
Haslemere, Galton at, IIIA 322, 326–336, 393–399, 425–427, 430–432, IIIB 583–585, 600–617
Haughton, Dr S., and the 'long drop', II 407
Havelock Ellis, his study of British genius, IIIA 261; and Eugenics, IIIA 296, 372
Haviland, Dr, advises Galton, I 180; on gout and asthma, I 185
Haweis, Mrs, association of words and visualised pictures in, II 243, 244; number form of, IIIB 469
Hawkins, Vaughan, on size of families of the rich and poor, IIIA 392
Hawks, in confinement, IIIA 380
Hawksley, and Galton's whistle, II 217; and Galton's 'registrator,' II 341
Head, measurements on, II 373, IIIA 54; size or shape of—and ability, II 94, 343, 387, 388, IIIA 247–249; and energy, II 150; and sturdiness of build, IIIA 248, 249; of scientists, II 149, 150; of Gladstone, II 379; of stablemen, IIIA 249; of Francis Galton, phrenologists on, I 157
Headlam, A. C., proposed for Galton Laboratory Committee, IIIA 385
Headmasters, and the needs of schoolboys, I 87, 88
Heald, Mr, reports death of Ali, IIIB 454
Health, and climate, II 36; and piety, II 101; of savants, II 141; inheritance of, II 151; and longevity, II 349; influence of, on delicacy of perception, II 308; and pigmentation, IIIB 476, 477
Heape, W., on Committee for Measurement of Plants and Animals, IIIA 127, 291
Hearing, of insects and animals, II 216, 217
Heat, Galton delighted in extreme, IIIB 447
Hebrew Race. See *Jews*
Height, of boys in town and country schools, II 125, 126, 337; absolute values of, at each rank, II 390; percentile values of, sitting, II 376. See also *Stature*
Heiresses, and fertility, II 95, 96
Heliostat, Galton's invention of a pocket, II 18–21, 50
Helm, H. T., 'American Trotter' of, IIIB 498
Helmholtz, a vice-president of section of physics at loan collection of scientific instruments, II 215
Helmingham Hall, Galton stays at, IIIA 323
Henry, E. R., becomes Chief Commissioner of Metropolitan Police, IIIA 151, 152; on finger-prints, IIIA 151, 152, 153, 158; on use of finger-prints in India, IIIA 157, 158, 187, 199, 254, IIIB 590; at Scotland Yard, IIIA 249, IIIB 572; the Kaffir police and, IIIB 571; dines with Galton, IIIB 513; visits Galton's Laboratory for instruction in finger-printing, IIIA 151
Henslow, Prof. J. S., the botanist, assists Galton, I 179
Henslow, Rev. G., number form of, II 242, IIIB 469; pictures seen with closed eyes by, II 244; his criticism of the law of ancestral inheritance, IIIA 42
'*Herbert Spencer*' *Lecture*, Galton's, IIIA 309, 311–313, 316–321, IIIB 583; translated into Hungarian, IIIB 585
Herbette, on value of anthropometric records, II 398; on personal identification, IIIA 188
Herbst, C., book of, IIIB 525
Herculaneum, Waldstein's views on, IIIB 589
Herd Instinct, in man, II 72, 73, 122; psychology of, II 73
Herdman, Prof. W. A., letter of, to Galton, asking him to be President of Brit. Assoc., IIIA 276; directs Weismann's attention to Galton's work, IIIA 340
'*Hereditary*' and 'Heredity,' history of words in English, IIIA 347
Hereditary, mannerism, II 63; savagery in man, II 85; liabilities, desire to screen, II 360; link, nature of, II 173; principle, and the Upper House, IIIA 33
'*Hereditary Genius*,' I 5, 7, II 70, 76, 87–115, 145, 147, 156; Charles Darwin on, I 6; reception of, II 88; Miss Shirreff on, II 132, 133; de Candolle on, II 134, 135, 136, 149; German translation of, IIIB 607
'*Hereditary Piety*,' by Frances Power Cobbe, II 160
Heredity, application of statistics to problems of, II 89, 92, 113, 357, IIIA 1–137, 264; early study of, II 131–210; Galton's theory of, II 182–190, IIIA 190, etc.; Law of Ancestral, II 84, 113, IIIA 21–23, 34–44, 47–49, 251; measurement of, II 92, 318, IIIA 240, 242; Galton's five constants of, IIIA 47, 48; theory of progressive evolution and, IIIA 94; Galton's critics on, IIIA 42; graphical representation of, IIIA 44–45; with law of regression, IIIA 46; assumptions of, IIIA 251; and Mendelism, IIIA 329; doctrine of, II 170; science of, IIIB 504; smallest unit transmissible by, IIIA 191; statoblasts in, IIIA 245; tradition and, IIIA 409–411; ancestral piety and, II 100, 101; the criminal class and, II 231; the blood and, II 177; knowledge required for study of, II 321, IIIA 70–73, 221; in man, II 72, 73, 75, 81–83, 86, 117, 118, 122; in twins, II 126–130, 181; in man, plants and animals, IIIA 266; insurance data and, IIIB 536–539; environment and, see *Nature and Nurture*; Galton reviews Ribot's book on, IIIB 463. Inheritance—of acquired characters, II 81, 147, 169, 170, 173, 174, 182–184, 186, IIIA 57, 59, 129–132, 374; of mental and moral characters, I 5–7, II 72, 75–78, 81–83, 86, 87, 89, 117, 118, 126, 128, 135, 146, 174; of physical and mental characters, II 230, 231, first determination of, IIIA 247; of ability, II 104–106, 141, 146, IIIA 102, 103, 108–121, 347, 348; of legal ability, II 93; of scientific ability, II 97; of mathematical ability, II 97; of artistic faculty, IIIA 68, 69; of talent, II 75–77; of physical and psychical characters, IIIA 69; of energy, II 251; of mental inertia, II 257; of memory, II 151; of perseverance, II 151; of piety, II 101; of independence of character, II 151; of character and temper, II 269, 271; of temper, II 271; of conscience, II 257; of vicious instincts, II 230; of visualising faculty, II 239, 242, 253; of number forms, II 252 Plate XXIV; of physical characters, II 76, 135; of muscularity, II 104; of constitution, II 151; of longevity, II 348, 349; of stature, IIIA 11–34; of cephalic index, IIIB 504; of finger-prints, IIIA 140, 168, 203, 189–193, IIIB 485, 488–491, 522–524; of fertility, II 95; of eye colour, IIIA 376; of eye characters, IIIA 345, 356, 368; of defective eyesight, IIIA 322, 368; of pathological states, II 80, IIIA 70–76; of sports, IIIA 120; of degeneracy, IIIA 373, 374; of insanity, IIIA 322; of tuberculous diathesis, II 292, IIIA 260, 326; in plants, IIIA 251; in bearded wheat, IIIA 314; in weight of seed, IIIA 3; in size of sweet-pea seeds, II 392; of colour in horses, IIIA 95–98; blended or alternative, II 397, IIIA 27, 29, 34, 59, 60, 76; natural, see *Natural Inheritance*. See also *Hereditary*
Herkomer, Galton meets, IIIB 513
Hermaphrodites, twins among, IIIA 359
Heroes, of Galton and Carlyle, II 94
Heron, D., Eugenics Research Fellowship of, IIIA 289, 330, 332; appointment of, IIIA 304, 305; work of, IIIA 324, 325, 327, 345, 355–360, 371, 376, 382, 387, 426, 427; on statistics of insanity, IIIA 311, 322; visits Galton, IIIA 225, 342, IIIB 593
Herschel, Sir J., and Quetelet's advice, II 418

Herschel, Sir W. J., work of, on finger-prints, III^A 140–149, 161, 166, 174–177, 187, 195, III^B 590, 591; letter of, on finger-printing, III^A 152; Galton's dedication to, III^A 199; illustrations from material on finger-prints of, III^A 182 Plate XVI, III^A 197 Plates XIX–XXII; forefinger-prints of, at interval of 54 years, III^A 439; contract of, with Rajyadhar Konai, first known Indian print for legal purposes, III^A 146 Plate V
Hertz, Miss, Galton congratulates her on her engagement, III^B 468, 469
Hertz, Mrs, founder of a scientific salon, Galton writes to, III^B 464, 476, 500, 503, 527; letter of, to Galton, III^B 500
Hewlett, Prof. R. T., notice by, of E. Wheler-Galton's work, III^B 579
Heydon Hall, home of the Earles in Norfolk, I 246 Plate LXIV
Hibbert, Mrs, Galton's nurse-housekeeper, III^B 582, 583
Hill, Elizabeth, Erasmus Darwin's mother, I 17; portrait of, I 16 Plate VII
Hill, Sir John, the botanist, I 17
Hill, Leonard, and the inheritance of ability, III^A 103, 229
Hill, Sir Rowland, size of head and stature of, II 150
Hill-James, Col., Galton meets at Biarritz, III^B 557
Hills, Mrs (née Grove), Galton's warm friendship for, III^B 596
Hindhead, Galton at, III^A 277, 323–326. See also *Haslemere* (with Grayshott)
Hindoos, finger-prints of, III^A 195
'*Histoire des Sciences et des Savants depuis deux Siècles*' of de Candolle, II 134, 135, 145, 207, 208
Historical Society, foundation of, at Cambridge, I 171, 172, 174, 178, 183
History, in educational scheme, II 155; of personal identification by finger-prints, III^A 138–154
Hobbes, on generic image of man, II 298
Hobhouse, L. T., on Eugenics, III^A 260
Hodgkin, on the Roman Empire, III^B 540, 541
Hodgson, Mrs Brian, her photograph of Galton, III^A 303
Hodgson, Prof. J., at birth of Galton, I 62; physician to Tertius Galton, I 192; medical sponsor to Galton, I 90; advises Galton, I 99, 103, 113, 126, 127, 130, 152, 187, III^B 452; Galton's recollections of, III^B 565
Hodgson, T. V., on finger-prints, III^B 486
Hodograms (isochronous curves), II 57
Holy Land, Galton's first impressions of the, I 203
Home (the medium), and spiritualism, II 62–66; and Mr Sludge, II 63, 64, 66
Homes, of the Galton and Darwin families and of their ancestors, I 28–30, 49–52, 62, 74, III^B 571, 580, I 30 Plate XXIII, I 48 Plates XXIX, XXX, I 49 Plate XXXI, I 50 Plate XXXII, I 74 Plate XLIII, I 246 Plate LXIV
Homogeneity, Galton on, in statistical data, III^A 303
Homoscedasticity, definition of, III^A 5; in sweet-pea experiments, III^A 7, 13
Homotyposis, reception of paper on, by Roy. Soc., III^A 241, 243; lecture on, III^A 247
Honours, concerning, III^A 236, 237; conferred on the young and on the old, III^A 249, 250
Hooker, J. D., and Galton, II 140; on farm for experimental breeding, III^A 134; on fertilisation of sweet-peas, III^A 325; medallist of Linnean Soc., III^A 340, Copley, III^A 431, III^B 614; size of head and stature of, II 150; death of, III^A 342
Hope-Pinker, models bust of Jowett, III^A 126; and bust of Weldon, III^A 281, 301, 302, 374; visits Galton, III^A 333
Hopkins, Alice, Biography of, III^A 357
Hopkins, W., Galton reads mathematics with, I 153, 162–167, 181; pupils of, I 171; not the ideal teacher for Galton, I 194; asks Galton to accept office of Gen. Sec. to Brit. Assoc., III^B 458
Hoppit, Mrs (bedmaker at Cambridge), lights Galton's fires, I 143; sevenfold offspring of, I 164
Horner, Leonard, and the child Francis Galton, I 65; attends a meeting of the Brit. Assoc., I 90; Galton visits his house in London, I 108, 111, III^B 453; praises Galton, I 143; gives Galton introductions at Cambridge, I 145
Hornets, specimens of, III^A 251
Horses, composite photographs of, II 288 Plate XXX; photographed for Galton, III^B 506, 507; standardised method of photographing, II 320, 321, 322; breeding of, II 322, 323; effect of continued selection in, II 399; prepotency in trotting, III^A 98–100; speed of American trotting, II 399, III^B 498; hereditary colour in, III^A 95–98; Galton's papers on, II 398–400
Horsley, Sir V., attacks work of Eugenics Laboratory, III^A 408, 430
Hospital, Galton admitted as pupil to Birmingham General, I 90; letter to Galton from his father on admission to, I 91; Galton's studies and experiences in, I 92, 99–104; at Frankfort, I 95
Hospitals, two purposes of, II 344
Hotham, at Cambridge, I 167, 171
Hottentot Ladies, study of the peculiar anatomy of, I 231, 232
Houghton, Lord, on epitaphs, III^B 562
House of Lords, reform of, III^A 33, 34, 405
Howard Ancestry of Charles Darwin, I 17, I 244–246
Huggins, Sir W., and dog 'Kepler,' II 66; speech on award of Darwin Medal to Galton, III^A 236
Hughes, friend of Galton at Cambridge, I 153
Hughes, Thomas (author of 'Tom Brown'), aids 'The Reader' newspaper, II 68
Huguenots, gain to England from, III^A 229
Human Faculties, measurement of, II 268–270
'*Human Faculty and its Development*,' Galton's 'Inquiries into,' I 5, II 87, 207, 212, 238, 241, 248–267, 361
Hume, on generic image of man, II 298
Humidity, register of, II 50
Hungarian, Galton's lecture translated into, III^B 585
Hunter, D., Galton calls upon, I 148
Hunting, and shooting, Galton's enthusiasm for, I 208, 209, 234
Hurst, Capt., on inheritance of eye colour, III 324, 376
Hutchinson, Jonathan, life register of, II 369; museum of, III^A 324
Hutchinson, W., and Renan's 'Antichrist,' III^B 572
Huxley, T. H., and Galton, II 62; on the principles of Charles Darwin, II 203; at Darwin's funeral, II 198; on Dr Erasmus Darwin, II 203, 204; London University and, III^A 289–291; Romanes lecture of, II 82; talents and characteristics of, II 178, III^A 333, III^B 506; on trickery of a medium, II 63; and spiritualism, II 67; and agnosticism, II 102; 'The Reader,' 'Nature' and, II 68, 69; size of head and stature of, II 150; chaffs Herbert Spencer, III^A 142; letters of, to Galton, in the *Galtoniana*, II 228
Huxley Lecture, given by Galton, III^A 226–235, 246
Huxley Medal, presented to Galton, III^A 226, 235
Hyacinthus candicans, named *Galtonia*, III^B 534; on Emma Galton's tombstone, III^B 533
Hybridisation, and variation, II 84; co-operation regarding observations on, III^A 135
Hybrids, Galton and Darwin on, II 169, 189; fertility of, III^A 130, 131
Hydrogen, use of, with Galton's whistle, II 217
Hydrophobia, through wolves and dogs, III^B 547

Hyères, Galton at, III^A 249
Hyperscope, invention of, II 20; use of, by Galton, III^B 618

'*Ibis*,' Galton's boat up the Nile, I 200
Ibrahim Pacha, obeys order not to drink wine by drinking brandy, III^B 572
Iceland Spar Compounder, Galton's, II 287
Ideals, and economics, II 254, 255
Ideas, association of, II 233–236, 256; limitation of, II 236; subconscious, and will, II 247; and consciousness, II 256; abstract, and nurture, II 255; selective action of brain in storing, II 256
Identification, personal, III^A 138–216; history and controversy regarding finger-prints, III^A 138–154; popularisation of finger-prints as method of, III^A 154–160; scientific papers and books on, III^A 161–216; by profiles, II 304, 306, 326; Bertillon's system of, II 304–306, 380, 383, etc.; anthropometric register and, II 398; American system of, III^A 188. See also *Finger-Prints*
Identification Offices, uses of, III^A 157, 158, 160
Idiocy, in children of drunkards, II 148; and cousin marriage, III^B 470
Idiosyncrasy, in Col. M., regarding mention of injuries, II 279
Idiots, in general population, II 90; mental capacity of, II 272
Illegitimacy, among the feeble-minded, III^A 366
Illnesses, grave, Galton's list of, II 366
Illustraria (Selenia moth), experiments with, IIIA 47, 49, III^B 484
Illustrious men, definition of, II 89
Images, optical combination of, II 285
Imagination, in science, II 98; in literary men, II 256; analytical power and, II 98; measurement of, II 307; aid of, in deafness, II 308; power of, III^B 492, 493
Imaginative Power, and lunacy, III^A 115
Impregnation, under effect of alcoholism, II 139
Impressions, and blended memories, II 296; general, frequent incorrectness of, II 296, 297, III^A 274, 364, 405; moral, in childhood, II 127
Imprisonment, statistics regarding terms of, II 406
Inaudi, mental sums of, II 275, 276; cited by Galton as a 'sport,' III^A 85
Inbreeding, following selection, III^A 94; Galton on, in man, III^A 270; intensive, and evolution, III^A 431
Incommensurable Motives, Galton on, II 246, 247
Independence of Character, in cattle, II 73, 74
Index, of finger-prints, III^A 140, 149–153, 159, 160, 164, 165, 167, 170–174, 186, 187, 197–214; of pedigrees, III^A 103, 104; of mistakability, II 329–333, depends on vision of operator, II 330, 331. See also *Finger-Prints*
Indexing, of profiles, II 304, 323, 324, III^A 325, 326; and numeralisation of portraits, II 323–328; diagrams, II 325, 327; and the Bertillon system, II 383, 398
India, queries regarding our treatment of, II 417; use of finger-prints in, III^A 146, 147, 151, 153–155, 157–159, 176, 187, 195, III^B 591; need for identification office in, III^A 157; finger-prints of criminals in, III^A 254
India Office, and marks for physical qualification for admission to Indian Civil Service, II 388, 394
Indian Civil Service, appointment in, as test of ability, III^A 247, 327
Indians, some characteristics of, III^A 157
Indifferentism, effect of, II 154
Individual, anthropometric description of the, II 380; and type, measure of difference between, II 311; life, control of, II 113, 114
Individual Difference Problem, Galton's paper on, II 411–414
Individuality, measurement of, II 303–306; of profiles, II 323; extreme and modal examples of, II 412; limitations of, II 413; of very gifted men, III^A 32; permanence of, III^A 279
Industrial Anthropometry, foreshadowing of, II 358; value of, II 382
Inebriates, report regarding, III^A 301
Infants, measurements on, III^B 496; finger-prints of, III^B 496–499
Infectious Illness, at the Birmingham Free School, I 81, 84, 85
Influences, effect of small but persistent, III^A 403, 404
Inge, Dean, and Eugenics Education Soc., III^A 339; preaches on Eugenics, III^B 612
Inheritance. See under *Heredity*
Innate Taste, for science, fostering of, II 153
Inquisition, and free-thinkers, II 111, 112
Insane, General Register of the, proposals regarding, III^A 365, 366
Insanity, and alcoholism, II 148; Galton's experiments as to, II 247, 248; in descendants of the very able, III^A 32; heredity and, III^A 277, 298, 322; convolutions of the brain in, III^A 298
'*Insectivorous Plants*,' Darwin's, II 181
Insects, hearing of, II 217
Inspiration, from an unseen world, II 260, 261
Instincts, Galton on vicious, II 230; observations regarding, III^A 135; of higher animals, III^A 130
Institute of Actuaries, Galton's appeal to, III^B 536–539
'*Instrumental Instructions for Mr Consul Petherick*,' II 27
Instrumentarium, of Galton, II 3
Instruments, or apparatus designed by Galton—anthropometric or psychometric, II 212, 213, 215–228, 370, 373; meteorological, II 44–49 (drill pantagraph), 59; a lock, I 148; a lamp, I 148, 149; a balance, I 149; a printing telegraph, I 212, 213; for compounding six objects, II 285; hyperscope, II 20, III^B 618; wave engine, II 51; heliostat, II 19; 'Tactor' machine, II 50; for optical combination of images, II 285; Iceland spar compounder, II 287; measurement of resemblance machine, II 332–333; instantaneous attitude snapper, III^B 629 (Addenda); spectacles for divers, II 34; whistles for high notes, II 212, 216; pocket registrator, II 341; for testing the perception of tint differences, II 219; for measuring rate of movement of a limb, II 220
Instruments, standardisation of, for Kew, II 59, 60, 223–226
Insurance Offices, eugenic data from, III^B 537–539, 542. See also *Institute of Actuaries*
Intellectual Ability, and the normal curve, II 90, 91; and eminence, II 91; and sensitivity, II 222
Intelligence, inheritance of, II 75–77, 104–106; and physical characters, II 77, 94, 388; and fertility, II 77–80, 94; and size of head, II 94, 343, 387, 388; of different races, II 106–109, 351, 352; normal curve and, II 104; dominance of, II 108; sign of high, II 236
Intermarriage, within the class, II 121, III^A 231
International Eugenics, III^A 219, 220
International Health Exhibition, Anthropometric Laboratory at, II 357, 362, 370–386
Interpolation, George Darwin's paper on, III^B 467
Intra-uterine Influence, III^A 59
Introspection, in women, II 242; free-will and, II 245–247
Iris, markings on, II 306
Iron Industry, and the Farmer family, I 39
Irresolution, and free-will, II 246

Irritability, effect of colour on, II 214
Island of St Paul, geographical model of, II 33 Plate IV
Isle of Wight, model of, II 34
Isochronic Passage Charts, construction of, II 35
'*Isochronous Curves*,' II 57, 58; machine for plotting, II 58
Isodic Curves, for sailing ships, II 56, 57
Isogens, for parental ages, II 408, 409, 410
Isograms, of stature and vital capacity, II 391, 392
Isoscope, definition of, II 332; Galton's use of, III^A 280
Italians, colour associations and number forms of, II 240
Italy, Galton visits, III^A 251, 276, III^B 474, 475, 521, 522, 540–543

J-Curves, examples of frequency distributions following, II 406
Jackson, Cyril, on Galton Laboratory Committee, III^A 386
Jacobs, J., and experiments on prehension, II 272; and composite portraits, II 290, 291, 293
Jaeger, on the nature of illness, III^B 594
Jahn, Otto, biographer of Mozart, III^B 500
James, Sir H., on maps, II 30
James, W., 'Varieties of Religious Experience' of, III^B 449
Japan, anthropology in, II 397; use of finger-prints in, III^A 145, 146, 148
Jastrow, composite photographs by, II 290
Jebb, Mrs, and twins, III^B 465
Jenkinsons, visit to the, II 130
Jennings, Mr, at Mrs Galton's funeral, III^B 503
Jephson, Dr, physician to Tertius Galton, I 185; advises Galton, II 53
Jeune, Dr, schoolmaster and bishop, I 81; Galton at school of, I 81–90; classical torments of, I 82; his reports on Galton, I 83, 87, 89; Galton's reports on, I 83–89; Galton visits, in Jersey, I 119; diary of Galton at school of, I 83–86; letters from Galton whilst at school of, I 82, 86–89, III^B 451, 452
Jevons, size of head and stature of, II 150
Jewish Type, composite portraits of, II 293, III^B 474, II 294 Plate XXXV; variation in, II 385
Jews, the, Galton and de Candolle on, II 209; finger-prints of, III^A 193, 194, III^B 485, 491
Johansen, pure line theory of, III^A 58
Johnson, Dr Alice, in Petrie's camp in Egypt, III^B 516; interested in the feeble-minded, III^B 599
Johnson, H. Vaughan, Galton shares rooms with, I 196
Johnson, Paddy, with Galton on visit to Fazakerly, III^B 456
Johnson, Samuel, and Erasmus Darwin, II 194
Johnson, W., of the Epigram Club, I 141; wins the Camden Medal, I 176
Johnston, Harry, and Speke memorial, III^B 588
Johnstone, Dr J., assists Galton, III^B 452
Jones, Augusta, acts as Secretary to Galton, III^A 398, III^B 603, 605, 612, 613
Jones, Dr Bence, experiments of, II 163
Jones, H. Gertrude, assistance of, I viii; work of, III^A 426
Jonker, promises to keep the peace in Damaraland, I 226, 227, I 226 Plate LVIII; Galton's opinion of, I 229
Jordan, Galton's journey down the, I 204
Journalism, and Galton, II 67–69
Journalists, and phantasmagoria, II 256
Jowett, B., sits for Hope-Pinker, III^A 126
Judges, fertility of, II 93–95; morality of, II 94, 95; judgment of, II 406, 407; personal equations of, II 406; in Utopia, II 408
Judgment, of physical efficiency on inspection, II 394; of weights, III^B 581; of weight of fat stock, II 403, 404; of democracy, II 403–405; of judges, II 406, 407; problem of fallible, III^B 501, 502
Judgment, measurement of time to form, II 359
Jukes family, II 231
Julia, daughter of Augustus, portraits of, and composite, II 296 Plate XLIII
Julia, daughter of Titus, portraits of, and composite, II 296 Plate XLIII
Julian Hill, visits to the Butler family at, II 11, 88, 161

Kahn, A., travelling fellowships of, II 152
Kant, ability of, II 107; his definition of religion, III^A 89, 90
'*Kantsaywhere*,' Galton's Utopia, III^A 411–424, 434, III^B 609, 611–616
Kay, E., on reading party with Galton, I 168, 169; and freemasonry, I 187, 189
Kay, J., the 'Cambridge idler,' I 141, 164, 166, 167, 169, 171; Galton visits, I 168; on reading party with Galton, I 168; and freemasonry, I 187, 189
Keats, quotation from, III^A 217; the 'Hyperion' of, III^A 319
Keltie, Dr J. Scott, assistance of, I 215
Kelvin, Lord, remarkable head of, III^A 248; letter of, to 'The Times,' III^B 515
'*Kepler*' (Sir W. Huggins' dog), and butchers, II 66, 148; Darwin on, II 173
Ker, W. Paton, assistance of, I viii, II vi, 298
Keswick, Galton joins reading party at, I 153, 155–162
Kew Gardens, and experimental work, III^A 131, 287
Kew Observatory, and Galton, II 49, 59, 60, 280, 282
Keynes, J. M., attacks work of Eugenics Laboratory, III^A 408
Khartoum, Galton at, I 200–202
Kidd, B., 'Social Evolution' of, III^A 88, 92; on Eugenics, III^A 260
Kimberley, Earl of, honours conferred on, III^B 494
King's College Chapel, Galton's sketch of, I 167 Plate LII
King's College, London, Galton at, I 105–127
Kinsfolk, number of, III^A 107, 114, 116, 117; of Fellows of Roy. Soc., III^A 107–121
Kinship, between parents and offspring, etc., II 169, 170; method of tracing, II 306; nearness of, III^A 59; arithmetic notation of, II 354, 355; nomenclature of, III^A 105, 106, 116, III^B 548, 549, 554–557
Kinsmen, Galton's method of measuring resemblance between, III^A 24
Kipling, R., quotation from, I 197
Kirby, Miss, Sec. of Inst. for the Feeble-minded, III^B 599
Kites, meteorological, III^B 535
Klaatsch, work of, on finger-prints, III^A 142
Klein, Dr, assistance of, in pangenesis experiments, II 167
Kleinwächter, as to definition of like twins, II 188
Knavery, a tinted map of, III^B 567
Knee, height of, III^A 54
Knowledge, increase of, and Charles Darwin, I 57, 58; scientific, Bateson and Karl Pearson on, III^A 288; and mental endowment, II 75; and science, II 347, 348
Knowles, J., Galton meets, III^B 520; 'Efficacy of Prayer' and, II 131
Koch, and Italian physicians, III^B 541
Kollmann, work of, on finger-prints, III^A 141, 161
Konai, Rajyadhar, signing of contract by finger-prints of, with Sir W. J. Herschel, III^A 146 Plate V
Körösi, data of, on fertility and age of marriage, II 408; work of, III^B 585

Krause, his Life of Dr Erasmus Darwin, II 192, 202, 203
Kreil, meteorological assistance of, II 39
Kubalees, atrocities of, in Africa, I 237
Kubo, researches of, on finger-prints, III^A 140, 194
Kuhn, P., letter of, with photographs showing Galton's name on rock in Ovampoland, III^B 616–617, III^B 617 Plate LIX

Labour, capacity for, and civic prosperity, III^A 401
Lagrange, Dr F., 'Physiologie des exercices du corps,' III^B 478
Lake Ngami, first explorer to reach, II 35
Lake Nyanza, and climate, II 36
Lake at Otchikoto, rock on, where Galton's name can still be read, III^B 617 Plate LIX; see also I 216 Plate LVII (Galton spells also 'Omutchikota')
Lalanne, anamorphic geometry of, applied to ogive curve, II 402
Lamarck, and transmission of acquired characters, II 147; Dr Erasmus Darwin and, II 202, 203
Lambeth Palace, Galton's inquiries at, III^B 578
Lamp, Galton's, I 148, 149
Lander, and the African memorial, II 25
Lane Fox, and anthropology, II 334
Languages, defects of, II 251, 252; modern, in educational scheme, II 155
Lankester, Sir E. Ray, on Committee for Measurements of Plants and Animals, III^A 127, 128, 291; on heredity and tradition, III^A 410, 411; eugenic ideas of, III^A 341; and experimental zoology, III^A 129; Galton and, III^B 605; Medallist of Linnean Soc., III^A 340; death of, III^A 342
Laplace, Galton introduced to theorem of, I 163; and application of mathematics to social phenomena, III^A 1
Laplace-Gaussian Curve. See *Normal Curve*
Larches, The, Galton's birth at, I 62; Darwin's recollections of, I 241; left by the Galtons, I 74; description of, I 75; plan and sketch of, I 75 Plate XLV; Dr Priestley and, I 51, 62, 75
Laski, H. J., Galton's interest in, III^B 606, 608; letter of Galton to, III^B 608, 609
Latency, Galton on, II 170, 171, 182, 364; and fertility, II 186
Latent Elements, III^A 76
Latin, in education, I 67
Law of Ancestral Heredity. See *Heredity*.
Law of Deviation from an Average, applied to discussions on heredity, II 89; Galton on, II 335. See also *Ogive, Normal Curve*
Lawrence, Sir T., albino orchids of, III^A 360, 370
Laws, Galton lays down, for Namaqualand natives, I 226–230, I 226 Plate LVIII
Leaders, amongst wild animals and men, II 73, 74
League of Nations, and racial boundaries, III^A 219; and international Eugenics, III^A 220
Leamington, Galton's home at, I 74; visits to, II 11, 53, 70, 161, 180
Least discernible Difference, Galton's method of measurement by, II 303, 304, 307; outline of Greek girl to illustrate, III^A ix, Extra Plate
Lecoq de Boisbaudran, and visualisation, II 253
Lee, Dr Alice, work of, III^A 248; recommends Ethel M. Elderton to Galton, III^A 258; resignation of post in Biometric Laboratory, III^A 328; degree of, III^B 513–515
Lee, Sidney, Galton consults, III^B 529
Leeds, Karl Pearson to lecture at, III^B 514
Lefevre, Sir J., large head of, III^A 248
Leffingwell, on illegitimacy, III^B 492
Left-handedness, and finger-prints, III^A 169
Legal Ability, inheritance of, II 93
Legislature, Upper House of, and inheritance of ability, III^A 32–34
Lengths, photographic measurement of, II 316, 317
Lennard, Sir J., death of, III^B 516
Lens, Galton's mounted watchmaker's, for comparing finger-prints, III^A 155, 178, and I Frontispiece
Le Pury, Galton rents house from, III^B 613
Lethbridge, Mrs (Millicent Bunbury), niece of Galton, I 193, III^A 281, 435; travels with Galton to Auvergne, I 93, 94; stays with Galton, III^A 393, III^B 574, 597, 603–605; children of, III^B 530, 532, 534, etc.; joins Eugenics Education Society, III^B 587; brailles Galton's 'Memories,' III^B 604; letters of Galton to, on prayer, III^B 471, III^A 412, III^B 472, 520–522, 528–536, 540–617; recollections of Galton by, III^B 446–449; portrait of, III^B 471 Plate XLVIII; assistance of, to biographer, I viii, III^A vii
Lethbridge, Amy ('Dim'), great-niece of Galton, III^B 521, etc.
Lethbridge, E. G. B., great-nephew of Galton, education of, III^B 473
Lethbridge, Frank, III^B 530, 532, etc.
Lethbridge, Guy, great-nephew of Galton, III^B 518, 520, 579, etc.; saves a life, III^B 544–546, 548; appointment of, III^B 586
Lethbridge, Hugh, great-nephew of Galton, a curate in Exeter, III^B 610, etc.
Lethbridge, R., great-nephew of Galton, III^B 518, 521
Letters, or numbers, associated with colours, II 253
Lewenz, work of, on correlation of phalanges, III^A 256
Lewes, G. H., and spiritualism, II 66; and 'The Reader' or 'Nature,' II 68, 69
Library, of Galton, II 12
License, poets and sexual, II 99
Lichfield Cathedral, monument in, to Dr Erasmus Darwin, II 202–204, III^B 476
Liebig, Galton proposes to work under, I 126, 127; laboratory of, I 128–130
Liebreich, Prof. Oscar, paper on 'Generic Images' sent to, III^B 469
Liége, Galton's sketch at, I 94, Plate XLIX
Life, expectation of, and heredity, III^B 536–539; after death, Galton on, III^A 91
Life-Histories of progenitors, value of, II 302, 303
'*Life-History Album*,' original scheme of, II 355, 362, 366; and 'Record of Family Faculties,' II 363–370
Lighthouse Signals, note on, by Galton, II 50
Limitation of Family, evils of, II 110–111, III^A 243
Lindley, Prof., lectures on Botany of, I 121, 123
Linnaeus, and strawberry cure for gout, III^A 124
Linnean Society, Samuel Galton and the, I 47; Darwin-Wallace celebration of, II 201, III^A 340–342; honours Galton, III^A 340–342
Linseer, Carl, memoir of, II 143
Lions, in Africa, I 222–224
Lippmann, von, pamphlet by, III^B 500, 503
Lister, Lord, remarkable head of, III^A 248; a Copley medallist, III^A 431, III^B 614
Lister, J. J., attacks Biometry, III^A 297–300; address of, at Brit. Assoc., III^B 579
Litchfield, Mrs, sends 'A Century of Family Letters,' by Mrs Darwin, to Galton, III^B 536; to stay with Galton, III^B 540
Literary Aptness of Galton as a child, I 63, 64
Literary Men, II 96; gifts of, II 256
Literature, in educational schemes, II 155
Livia, wife of Augustus, portrait of, and composite, II 296 Plate XLIII
Livingstone, travels of, in Africa, I 214–216, II 30, 35; and African memorial, II 25; writes first of the Zambezi, III^B 548

Lock, C. S., his opinion of Eugenics, III^{A} 260
Lock, R. H., on heredity, III^{A} 303
Lock, invented by Galton, I 148
Locke, on generic image of man, II 298
Lockyer, Norman, and journalism, II 67–69
Logic, in educational scheme, II 155
Lombroso, and modern criminology, II 232; on inheritance of ability, III^{A} 102
London, Galton's early stay in, I 91
London, Jack, 'Call of the Wild' by, III^{B} 551
London County Council, and the Galton Laboratory, III^{A} 322; material provided by, III^{A} 386
Longevity, of various classes, II 116; and health, II 266, 349; inheritance of, II 348, 349; inquiry concerning, II 349, 350; of the Galtons, III^{A} 347, III^{B} 594; source of, in the Galton family, I 26
Loops, in finger-prints, types of, III^{A} 209, III^{A} 213 Plates XXIV, XXVI; sheet of types, III^{A}, pocket in case; counting ridges of, III^{A} 201–202
Loti, Pierre, at Ascain, III^{B} 561
Louping-ill, E. Wheler-Galton's report on, III^{B} 564–566
Lourdes, Galton at, III^{B} 553
Lovelace, Lady, presents collection of casts to Galton Laboratory, I 180
Lowe, E. J., on Committee for Measurement of Plants and Animals, III^{A} 127
Lowell, J. R., pall-bearer at Darwin's funeral, III^{B} 471
Lowell Institution, and Galton, II 361
Loxton, home of Erasmus Galton, sketch of, I 48 Plate XXIX; Erasmus Galton farms at, I 125, III^{B} 528, 543; improvements in, III^{B} 595; cavern at, III^{B} 606; Galton visits Erasmus at, I 111, 147, III^{B} 527, 528. See also III^{B} 601
Loyola, III^{B} 567, 568
Lubbock, Sir J., and anthropology, II 334; and care in choice of words, III^{B} 506; peerage of, III^{B} 516. See also *Avebury, Lord*
Lunar Society, and Samuel Galton, I 46, 47; Dr Erasmus Darwin a member of, I 61; meeting place of, I 48, 49 Plate XXXI
Luschan, von, skin colour scale of, II 226
Lushington, Sir V., pedigree of, III^{A} 343, 345, 356
Lutz, F. E., paper of, in 'Biometrika,' III^{A} 251
Lyall, A., on the Indian Civil Service, III^{A} 327
Lying, Salonika centre of gravity of, in Europe, II 341
Lynch, surveys the Jordan and Dead Sea, I 205
Lyndhurst, Galton at, II 130, III^{A} 374–377, III^{B} 593, 594
Lyndon, Dr, attends Galton, III^{B} 613, 617, 618
Lyons, Capt. (now Sir H. G.), visits Galton, III^{B} 608
Lyrics, of Sir Charles Sedley, I 20

Macalister, Prof. A., on Committee for Measurement of Plants and Animals, III^{A} 126; on Evolution Committee, III^{A} 290; meets Galton in Egypt, III^{B} 517, 518
MacAlister, Sir D., and Galton-MacAlister curve, II 228; interest of, in Eugenics, III^{A} 430; on law of geometric mean, III^{B} 468
McCaskie, Dr, attends Galton, III^{B} 596, 597
Macdonell, Mr, becomes engaged to Miss Hertz, III^{B} 469
Macdonell, W. R., correlation coefficients of, compared with Galton's values, III^{A} 53, 54; on correlation of Bertillon measurements, II 305, 383, III^{A} 188; on criminal anthropometry, III^{A} 247; paper of, for 'Biometrika,' III^{A} 248, 249; one of original guarantors for 'Biometrika,' III^{A} 250
McFadyean, Mrs, on limitation of families, III^{A} 322
Machine inventing, Galton takes to, II 51
Macintyre, Violet, visits Galton, III^{A} 324, III^{B} 583, 584
Mackenzie, W. L., on nature and nurture, III^{A} 260
Mackinder, H. J., and the Fellowship in National Eugenics, III^{A} 222, 223
McLennan, D., Galton visits, III^{B} 470
MacMahon, P. A., and technical scientific terms, III^{A} 334
Macmillan, on 'Finger Print Directory,' III^{A} 151
Macnaghten, Lord, elected to Trinity College, Cambridge, Hon. Fellowship, III^{A} 236, 238; meets Galton at Trinity College, III^{B} 574
Macnaghten, M. L., and identification of criminals, III^{A} 149
Macrobius, III^{A} 290, 291
Magazines, mid-Victorian, II 117
'*Magnum Opus*,' of Roger Bacon, possible purchasers for, III^{B} 527
Magnus, Sir Philip, and Galton Committee, III^{A} 386
Mahan, Capt. A. T., honour bestowed on, III^{B} 494
Mahomed, Dr F. A., and composite portraits of the phthisical, II 290–292; and the 'Life-History Album,' II 366, 367
Maine, Sir H., at Cambridge, I 141, 153, 164; obtains Chancellor's English Medal, I 167
Mair, D., to be consulted regarding plans for eugenic certificates, III^{A} 296
Mailland, Prof., elected to Trinity College, Cambridge, Hon. Fellowship, III^{A} 236, 238
Makarius, Galton meets in Egypt, III^{B} 519
Mallock, W. H., novel by, III^{B} 595, 596
Malthus, Galton on maxims of, II 110, 265
Malthusian League, III^{A} 304
Mammals, need of work in their case, III^{A} 401
Man, Darwin and Galton on, II 86; ancestry of, II 85, 274, 275; source of slavish aptitudes in, II 72, 73; herd instinct in, II 72, 73; domestication of, III^{A} 219, 220; and his environment, III^{A} 218, 219; nature and nurture of, II 254; and heredity, II 117; and civilisation, II 85; duty of, regarding evolution, II 263; future of, III^{A} 217–220; inequalities in, III^{A} 347, 348; on earth, III^{A} 318; Galton, on salvation of, II 109, 110; generalisations regarding, II 298; and woman, artistic gifts in, III^{A} 66; sensitivity of, II 221–222
Manhood, forecast of, from youth, III^{A} 232; associations of, II 235
Mann, Dr, on drawing of Bushmen, II 239–240
Manourrier, his lecture on 'bertillonage,' III^{A} 148
Map Making, and measurements in Africa, I 232, 234
Maps, II 21, 22; proposals for, II 29, 30, 33, 34, III^{B} 461; stereoscopic, II 283; meteorological, II 36–43, II 36 Plates V, VI, II 38 Plate VII; portable frames for maps of world, III^{B} 462
Marking, system of, for bodily efficiency, II 382, 386–396
Marks, desirable tests carrying, II 388
Marlborough College, statistics from, II 343, 396
Marriage, early, and fertility, II 110, 123–125, 265, 266, 408, 409; factors influencing, II 112, 149, III^{A} 265, 270, 273; customs regarding, III^{A} 233, 234, 267–270; among primitive peoples, III^{A} 267–270; German and Austrian laws regarding, III^{A} 268; interference with freedom of, III^{A} 269; and religion, III^{A} 269; and civic worth, III^{A} 233, 242; within the caste, II 121; into tainted stock, II 132; of cousins, II 188, III^{B} 470; encouragement of, in good stock, II 139; the modern woman and, II 132–134; Bernard Shaw on conventions regarding, III^{A} 260
Mars, signals from, II 279, 280
Marshall, Prof. Alfred, criticises work of Eugenics Laboratory, III^{A} 408, 430
Martin, Prof. R., eye scale of, II 226
Martin, Sophia, poem of, on Eugenics, III^{A} 357
Mason, Mr, teaches Galton mathematics, I 88, 99

Masters, Maxwell. T., on Committee for Measurement of Plants and Animals, III^A 127, 291
Maternal Impressions, II 279; Galton's views on influence of, II 146, 209; de Candolle on, II 209, 210
Mathematical Analysis, Galton's, and coefficient of reversion, III^A 7–9; aid from Dickson in, III^A 12, 13
Mathematical Studies, of Galton, I 88–90, 99–101, 107; advice from Darwin and Bowman regarding, I 110; advice from Hodgson on, I 113; at Cambridge, I 115, 140–195
Mathematicians, and 'Biometrika,' III^A 256
Mathematics, inherited ability in, II 97; success in life and, II 97, 98; in educational scheme, II 155, III^A 302; mental imagery and, II 243; application of, to social phenomena, III^A 1; and science of heredity, III^B 504; need of, in statistical work, III^A 302, 303
Mathison, Tutor of Trinity College, I 153, 155, 156, 158–160, 162, 163
Mating, and parentage, instincts of, III^A 218
Maudsley, Dr Henry, on Eugenics, III^A 259
du Maurier, G., drawing in 'Punch' of, III^A 375
Maxwell-Masters, see *Masters*
Maxwell, Clerk, size of head and stature of, II 150
Mean, Galton's method of determining, III^A 24, 25; tables, giving some values of, for bodily characters with standard deviations, III^A 54
Measurement, Galton's passion for, III^B 458; value of habit of, II 382; need of, in anthropology, II 334; advantages of, to the measured, II 381; diurnal changes in, II 380; choice of, II 372, 373; by photography, II 316–323; of resemblance, II 329–333; III^B 562, 564, 565, 569, 576; by least discernible difference, II 303; anthropometric, II 336, 337, 370, 373; of individuality, II 303–306; of difference between individual and type, II 311–315; of physical powers, II 336; of bodily efficiency, II 387, 388; of acuity of vision, II 222, 223, 336; of sight differences, difficulties of, II 303; of imagination and sensation, II 307, III^B 493; of pain, II 408; of infants, III^B 496; of animals, II 317, 318, 320–323; of correlation, III^A 50–57; and map making in Africa, I 232
Measurer, fallibility of, II 380
Mechanical Aptitude, in scientists, II 151
Mechanical Ingenuity, in Galton and his ancestors, I 16, 60, 148–150, 181, 212, 213, II 3, 19, 35, 49–53, 59, 274. See also *Galton, Francis*, mechanical ingenuity of
Mechanical Manipulation, in educational scheme, II 155
'*Mechanical Selector*,' for border-line cases, II 305, 306
Mechanics, Hopkins compliments Galton on his, I 166; mental imagery and, II 243
Medallions, of Erasmus Darwin, by Onslow Ford, II 202, II 204 Plate XX; by Fassie, III^B 473 Plate L; Galton's annual, specimen of, III^B, below list of illustrations
Medals awarded to Galton, Royal Geographical, I 239; French Geographical Silver, I 239, III^A 236; Huxley, of Anthropological Institute, III^A 226, 235; Royal Gold Medal, II 201, III^B 476; Darwin, of Royal Society, III^A 236, 237; Darwin-Wallace Medal of Linnean Society, III^A 340–342; Copley Medal of Royal Society, III^A 400, III^B 611, 614, 615
Median Value, Galton's definition of, II 338
Medians, and means, II 344; Galton's use of, III^A 51, 54, 422; papers concerned with, II 400–405
Medical Life History, value of, II 359; Galton's interest in, II 360; difficulty in obtaining, II 360; scheme for obtaining, falls through, II 361
Medical Men, visits to, in Heidelberg, I 95; indices of capacity for, II 407
Medical Research Council, grants of, III^A 361
Medical Research Fellowships, need for, II 153
Medical Studies, of Galton—in Birmingham, I 92, 99–104; at King's College, London, I 105–128; at Cambridge, I 180, 181, 184–187; at St George's Hospital, I 190, 191; Galton gives up his, I 193, 194, 196
Medicine, Galton dislikes the idea of practising, I 199; Sir Francis S. Darwin and, I 22, 23
Medico-metric Laboratory, II 359
Mediocrity, Galton on, II 384, 385
Meldola, Prof., on Committee for Measurement of Plants and Animals, III^A 126, 133, 290, 291; on a farm for experimental breeding, III^A 134; suggests use of the word 'phylometry,' III^B 500; Galton congratulates, III^B 614; letters of Galton to, III^B 500, 501, 598, 614
Melville, Col., interest of, in Eugenics, III^A 392; calls on Galton, III^B 593; praises Lord Haldane, III^B 603
Memoirs, Papers and Letters to Journals of Francis Galton:
(1) 'The Telotype,' a printing Electric Telegraph (1849), I 212
(2) 'Recent Expedition into the Interior of South-Western Africa' (1852), I 215
(3) 'Modern Geography,' Cambridge Essays (J. W. Parker) (1855), II 21
(4) 'Ways and Means of Campaigning' (1855), II 14, 15, 16, 17, 18
(5) 'Course of Public Lectures in Camp at Aldershot' (Privately printed) (1856), II 15
(6) 'Catalogue of Models, illustrating Camp Life' (Privately printed) (1858), II 18
(7) 'The Exploration of Arid Countries' (1858), II 24, 25
(8) 'Sun Signals for the Use of Travellers' (1859), II 21
(9) 'Table for Rough Triangulation, etc.' (1860), II 23, 24
(10) 'On a New Principle for the Protection of Riflemen' (1861), II 18
(11) 'Additional Instrumental Instructions for Mr Consul Petherick' (1861), II 27
(12) 'Zanzibar,' a Lecture at the C.M.S. (1861), II 28
(13) 'Weather Map of the British Isles for Tuesday, September 3, 9 a.m.' (1861?), II 36 Plate VI
(14) 'Synchronous Weather Chart of England, January 16, 1861, 9 a.m.' (1861), II 36, 38 Plate V
(15) 'Circular asking for Synchronous Observations during one month three times daily, with Map' (Privately printed) (1861), II 37
(16) 'English Weather Data, February 9, 1861, 9 a.m.' (1861), II 37
(17) 'Meteorological Charts' (1861), II 37
(18) 'Recent Discoveries in Australia' (1862), II 24
(19) 'Report on African Explorations' (1862), II 27
(20) 'A Development of the Theory of Cyclones' (1862), II 39
(21) 'Hereditary Talent and Character.' (Written 1864), II 75, 88, 92. (Published 1865), II 70
(22) 'On Stereoscopic Maps taken from Models of Mountainous Countries' (1865), II 33, 34
(23) 'Spectacles for Divers and the Vision of Amphibious Animals' (1865), II 34
(24) 'The first Steps towards the Domestication of Animals' (1865), II 70
(25) 'Domestication of Animals' (1865), II 258
(26) 'On an Error in the usual Method of obtaining Meteorological Statistics of the Ocean' (1866), II 53, 54
(27) 'On the Conversion of Wind Charts into Passage Charts' (1866), II 55, 56
(28) 'Drill Pantagraph, reducing horizontally and vertically to different Scales. Also a mechanical Computer of Vapour Tension. Report of Meteorological Council' (1869), II 45–49

(29) 'Barometric Prediction of Weather' (1870), II 54, 55
(30) 'Experiments in Pangenesis by Breeding from Rabbits of a pure Variety, into whose Circulation Blood taken from other Varieties had previously been largely transfused' (1871), II 156
(31) 'Address to Geographical Section of the British Association at Brighton' (1872), II 28, 29
(32) 'Gregariousness in Cattle and in Men' (1872), II 72
(33) 'Statistical Inquiries into the Efficacy of Prayer' (1872), II 115–117
(34) 'Blood Relationship' (1872), II 169, 170, 184; reference to, II 146
(35) 'Africa for the Chinese' (1873), II 32–33
(36) 'On the Employment of Meteorological Statistics in determining the best Course for a Ship whose sailing Qualities are known' (1873), II 57
(37) 'Hereditary Improvement' (1873), II 117–122; quotation from, II 131
(38) 'The Relative Supplies from Town and Country Families to the Population of Future Generations' (1873), II 123–125
(39) 'On the Causes which operate to create Scientific Men' (1873), II 146
(40) 'English Men of Science, their Nature and Nurture' (Lecture R.I.) (1874), II 145
(41) 'Proposal to apply for Anthropological Statistics from Schools' (1874), II 336
(42) 'Excess of Females in the West Indian Islands from Documents communicated to the Anthropological Institute by the Colonial Office' (1874), II 337
(43) 'Proposed Statistical Scale' (1874), II 337
(44) 'Notes on the Marlborough School Statistics' (1874), II 343
(45) 'The History of Twins, as a Criterion of the Relative Powers of Nature and Nurture' (1875), II 126–128; reference to, I 8
(46) 'A Theory of Heredity' (1875), II 177, 182–188
(47) 'Statistics by Intercomparison with Remarks on the Law of Frequency of Error' (1875), II 338
(48) 'On the Probability of the Extinction of Families' (with H. W. Watson) (1875), II 341–3
(49) 'On the Height and Weight of Boys aged 14 in Town and Country Public Schools' (1876), II 125–6
(50) 'Short Notes on Heredity in Twins' (1876), II 128–130
(51) 'Whistles for determining the upper Limits of audible Sound in different Persons' (1876), II 215
(52) 'Apparatus for the Rapid Verification of Thermometers; now in Use at Kew Observatory' (1877), II 59
(53) 'Whistles' (1877), II 212
(54) 'Address to the Anthropological Department of the British Association' (1877), II 228. (Anthropology, a department under Biology, founded 1866)
(55) 'Typical Laws of Heredity' (1877), III^{A} 6
(56) 'Hints to Travellers' (1878), II 4, 23
(57) 'Review of Letters of H. M. Stanley from Equatorial Africa' (1878), II 30
(58) 'On Means of combining various Data in Maps and Diagrams' (1878), II 284
(59) 'Composite Portraits made by combining those of many different Persons into a single resultant Figure' (1878), II 285
(60) 'Composite Portraits' (1878), II 286; reference to, II 212
(61) 'The Average Flush of Excitement' (1879), III^{A} 124
(62) 'Psychometric Inquiries' (1879), II 212
(63) 'The Geometric Mean in Vital and Social Statistics' (1879), II 227
(64) 'The Law of the Geometric Mean' (1879), II 228
(65) 'Psychometric Facts' (1879), II 233
(66) 'Psychometric Experiments' (1879), II 233
(67) 'Generic Images' (Lecture R.I.) (1879), II 295
(68) 'Generic Images' (*Nineteenth Century*) (1879), II 297; reference to, II 212
(69) 'On Determining the Heights and Distances of Clouds by their Reflexions in a low Pond of Water and in a mercurial Horizon' (1880), II 61
(70) 'Visualised Numerals' (*Nature*) (1880), II 242; reference to, II 195
(71) 'Visualised Numerals' (*Journal Anthrop. Inst.*) (1880), II 242
(72) 'Statistics of Mental Imagery' (1880), II 236
(73) 'Pocket Registrator for Anthropological Purposes' (1880), II 341
(74) 'Opportunities of Science Masters at Schools' (1880), II 344
(75) 'Galtonia (*Hyacinthus candicans*)' (1880), III^{B} 533–534
(76) 'The Equipment of Exploring Expeditions, now and fifty years ago' (1881), II 34
(77) 'On the Construction of Isochronic Passage Charts' (1881), II 35
(78) 'The Visions of Sane Persons' (1881), II 243
(79) 'On the Application of Composite Portraiture to Anthropological Purposes' (1881), II 288
(80) 'An Apparatus for testing the Delicacy of the Muscular and other Senses in different Persons' (1882), II 217
(81) 'A Rapid View Instrument for Momentary Attitudes' (1882). Addenda III^{B} 629
(82) 'An Inquiry into the Physiognomy of Phthisis by the Method of Composite Portraiture' (1882), II 291
(83) 'Photographic Portraits from Childhood to Old Age' (1882), II 302
(84) 'The Anthropometric Laboratory' (1882), II 358
(85) 'Conventional Representation of the Horse in Motion' (1882), II 399
(86) 'Hydrogen Whistles' (1883), II 216
(87) 'Arithmetic Notation of Kinship' (1883), II 354
(88) 'Medical Family Registers' (1883), II 359
(89) 'Outfit for an Anthropometric Laboratory (Privately printed) (1883), II 370
(90) 'The American Trotting Horse' (1883), II 399
(91) 'Table of Observations [on physical characters of 400 Persons]' (1889), II 378
(92) 'The Weights of British Noblemen during the last three Generations' (1884), III^{A} 136, 137
(93) 'Free-will, Observations and Inferences' (1884), II 245
(94) 'Measurement of Character' (1884), II 268
(95) 'Anthropometric Laboratory arranged by Francis Galton, F.R.S.... International Health Exhibition, 1884,' II 370
(96) 'Address to Anthropological Section B.A.' [On Inheritance and Regression] (1885), III^{A} 11
(97) 'Regression towards Mediocrity in Hereditary Stature' (*Journal Anthrop. Inst.*) (1885), III^{A} 12
(98) 'On the Anthropometric Laboratory at the late International Health Exhibition' (1885), II 371; reference to, 221
(99) 'Measure of Fidget' (1885), II 277
(100) 'Photographic Composites' (1885), II 293
(101) 'Some Results of the Anthropometric Laboratory' (1885), II 374

(102) 'The Application of a Graphic Method to Fallible Measures' (1885), II 375
(103) 'Anthropometric Percentiles' (1885), II 375–376
(104) 'A Common Error in Statistics' (1885), II 377
(105) 'Presidential Address, Anthropological Institute' (1886), III^A 12
(106) 'Notes on Permanent Colour Types in Mosaic' (1886), II 225
(107) 'Anniversary Meeting of Royal Society (Galton's Speech at the dinner)' (1886), II 201
(108) 'On Recent Designs for Anthropometric Instruments' (1886), II 226
(109) 'The Origin of Varieties (Curve of Attractiveness)' (1886), II 272–273
(110) 'Chance and its Bearing on Heredity' (Birmingham Lecture) (1886), III^A 12, 29
(111) 'Family Likeness in Stature' (1886), III^A 12
(112) 'Family Likeness in Eye Colour' (1886), III^A 34
(113) 'Good and Bad Temper in English Families' (1887), II 271; reference to, III^A 69
(114) 'Thoughts without Words' (1887), II 274–275
(115) 'Presidential Address, Anthropological Institute' (1887), II 396–397
(116) 'The Proposed Imperial Institute, Geography and Anthropology' (1887), II 411
(117) 'North American Pictographs' (1887), II 411
(118) 'Pedigree Moth-breeding as a means of verifying certain important Constants in the General Theory of Heredity' (1887), III^A 47
(119) 'Remarks on Replies by Teachers to Questions respecting Mental Fatigue' (1888), II 276
(120) 'Presidential Address, Anthropological Institute' (1888), II 397
(121) 'Personal Identification and Description' (Lecture 1888, publication 1889), II 303–306 reference to, III^A 141
(122) 'Co-relations and their Measurement, chiefly from Anthropometric Data' (1888), III^A 50
(123) 'Instrument for testing the Perception of differences of Tint' (1889), II 219
(124) 'Presidential Address, Anthropological Institute' (1889), 'Human Variety,' II 383
(125) 'Notes on Australian Marriage System' (1889), Addenda III^B 629
(126) 'On the Advisability of Assigning Marks for Bodily Efficiency in the Examination of Candidates for those Public Services in which Bodily Efficiency is of Importance' (1889), II 386
(127) 'Head Growth of Students at the University of Cambridge' (1889), II 387
(128) 'On the Principle and Methods of Assigning Marks for Bodily Efficiency' (1889), II 388
(129) 'The Sacrifice of Education' (Tests of Physical Capacity) (1889), II 393
(130) 'Feasible Experiments on the Possibility of transmitting Acquired Habits by Means of Inheritance' (1889), III^A 57
(131) 'Tests and Certificates of the Kew Observatory, issued by the Kew Committee of the Royal Society' (1890), II 59, 60
(132) 'A new Instrument for measuring the Rate of Movement of the various Limbs' (1890), II 220
(133) 'Anthropometric Laboratory, Notes and Memoirs No. 1' (1890), II 381
(134) 'Why do we measure Mankind?' Section of (133), II 381
(135) 'Human Variety,' Section of (133), II 383
(136) 'Variety,' Section of (133), II 384, 385
(137) 'The Measurement of Variety,' Section of (133), II 385
(138) 'Physical Tests in Examinations' (1890), II 394
(139) 'Dice for Statistical Experiments' (1890), II 405
(140) 'The Patterns in Thumb and Finger Marks; on their arrangement into naturally distinct classes, the permanence of the papillary ridges that make them, and the resemblance of their classes to ordinary genera' (1890), III^A 161
(141) 'Sexual Generation and Cross Fertilisation' (1890), III^A 318
(142) 'Decrease of Mortality by Smallpox, 1838–1887' (1890), III^B 482
(143) 'Method of Indexing Finger-Marks' (1891), III^A 170
(144) 'Identification by Finger Tips' (1891), III^A 154
(145) 'Address to the Demographers' (1891), III^A 218
(146) 'Retrospect of Work done at my Anthropometric Laboratory at South Kensington' (1891), II 378
(147) 'Galton's Pantagraph and Vapour Tension Computer (German)' (1892), II 47
(148) 'Identification' (1893), III^A 155
(149) 'The Just-Perceptible Difference' (1893), II 307, III^A viii, Extra Plate
(150) 'Enlarged Finger Prints' (1893), III^A 155
(151) 'Discontinuity in Evolution' (*Mind*, 1894), III^A 84–86
(152) 'The Part of Religion in Human Evolution' (1894), III^A 88; reference to, II 102
(153) 'The Relative Sensitivity of Men and Women' (1894), II 222
(154) 'Review of A. Binet's "Psychologie des Grands Calculateurs et Joueurs d'Échecs"' (1894), II 275
(155) 'Arithmetic by Smell' (1894), II 275
(156) 'A plausible Paradox in Chances' (1894), II 405
(157) 'Results derived from the Natality Table of Kőrösi by employing the Method of Contours or Isogens' (1894), II 408
(158) 'Physical Index to 100 Persons on their Measures and Finger-Prints' (1894), III^A 197
(159) 'Terms of Imprisonment' (1895), II 406
(160) 'A New Step in Statistical Science' (1895), II 411
(161) 'The Wonders of a Finger Print' (1895), III^A 159
(162) 'Intelligible Signals between Neighbouring Stars' (1896), II 279
(163) 'A curious Idiosyncrasy' (1896), II 279
(164) 'On Bertillon's System of Identification' (1896), II 411, III^A 144
(165) 'Three Generations of Lunatic Cats,' Letter only (1896), III^A 87
(166) 'Scheme for further accurate Observations on Variation, Heredity, Hybridism, and other Phenomena that would elucidate the Evolution of Plants and Animals' (1896), III^A 135
(167) 'Prints (Finger) of Scars' (1896), III^A 154
(168) 'Les empreintes digitales' (1896), III^A 159
(169) 'The Average Contribution of each several Ancestor to the total Heritage of the Offspring,' *Roy. Soc. Proc.* (Basset Hounds) (1897), III^A 40
(170) 'Relation between the Individual and Racial Variability' (1897), III^A 95
(171) 'Dr Haughton and the Long Drop' (1897), II 407
(172) 'Rate of Racial Change that accompanies Different Degrees of Severity in Selection' (1897), III^A 93
(173) 'Retrograde Selection' (1897), III^A 95
(174) 'Hereditary Colour in Horses' (1897), III^A 95
(175) 'Temporary Flooring in Westminster Abbey for Ceremonial Processions' (1898), II 198
(176) 'Photographic measurement of Horses and other Animals' (1898), II 320
(177) 'Photographic Record of Pedigree Stock' (1898), II 321–322

(178) 'An Examination into the Registered Speeds of American Trotting Horses, with Remarks on their Value as Hereditary Data' (1898), II 400
(179) 'A Diagram of Heredity (illustrating the Ancestral Law)' (1898), III^A 44–45
(180) 'The Distribution of Prepotency' (in Trotting Horses) (1898), III^A 98
(181) 'The Median Estimate' (1899), II 401
(182) 'A Geometric Determination of the Median Value of a System of Normal Variants, from Two of its Centiles' (1899), II 402
(183) 'Linnaeus' Strawberry Cure for Gout' (1899), III^A 124
(184) 'William Cotton Oswell, Hunter and Explorer' (1900), Preface to, by Galton, II 35
(185) 'Analytical Photography' (1900), II 311–316
(186) 'Identification Offices in India and Egypt' (1900), III^A 157
(187) 'Souvenirs d'Égypte' (1900), III^A 158, III^B 455
(188) 'Biometry' (1901), III^A 100
(189) 'On the Probability that the Son of a very highly gifted Father will be no less gifted' (1901), III^A 102
(190) 'The possible Improvement of the Human Breed under the existing Conditions of Law and Sentiment' (1901), III^A 226
(191) 'The most suitable Proportion between the Values of First and Second Prizes' (1902), II 411
(192) 'Finger-Print Evidence' (1902), III^A 160
(193) 'Pedigrees' (1903), III^A 103
(194) 'Sir Edward Fry and Natural Selection' (1903), III^A 122
(195) 'Our National Physique—Prospects of the British Race—Are we Degenerating?' (1903), III^A 252
(196) 'Nomenclature and Tables of Kinship' (1904), III^A 105
(197) 'Average Number of Kinsfolk in each Degree' (1904), III^A 107
(198) 'Distribution of Successes and of Natural Ability among the Kinsfolk of Fellows of the Royal Society' (1904), III^A 108
(199) 'Number of Strokes of the Brush in a Picture' (1905), III^A 125
(200) 'On Dr Faulds' "Guide to Finger Print Identification"' (1905), III^A 147–148
(201) 'Eugenics, its Definition, Scope and Aims' (1905), III^A 262
(202) 'A Eugenic Investigation, Index to Achievements of Near Kinsfolk of some of the Fellows of the Royal Society' (1905), III^A 266
(203) 'Restrictions in Marriage' (1905), III^A 266
(204) 'Eugenics as a Factor in Religion' (1905), III^A 273
(205) 'Studies in National Eugenics' (1905), III^A 274
(206) 'Cutting a Round Cake on Scientific Principles' (1906), III^A 124
(207) 'Measurement of Resemblance' (1906), II 329–333; reference to, III^A 279
(208) 'The Measurement of Visual Resemblance' (1906), II 331
(209) 'Anthropometry at Schools' (1906), II 345
(210) 'Classification of Portraits' (1907), II 325
(211) 'One Vote, one Value' (1907), II 400
(212) 'Grades and Deviates' (1907), II 401
(213) 'Vox populi' (1907), II 403
(214) 'Probability—the Foundation of Eugenics' (1907), III^A 317–321; quoted, II 415, III^A 309
(215) 'Suggestion for improving the Literary Style of Scientific Memoirs' (1908), III^A 336
(216) 'Address on Eugenics' (1908), III^A 346
(217) 'Local Associations for promoting Eugenics' (1908), III^A 350
(218) 'Sequestrated Church Property' (1909), II 410, III^A 367
(219) 'Segregation (of the Feeble-Minded)' (1909), III^A 365
(220) 'Numeralised Profiles for Classification and Recognition' (1910), II 326
(221) 'Eugenic Qualities of Primary Importance' (1910), III^A 401
(222) 'Note on the Effects of small and persistent Influences' (1910), III^A 403
(223) 'The Eugenic College of Kantsaywhere' (1910), III^A 411
Memories, blended, and general impressions, II 296
'*Memories*,' Galton's, III^A 330, 333, 335, 339, 342, 343, 345, 346, 354, 355, III^B 585–588, 613
Memory, inheritance of, II 151; physiological basis of, II 296; of form and number, tests for, II 359; of form in men and women, II 376; a deceitful guide, II 367; will and, II 241
Men, inequalities of, I 61, II 89–92, 121, 122, 135, 137, III^A 252; and women, sensitivity of, II 221, 222
Mendel, G., and Galton, born in same year, III^A 335; Galton's appreciation of, III^B 542
Mendeleef, Prof., honour conferred on, III^B 494
Mendelians, and eye colour, III^A 36, 37, 324; and albinism, III^A 388; on Galton's Ancestral Law, III^A 41
Mendelism, and Law of Ancestral Inheritance, II 84, III^A 329; Galton's approach to, II 190: some modifications in, III^A 81; biometry and, III^A 287, 288, 357, 358; bearded wheat and, III^A 314; and white man and albino negress giving mulattoes, III^A 370
Mental, character—of Francis Galton, I 56–60, II 1, 4, 10, 233–236, 308, 317, 355 (see also *Galton, Francis*, mentality of); inheritance of, II 72, 75–77, 81, 82, 86–89, 126, 128, 135, 146, 174; and physiognomic character, II 231–233; and anthropometric characters, II 229, 232, 233; resemblance of, in twins, II 127; correlation of, with temperament, etc., II 229; source of, II 72, 73; of different races, II 31, 88; of criminals, II 230; measurement and classification of, II 229
Mental Capacity, tests for, II 272; of school children, and physique, III^A 356
Mental Defect, in the general population, III^A 365–367; plans for work on, III^B 541, 542, 544
Mental Discomforts, weariness in Galton, I 56, 58, 69, 70; fatigue, signs and warning of, III^B 478; overwork, circular concerning, II 351, 352; and idiosyncrasy, II 279
Mental Imagery, in Dr Erasmus Darwin, II 196; in Charles Darwin, II 207; Galton and, II 236, 238; in scientists, II 237, 243; in different races, II 239, 240; in artists, etc., II 243; in women, II 242; accounts of, II 240; desirable power of, II 241; and ability, II 243
Mental Impressions, source of error in, II 296, 297
Mental Processes, and imaginary smells, II 275, 276
Mental Types, distinguished by portraiture, II 301, 303
Menzies, Mrs, Galton sees, in Cape Town, I 219
Menzies, Sir N. and Lady, hospitality of, I 168, 169
Mercier, C., on Eugenics, III^A 259
Meredith, George, and London University, III^A 289
Merrifield, G. W., and Galton's wave-machine, II 52, 53
Merrifield, F., cooperates with Galton in breeding experiments on moths, III^A 46, 47, 49, 130, III^B 484
Mesmerism, Galton's experiments in, II 197
Meston, A. J., graphical representation of Galton's ancestral law by, III^A 44
Metaphysicians, mental imagery in, II 243

Meteor, seen by Galton, III^B 579, 580
'*Meteorographica*,' work by Galton, II 38, 40–43, II 38 Plate VII
Meteorological, data, appeal for, II 37; predictions, III^B 475, 481; instructions for travellers, II 44
Meteorological Council, II 43, 44, 56, 57
Meteorologists, need for co-operation among, II 38, 39; biometry and, III^A 282, 283
Meteorology, and Galton, II 7, 13, 35–49, 53–62; on how to measure data regarding, III^B 466–468, 471–473
Methuen, asks Galton for an autobiography, III^A 329, III^B 585; considers 'Kantsaywhere,' III^B 615
Meyer, Dr, Galton travels to Frankfort with, I 132
Meyrick, on experiments at Marlborough, II 396
Mice, experiments and work on, II 139, 140, III^A 251, 315, III^B 542
Middle Classes, lower, Galton on, III^A 252
Middleton, 'Biographia Evangelica' of, II 100–102
Middleton, J. H., honour conferred on, III^B 494
Midparent, Galton's definition of, III^A 8, 15
Mill, John Stuart, definition of 'religion' of, III^A 89, 90, 93
Millais, Sir E., Reports on Basset hounds of, III^A 40
Miller, W. A., works in Liebig's laboratory, I 126, 130; his advice to Galton, I 129; chemical investigations of, I 130
Mimetic Insects, and natural selection, III^A 122, 123
Mimicry, argument from, III^A 370
Mind, and body, exercise and fatigue of, III^B 478, 479; over-concentration of the, II 241; of others, as a field for exploration, II 243
Ministers, Welsh, composite portrait of, II 288 Plate XXXIII
Minto, Lord, on permanence of individuality, III^A 279
Minutiae, of finger-prints, III^A 181, 183; definition of, III^A 178; persistence of, III^A 181, 195, 196; resemblance of, in twins, III^A 191. See also under *Finger-prints*
Misery, and fertility, III^A 218
Missionary Enterprise, II 32
Models, geographical, II 33, 34, II 33 Plate IV; to illustrate action of Weber's Law, II 307
Mohammedanism, Galton's interest in, and respect for, I 207, III^B 449; and missionary enterprise, II 32; Bosworth Smith on, I 207. See also *Moslems*
Moilliet, Grace, writes to Galton, III^B 579
Moilliet, J. K., brother-in-law to Galton, III^B 473; death of, III^B 580
Moilliet, Lucy, sister to Galton, visits to, at Selby Hall, I 155; messages to, III^B 453
Monkeys, Galton's pets in Egypt and Syria, I 202, 203
Monogamy, from the eugenic standpoint, III^A 86
Mont Blanc district, model of, II 34
Montford, bust of Darwin by, III^A 374
Monumental Tablet, to Erasmus Darwin in Lichfield Cathedral, II 204 Plate XX
Moral Character, source of, II 72, 73; inheritance of, II 72, 75, 81, 82, 83, 86, 87, 89; and natural selection, II 83
Moral Lapses, and piety, II 102
Morality, relative nature of, III^A 263
Morals, and the legal profession, II 95; and the Greeks, II 108
Morley, John, quotation from, III^B 441; anecdote of, and Lord Rosebery, III^B 615
Mosaic Code, and Eugenics, III^A 223, 224
Mosaics, as colour standards, III^B 474
Moslems, civilising effect of, II 28, 258
Motherhood, endowment of, and Eugenics, II 134
Mothers, of scientists, II 96, 97, 102; of divines, II 102; transmission of talent by, II 76, 77; mental influence on, III^A 135; and mothers' ancestry, influence of, III^A 102; fathers and offspring, finger-prints of, III^A 192
Moths, breeding experiments with, III^A 40, 45–47, 49, III^B 484
Motives, valuation of, II 351; material and immaterial, III^A 270, 271; and temptations, changing character of, III^B 492
Mott, F. W., on segregation of defective children, III^A 268; on internal pigment, III^A 372; material of, III^A 298
Mount Carmel, Galton's experiences at, I 204
Mountaineering, and Galton, II 6
Mozart, extraordinary powers of, III^B 500
Mügge, paper of, III^B 600
Mulattoes and Mendelism, III^A 370
Mules, on Galton's African journey, I 222–225
Müller, Prof. Max, on thought without words, II 274, 275; Prof. Sayce and, III^B 517
Mungo Park, and African memorial, II 25
Munro, R., and Kew Observatory, II 59
Murchison, Sir R., and Galton, II 61; on Committee of Brit. Assoc., III^B 458
Murie, and experiments on pangenesis, II 159
Murray, travels of, in Africa, I 214
Murray, Prof. Gilbert, on religion of Pagan Greeks, III^A 272
Murray, John, and publication of eugenic matter, III^A 277, III^B 544, 545; dines with Galton, III^B 530; advice of, to writers, III^B 564
Muscle, hereditary character of, II 104
Muscular, power and training, II 91; co-ordination, measure of, II 358, 359; sense, testing of, II 217, 218
Musical Box, facility in identifying perforated discs of, III^B 505
Musical Sense, lack of, in Galton, III^B 441, 448
Musicians, fertility of, II 96
Mutations, and origin of species, II 84; evolution by, III^A 31, 32. See also *Sports*
Mutiny, among men on Galton's African journey, I 223–225
Muybridge, his photographs of the horse in motion, II 399
Myners, Miss, Galton's interest in, III^B 453
Mytton, Jack, extravagant exploits of, I 211

Nail Mark, on Chinese coin, III^A 174
Namaquas, behaviour of the, I 224–227, 229, 230, 232, 233, 235; Galton calls on, and gives laws to Jonker, the captain of the, I 226; and rewards him for keeping the peace, I 235
Name, change of, and loss of hereditary record, III^A 119
Nangoro, African chief, crowned by Galton, I 233, 235; death of, I 240; sketch of, I 59 Plate XXXVIII, I 237 Plate LIX
Napoleon, composite portrait of, II 295, II 296 Plate XLIV; on undesirability of mental imagery in commanders, II 253
Naqada Crania, measurements of, III^A 247
Nash, Mrs Vaughan, gift of, to the Galton Laboratory, of Florence Nightingale's copies of Quetelet, II 414
Nasmyth, and picture of a meteor, III^B 579
National Eras of Progress, Galton on, II 254, 255
National Physical Laboratory, Galton on deputation regarding, III^A 248
Nations, modern, nature of, II 118; influences affecting natural ability of, II 109–115
Natural Ability, among kinsfolk of scientific men, I 6
Natural Children, of Erasmus Darwin, I 18, 19; of Col. E. S. Pole, I 19
Natural Equality, I 61, II 90–92, 121, 122

'*Natural Inheritance*,' work by Galton, I 2, 5, II 84, 87, IIIA 57–82; Weismann's appreciation of, IIIA 340, 341
Natural Selection, Galton introduces idea of, into his work, IIIA 9; quincunx, to illustrate action of, IIIA 10; Galton on, IIIA 122; Sir E. Fry on, IIIA 122, 123; action of, IIIA 81, 82; survival and, II 74; and man, II 74, 75, 79, 82, 83, 86; and evolution, II 79, 171; and civilisation, II 83; and original sin, II 85; Eugenics and, IIIA 355; popular lecture on, IIIA 247
Nature, carelessness of, II 109; requirements of, II 119; and the improvement of man, IIIA 218–220; and man, IIIA 318
'*Nature*,' origin of, II 67, 69; letters of Galton and Darwin in, on pangenesis experiments, II 162–165. For Galton's many letters and papers in, see under *Memoirs*
Nature and Nurture, I 7, 8, 60, II 81, 118, 254–258, 358, 397, IIIA 260, 374, 384, 390, 392, 400; and measure of heredity in twins, II 126–130; of scientific men, II 177–179; de Candolle on, IIIB 483
Navigation, II 55–58
Navy, characteristics of boys in, II 120
Naworth Park, the birthplace of the coefficient of correlation, II 393, IIIA 5, 50
Nebula, shedding a satellite, IIIB 464
Negro, The, Galton on, II 28; characteristics of, II 31–33, 81, 106, 107; and Africa, II 32, 33, 264, 265; profiles of, II 324; finger-prints of, IIIA 193; in America, IIIA 219, 252; in Africa, IIIA 252
Neolithic Man, and domestication, II 72
Neo-Malthusianism, II 111, 132–134, IIIA 243, 302, 304. See also *Malthus* and *Limitation of Families*
Nephews and grandsons, relative nearness in kinship of, IIIA 33
Nero, composite portrait of, II 295, II 296 Plate XLI; the Beast of the Apocalypse, IIIB 572
Nervous Disorders, in parents of the sane and insane, IIIA 277
Nettleship, E., meets Galton, IIIA 327, 330, 393; albino dogs of, IIIA 356, 357, 393
Neurath, Drs Otto and Anna, translate 'Hereditary Genius,' IIIB 607
Newcastle, lecture at, IIIA 394
Newnham Grange (Sir George Darwin's home), Darwin portraits at, I 243
New Year's Greeting, to Galton, IIIA 248
New Zealand, Eugenics in, IIIB 612
Ngami, Lake, and Galton's projects in Africa, I 214, 215, 219
Nietzsche, and Galton, II 109, 119
Nightingale, Florence, and the Crimean War, II 13; and statistical inquiry, II 156, 250; on treatment of the wounded, II 416; 'The passionate statistician,' II 414–424; correspondence of, with Galton, II 416–424; on Quetelet, II 414, 418
Nile, The, Galton's impressions of, I 200–203
Nobility, ability of the, II 93
Noel, Captain, Galton and his sister stay with, I 179, 180; collection of casts from living heads, I 180
Normal Curve, Galton's use of, II 89, 90, IIIA 5, 7–10, 30, 31; and Galton's ogive, II 338, etc.; and the binomial, II 338, etc.; intelligence and the, II 104; physical measurements and the, II 384, 386; finger-prints and the, IIIA 167, 168
Normal Surface, IIIA 63
Norris, Major, and 'Daddy Tin Whisker,' IIIB 603
Northampton, Marquess of, Galton meets, in Egypt, IIIB 516–518; invalid wife of, IIIB 517
Northbrook, Lord, characteristics of, IIIA 113
Norths, The, and Galton, II 11, 53, 70, 88
Norway, Galton plans a tour in, I 122, 123, and gives it up, I 124, 125; Eugenics in, II 267
Noteworthies, proportion of, in population, IIIA 116, 118, 120; percentage amongst kinsmen of Fellows of Roy. Soc., IIIA 119
Noteworthiness, Galton's measure of, IIIA 108, 114, 115, 118, 119, 121; of kinsmen of Fellows of Roy. Soc., IIIA 108–121; register of, IIIA 112; ability and, IIIA 116
'*Noteworthy Families*,' work by Galton and Schuster, IIIA 113–121; publication of, IIIB 548, 550, 564, 567, 568
Novel, Galton's. See '*Kantsaywhere*'
Number, associated with colour, II 214, 253
Number-Forms, Galton on, II 205, 206, 240, 243, 253, IIIB 486; types and associations of, II 241; Galton's collection of, II 241, 242; examples of, II 252 Plate XXIV; heredity and, II 242; frequency of, II 242; in the sexes, II 242; of de Candolle, II 207, 208
Nurture, of Galton, I 11, 12; weeds and, IIIA 220; problems regarding State aid and, IIIA 253, 254; limited power of, II 118, 127, 128; dependence of abstract ideas on, II 255; early sentiments and, II 256, 257; conscience and, II 257; domestication and, II 258; and nature, see *Nature and Nurture*

Oakley, his portrait of Tertius Galton, IIIB 570, I 52 Plate XXXIII; his portrait of Francis Galton I 93 Plate XLVIII
Obituary Notices, written by Galton, II 396
O'Brien, tutor to Galton at Cambridge, I 143–146, 149, 153
Observation, in Galton's educational scheme, II 155
Occupation, effect of, on finger-prints, IIIA 155; of parents and health of children, IIIA 356
Ocean, meteorological statistics of, II 53, 54, 58; determination of currents in, IIIB 579
Octogenarians, inquiries concerning, II 349, 350
Odell, Maud Gardiner, takes finger-prints of her child, sends to biographer copies of Galton's letters to her, IIIB 496–499
Ogive, exponential curve of Galton, II 191, 239, 335, 338–340, 402–404, IIIA 62, 167, 168, 422, IIIB 463, IIIA 31 Plate II
Ogle, Dr, case of crooked finger in twins, II 181; on medical life-histories, II 359
Oldenshaw, Mrs, to have medallion of Erasmus Darwin, IIIB 473
Omabonde Lake, Galton reaches, I 231, 233, 235
Omutchikota Lake. See *Otchikoto Lake*
Opinion, necessity to test, II 296
Optical Continuity and just perceptible difference, II 308, 309, IIIA Extra Plate facing Table of Contents
Orators, mentally reading manuscript, II 252; gifts of, II 256
Ord, Dr W. M., on medical life-histories, II 359
Organisation, spirit of, in the Galton blood, I 124
Oriental Subjects, and political power, IIIA 279
'*Origin of Species*,' and Galton, I 207, 208, II 4, 5, 70, 82, 201, 206; letter of Galton to Darwin on, II 200 Plate XVIII
Ortler Spitze, geographical model of, II 34, II 33 Plate IV
Osborn, H. F., on palaeontological evidence for evolution, IIIA 82
Osborn, Admiral S., large head of, IIIA 249
Osler, Prof., assistance of, IIIA 343
Ostwald, erroneous statement as to priority of de Candolle and Galton, II 145
Oswell, W. E., Galton on, II 35; travels of, in Africa, I 214, 215

Otchikoto Lake, Galton's name on rock by side of, III^B 617, Plate LIX [In 'Tropical South Africa,' 1st ed. this lake on pp. 200 and 238 is termed 'Otchikoto.' Galton was there the second time on June 25, 1851; in his sketch of that date (see I 216 Plate LVII) it is termed 'Omutchikota'; the modern Staff map gives no aid]
Ovahereros, atrocities of the, I 236
Ovampo (Karupi), a sketch of, by Galton, I 216 Plate LVII
Ovampos, The, Galton crowns the king of, I 233, 235, 237 Plate LIX; country and welfare of, I 233, 235, 236; fowls of, I 233; subsequent difficulties of, I 240
Overwork, of Galton, I 101–103, 108, 146, 154, 166–172, 194; of Cambridge students, I 171, 194; victims of, II 278
Oxen, on African journey, I 222–226, 231–236, 239, I 237 Plate LIX
Oxford Philosophical Club, Karl Pearson lectures to the, III^A 315
Oxford University, and Weldon memorial, III^A 286, 287; anthropometric laboratory for students of, III^A 328; honours Galton, III^B 493, 494
Owen, R., size of head and stature of, II 150; asks for specimens from Africa, I 216

Paestum, Galton sees ruins at, III^B 475
Paget, Sir J., singular case of, II 163; analyses the successes of his pupils, III^A 62, 63; advises Mrs Galton, III^B 464
Pain, sensitivity to, in criminals, II 408; relation of degree felt to force of blow, II 408; measurement of unit of, II 408
Paine, Thomas, and metal bridges, III^A 148
Painters, fertility of, II 96, 99; Galton on, II 99
'Pairs of Things,' Arab view of, III^B 535
Palaeolithic Man, and domestication, II 71; drawings of, and mental imagery, II 240
Palmer, A., honour conferred on, III^B 494
Palmer, Winifred, finger-prints of, III^B 498, 499
Pangenesis, adopted by Galton, II 113; experiments on, II 156–177, 181–183, III^A 129; rejected by Galton, II 163, 164, 182–184; letters of Darwin and Galton to 'Nature' on, II 163, 164; danger of theory of, II 173
Panmixia, Galton on observations regarding, III^A 135
Pantagraph, double, Galton designs, II 42, 45–47
Pantheism, of Galton, II 117, 119
Paper, modern, good and bad, III^B 575
Parentage, Galton on, II 110; and marriage, II 134; and mating, instincts of, III^A 218; in complex organisations, II 182, 185; sex prepotency in, III^A 17, 18; exceptional, and regression, III^A 32; of able men, III^A 27
Parental Difference, effect of, on offspring, III^A 275
Parents, and offspring, resemblance of, II 169, 170, 185, 186, III^A 2–4, 17–19, 229, 242; relationship of, II 172–174; finger-prints of, III^A 192; latent elements in, II 170, etc.; age of, and fertility, II 408–410; age of, and vigour of offspring, II 348, 349; modern youth and, II 257; and training of children, II 127
Paris, Galton's holiday visit to, I 118, 119, 147
Parker, Gilbert, novel of, III^B 555
Parker, Sir H., Galton meets in Cape Town, I 219; yachting with, I 241
Parker, Lord, of Waddington, in the Temple, III^A 302; 'Biometrika' and, III^A 250; to meet Galton, III^A 326, 327
Parker, The Misses, natural daughters of Erasmus Darwin, I 18 Plate X, I 17, III^B 462
Parkes, Dr E. A., on medical hygiene, III^B 593, 594
Parkyns, Mansfield, Galton meets at Khartoum, I 201, 202
Parry, Major, Galton on the Danube with, I 133, 134
Parthenogenesis, observations regarding, III^A 135
Partridge, Prof. R., Galton in the house of, I 105, 106, 113, III^B 565; Galton studies anatomy under, I 105; diagnoses a carbuncle on Galton, I 108
Passports, Galton's early, I 93, 114
Patency, and latency, II 172, 173
Paternity, finger-prints in determination of, II 307
Pathological States, inheritance of, II 80; resemblance of, in twins, II 127
Patronage of landowners might aid Eugenics, III^A 234
Pau, Galton at, III^B 551–553
Paupers, marriage restrictions about, III^A 268
Pearl, R., work of, III^A 103, 279, 283, 297, 383; post held by, III^A 381
Pearson, E. S., assistance of, II vii, 311
Pearson, Prof. H. H. W., on Galton's name on rock at lake in Ovampoland, I 216 Plate LVII; hears of Galton in Rehoboth, I 223. See also *Otchikoto Lake*
Pearson, Karl, some personal recollections of Galton, III^A 239, 262, 432–436; correspondence of, with Galton, see *Galton, Francis*, correspondence of; views on Galton's School, III^A 57; Eugenic Record Office and, III^A 296–299, 303, 304; early plans of, for the Galton Laboratory, III^A 304–306; directs the Galton Laboratory, III^A 299–302, 304, 322, 332, 333, etc.; Fellowship in National Eugenics and, III^A 222, 223, 300, 301, etc.; higher mathematics in statistical work of, III^A 249, 301, 302; on Committee for Measurement of Plants and Animals (Evolution Committee), III^A 126, 127, 128, 286–291, III^B 501; resigns from Evolution Committee, III^A 291; farm for experimental breeding, III^A 134, 135, 251, 287; fellowship of Royal Society of, III^A 282, 283; University of London and, III^A 289, 290; Weldon memorial and, III^A 281–287, 291, 297, 301; Chairman at Galton's lecture to the Sociological Society, III^A 261, 262; Boyle lecture of, III^A 309, 314, 315; lectures to Oxford Philosophical Club, III^A 315; Darwin Medal awarded to, III^B 505; 'Chances of Death' by, received by Galton, III^B 502; death and characteristics of Father of, III^A 327, 328; last visit of, to Galton, III^A 432; portrait of, with Galton, III^A 353 Plate XXXVI. See also under '*Biometrika*'
Pearson, M. V., assistance of, III^A viii
Pedigree, showing direct ancestors of Galton, I 10; of immediate ancestry and collaterals of Galton, Plate A in pocket of Vol. I; showing connections of Barclays with noteworthy ancestors, Plate B in pocket of Vol. I; showing relationships of Freames, Barclays and Galtons, Plate C in pocket of Vol. I; of Abrahams, Farmers and Galtons, Plate D in pocket of Vol. I; showing connection of Charles Darwin with noteworthy ancestors, Plate E in pocket of Vol. I, and Plate F in pocket of Vol. III^A; of Darwins and Wedgwoods, III^A 313, 314; of Darwin ancestry, I 244–246; of polydactyly and twinning, II 129; of multiple births, II 129; of criminality (the 'Jukes'), II 231
Pedigrees, indexing of, III^A 103–105; classification in, III^A 343, 344; schedule for use in collecting, III^A 105, 106; difficulties in working out, I 9; of able families, III^A 343–345, 356
Pedigree Stock, collection of records of, II 321–323; of moths, Galton's circular regarding, III^A 45, 46
Pedler, Sir A., on use of finger-prints in India, III^A 153
Peek, A. E., and Dr Faulds' letter on finger-prints, III^A 143
Peerages, extinction of, II 95, 96, 341, 343
Pekingese, and Pomeranians, cross-breeding of, II 232

Pelly, Lady, III^A 400, 433, III^B 530, 587
Pember, E. H., delivers Galton's lecture before Roy. Soc. of Literature, III^A 336
Penn, Margaret, married Thomas Freame, I 32
Pensioners, use of finger-prints of, III^A 176
Peppard, biometricians at, III^A 135, 279
Percentiles, Galton's method of, II 125, 236, 374–376, 381, 401–404; table of anthropometric, II 376. See also *Normal Curve* and *Ogive*
Perception, delicacy of, under varying conditions, II 308
Perry, mathematical lectures of, at Cambridge, I 142, 153
Perry, Sir E. Cooper, in the Temple with Karl Pearson, III^A 302
Perry Coste, F. H., takes finger-prints in Polperro for Galton, III^B 522–524; correspondence of, with Galton, III^B 522–524, 616; colour faculty in son of, III^B 616
Perseverance, heredity of, II 151
Pertz, Miss, Galton meets, III^B 513
Pet Animals, II 70–72
Petit, Dr, attends Mrs Francis Galton, III^B 502
Petrie, Flinders, Galton stays with in Egypt, III^A 240, III^B 515–517
Petworth, biometricians on holiday at, III^A 322–326
Phantasmagoria, II 244; in journalists, II 256
Phidias, ability of, II 107
Philanthropy, views of, III^A 402, 403. See also *Charities*
Philistis, Queen of Sicily, portraits of, and composite, II 296 Plate XLII
Phillimore, Sir W. and Lady, white ducks of, III^A 360, 361; daughter of, killed in accident, III^B 530
Phillips, Emma, Galton's appreciation of, III^B 532
Phillips, Prof. J., retires from office of Secretary to Brit. Assoc., III^B 458
Phillips, Mary, marries Darwin Galton, III^B 454
Phillips, Stephen, 'Nero' of, III^B 572
Philosophers, and mental imagery, II 240
Photographic Researches of Galton, and portraiture, II 283–333; indexing of profiles, etc., II 298–311; method and study of mental characters, II 297; for family records, importance of, II 302; reduction of circle to ellipse, II 299, 300; bi-projection, II 300; measurement of distance by, II 316, 317; measurement of animals by, II 317, 318, 320–323; records of pedigree stock by, II 321; lecture on composite portraiture to Photographic Society, II 288
Photography, air plane, colour and stereoscopic, II 29; Galton turns to, II 229; composite, II 192, 204; Galton's first announcement on, II 229; composite, as method of measuring association of mental and physical characters, II 229; analytical, II 311–316; photograph of a smile, II 312; measurements by, II 316–323; finger-prints and, III^A 187, 196, 197
Phrenology, I 157, III^B 577; Captain Noel's interest in, I 180
Phthia, Queen of Epirus, portraits of, and composite, II 296 Plate XLII
Phthisical, diathesis and physiognomy, II 290–293, composite portrait representing the, II 291 Plate XXXIV
Phthisis, inheritance of, II 292, III^A 73–76, 260, 326, 399
'*Phylometry*' and 'Biometry,' III^B 500
Physical Characters, source of Galton's, I 55, 56; of the Negro, II 31; inheritance of, II 76, 82, 135 (see also under *Heredity*); resemblance of, in twins, II 127; psychological significance of, II 212; correlation of, II 390; correlation of, with mental qualities, II 128, 229, 232, 388
Physical Efficiency, marks for, II 382, 386–396, III^B 480; measurement of, II 90, 336; ability and, II 94; intelligence and, II 77
Physical Overstrain, and mental overwork, II 351, 352
Physicists, mechanical aptitude in, II 151
Physiognomic Characters, and mental traits, II 231–233, 283; weight given to, II 301; and phthisical tendency, II 291–293, II 291 Plate XXXIV
Physiognomy, Charles Darwin on, II 335
Physique, correlation of, with ability, II 128; of the British, II 336, III^A 252, 253; of school-children, and mental capacity, III^A 356
Pictographs, of American Indians, II 411
Pictures, seen with closed eyes, II 244, 247; visualised, associated with words, II 243; signalling of, II 280
Piety, and health, II 101; hereditary nature of, II 101, 103; and the sceptic, II 102; and the sense of sin, II 102; and moral oscillations, II 102; and instability of disposition, II 103
Pigmentation, and relative healthfulness, III^B 476, 477
Pigments, internal and external, III^A 372; in hair, analysis of, III^A 97, 98; painting of trees in, III^A 97 Plates III, IV
Pigs, Galton's interest in, III^A 390, 391
Pincushion Doll, legend regarding lady and doctor, III^B 553
Pinsent, Mrs Hume, on segregation of the feeble-minded, III^A 374
Pitch, of musical note, measure of sensitivity to, II 226, 227
Pitfalls, in experiments on theory of chance, II 405, 406
Plague, and Sir Francis S. Darwin, I 23
Planimeter, Amsler's, use of, II 144
Plants, laws of inheritance in, III^A 251; in windy and still air, III^A 131–133
Plato, ability of, II 107; eugenic passages in works of, III^A 312
Playfair, Dr Lyon (afterwards Lord), Galton introduced to, at Liebig's laboratory, I 130; on Catholic control and education, II 139
Ploetz, A., interest of, in Eugenics, III^A 388, 389; 'Rassen-Hygiene' of, III^B 599; visits Galton, III^A 429, III^B 606; letter of, to Galton, III^B 545, 546
Plymouth, Galton at, III^A 296, 297–310, III^B 579–581
Poem, of Dr Erasmus Darwin to Mrs Pole, I 18 Plate XI; of Galton on birth of Prince of Wales, I 167. See also *Verses, Epigram, Ardennes*
Poetry, and Science, III^A 337, 338, 343; in educational scheme, II 155
Poets, Galton on, II 99; fertility of, II 96, 99; and expression of ideas, III^A 337, 338
Pole, Mrs Chandos. See *Collier, Elizabeth*
Pole, W., on Roy. Soc. Committee on Colour-blindness, II 227
Pollaky, and finger-print work, III^B 590
Pollocks, pedigree of, III^A 343, 345
Polperro, finger-prints from, III^B 522–524
Polydactyly, in siblings of twins, II 129
Ponies, hearing of, II 216
Poole, W. H., material on visualisation provided by, II 237, 344
Poor-Law, effect of working of, II 417
Poppy, Shirley, work on inheritance in, III^A 251, 279, 280, III^B 515
Popular Lectures, difficulties of, III^A 246, 247
Population, Galton on, II 335; error in treating statistics of, II 147, 265, 266; of distinguished scientists in British Isles, II 151; prediction of, from measurement on small sample, II 179; and Malthus' conclusions, II 265; effect of town life on, II 123–126; stability of, III^A 58, 61, 65; reproduction of a stable, III^A 28, 29; limitation of, III^A 322; over-growth of, III^A 218; sudden advances in, III^A 426, 427; descent of qualities in, III^A 229; from early and late marriages, II 266

Portmanteau words, III^A 337
Portmore, Lord. See *Colyear, Charles*
Portraits, of the Darwin family, I 243, 244; indexing and numeralisation of, II 283, 323–328; telegraphy of, II 283, 309, 324; measure of resemblance between, II 329–331; composite family, II 356. See individual names for reference to portraits of all members of Galton and Darwin families, except Francis Galton
Portraits of Francis Galton:
(1) Aged 8, silhouette, I 63 Plate XXXIX
(2) Aged 18, from water-colour by Oakley, I 93 Plate XLVIII
(3) Aged *c*. 26, from a water-colour sketch, III^B 456 Plate XLVII
(4) Aged *c*. 28, from a daguerreotype, II 1 Plate I
(5) Aged 30 (?), from photographs on glass, I 211 Plate LV
(6) Aged 31, with Louisa Galton, from a photograph, I 241 Plate LX
(7) Aged 33, from a photograph, II 2 Plate II
(8) Aged 38, from a photograph, III^B 531 Plate LV (ii)
(9) Aged 38, from a photograph, II 40 Plate VIII
(10) Aged 42, from photographs, II 67 Plate IX
(11) Aged 45 (?), with Louisa Galton, I 242 Plate LXI
(12) Aged 50, compared with Charles Darwin, aged 51, from photographs, I 56 Plate XXXVII
(13) Aged about 50, from a photograph, II 131 Plate XIII
(14) Aged 55, from a photograph, II 11 Plate III
(15) Aged 56, in holiday garb, from a photograph, II 226 Plate XXII
(16) Aged 60, from the oil-painting by Graef, II 99 Plate XI
(17) Aged about 60, from Galton's standard photographs of himself, II 356 Plate XLIX
(18) Aged about 60, from photographs, II 211 Plate XXI
(19) Aged about 65, from a photograph, II 283 Plate XXVII
(20) Aged 65, silhouette, compared with that of age 8, II 310
(21) Aged about 65, from a photograph, III^B 472 Plate XLIX
(22) Aged 66, copper-engraving from 'Biometrika,' III^A Frontispiece
(23) Aged about 68, from a photograph, II 334 Plate XLVIII
(24) Aged 71, photographed in two aspects as criminal by Bertillon, II 383 Plate LII
(25) Aged 73, from a photograph, II Frontispiece
(26) Aged about 74, from a photograph, II 270 Plate XXV
(27) Aged 75, from a photograph, II 281 Plate XXVI
(28) Aged about 75, from a photograph, III^B Frontispiece
(29) Aged 75, in holiday garb, July at Royat, III^B 502 Plate LII
(30) Aged 80, from a photograph by Dew Smith, III^A 217 Plate XXXI
(31) Aged about 80, from a photograph, III^A 249 Plate XXXII
(32) Aged about 80, a reverie 'caught when the spirit was not there,' III^A 354 Plate XXXVII
(33) Aged 81, from the oil-painting by Furse, I Frontispiece
(34) Aged 82, from a photograph, III^A 259 Plate XXXIV
(35) Aged 83, from a photograph by W. F. R. Weldon, II 332 Plate XLVII
(36) Aged 84, with his great-niece Eva Biggs at Bridge End, III^B 576 Plate LVIII
(37) Aged 85, from a sketch by Eva Biggs, II viii
(38) Aged 85, in donkey-chair, from a photograph, II 415 Plate LIII
(39) Aged 87, with his biographer on the Stoep at Fox Holm, III^A 353 Plate XXXVI
(40) Aged 87, with the faithful Gifi and Wee Ling on the Stoep at Fox Holm, III^A 390 Plate XXXVIII
(41) Aged 88, from a sketch by Frank Carter, II xi, tailpiece to 'Contents'
(42) Aged 88, from a sketch by Frank Carter, III^A 432 Plate XL
(43) Aged 88, from a sketch by Eva Biggs, II 425 Plate LIV
(44) Aged 89 all but a month, from a photograph taken after death, III^A 433 Plate XLI
Portraiture, and photographic researches, II 283–333; used to distinguish racial and mental types, II 301, 303
Portugal Street, Galton living in, II 11
Post-mortem examination, Galton sees his first, I 100
Poulton, Prof. E. B., co-operation of, II 321; on Committee for Measurement of Plants and Animals, III^A 126; on Evolution Committee, III^A 290; calls attention in his translation of Weismann to Galton's ideas, III^A 341
Practice and proficiency, II 372, 373
Prayer, nature of, II 115; efficacy of, II 115–117, 131, 250, 259, 260; subjective value of, II 117; effect of Galton's article on, II 175, 250, 258; Galton on, in letters to his nieces Millicent Lethbridge and Evelyne Biggs, III^A 271, 272; chapter on, in 'Inquiries into Human Faculty,' III^B 448, 449, 471, 472; Galton's, on writing for publication, III^B 449
Prediction, of character of population from knowledge of a small sample, II 179; of extinction of families, II 360; of speed of fastest horse, II 399; of fertility, from ages of parents, II 408
Prepotency, and sports, III^A 98, 99; in sires, III^A 44, 99, 100; co-operation regarding observations on, III^A 135
Price, Bartholomew, proposes that Hon. D.C.L. be conferred upon Galton at Oxford, III^B 493, 494
Price, Bedford, pedigree from, III^A 371
Price, Bonamy, and Wordsworth, III^A 115
Pridham, Miss, helps with experimental breeding, III^B 484
Priestley, Dr, and Samuel Galton, I 44, 45; house of, I 51, 62
Primates, finger-prints of, III^A 143
'*Primer of Statistics*,' by W. P. and E. M. Elderton, III^A 316, 317, 320, 363, 364, 376–379, 385, 386, III^B 601
Primogeniture, and heredity, III^A 405
Prince of Wales, plot to kidnap, III^A 159; verses by Galton on birth of, I 167
'*Principia botanica*,' of Robert Waring Darwin, I 15
Pritchard, Dr, physician to Tertius Galton, I 185
Prizeman, Galton, in Anatomy and Chemistry, I 119
Prizes, offered by Galton for family histories, II 359–363, 370; method of proportioning first and second, II 411–414
Probability, and Eugenics, III^A 274, 320; the basis of knowledge, III^A 314
Probable Error, Galton's determination of, III^A 63
Procter, Mrs, meets Galton in Egypt, III^B 519
Professorship of Eugenics, proposals and plans for, III^A 224, 225, 300, 381–384, 437, 438
Professorship of Applied Statistics, proposals for, by Florence Nightingale, II 414–424

Proficiency, and practice, II 372, 373
Profiles, identification by, II 304, 306, 324, 326; of Greek girl from a gem, II 309; formula for, II 309; indexing of, II 304, 323–328, III^{A} 325, 326; Galton's collection of, II 323, 324; racial, II 324; resemblance of, in relatives, II 309, 310, 324; Galton working on, III^{B} 578, 602, 605. See also *Silhouettes*
Progenitors, value of life-histories of, II 302, 303; knowledge of, in 'Kantsaywhere,' III^{A} 415, 422
Propaganda, and scientific research, III^{A} 371, 372, 379; needs and difficulties of, III^{A} 407, 408; Galton's methods of, III^{A} 412
Pryor, collection of silhouettes of, III^{B} 580
Psychical, measurements, and normal curve, II 90, 91; in schools, II 344; characters, ranking of, II 351; measure of strength of resemblance in, III^{A} 69
Psychological, investigations of Galton, II 211–282; papers, English and Continental, II 213
Psychologists, and the correlation of ranks, II 393
Psychology, experimental, and Galton, II 211–213; of the herd, II 73, 74
Psychometric, instruments devised by Galton, II 215–228; observations and experiments of Galton, II 228–248
Psychometry, definition of, II 233
Psycho-Physics, II 229; science of, II 307
Public Opinion, power of, III^{A} 352, 367; instability of, III^{A} 321, 404
Pull, strength of, II 374, 376
'*Punch*,' drawing in, III^{A} 335, 375, 376; on 'the squeeze of 86,' II 375
Punishments, legal, result of Florence Nightingale's queries on, II 417
Pure Breed, III^{A} 76, 77
'*Pure-Line*' hypothesis, II 171, III^{A} 10, 11, 58, 221
Purkenje, work of, on finger-prints, III^{A} 141, 143, 161, 174, 179, 184; standard patterns of, III^{A} 179 Plate IX
Pursuits, and early interests of Galton, I 68, 69, 71, 76, 77, 78, 80, 82, 83, 85, 90, 101, 109, 111

Quagga taint, II 159
Quakers, and Galton ancestry, I 9, 11, 27–49, 59, 74; persecutions of, I 31, 32, 34–38; and commercial enterprise, I 32–40, 46; Samuel Galton disowned by, I 45
Quartiles, Galton's definition of, II 338; Galton's use of, III^{A} 51, 54, 55, 422; and standard deviations, II 339
'*Quedley*,' meaning of, III^{A} 325, 326
'*Questions of the Day and of the Fray*,' II 27
Quetelet, and Galton, II 12, 39; and Florence Nightingale, II 414, 418; 'mean man' of, II 295; and composite portraits, II 297; and application of mathematics to social phenomena, III^{A} 1; and the normal curve, II 90; on expectation and realisation, etc., II 418; achievements of, II 420; 'Letters on Probabilities' of, II 89, 335
Quincunx, Galton's, III^{A} 9, 10, 63

Rabbits, and experiments on pangenesis, II 156–169, 174–177
Race, Galton's definition of, III^{A} 84, 85; improvement of, III^{A} 233–235, 241, 242, 252, 253, 355 (see also under *Eugenics*); survival of, II 82, 83; extinction of, II 264; selection and, II 120, 263; influence of man upon, II 264
Races, inequalities of, II 32, 33, 106–109, III^{A} 252; mental peculiarities of, II 88; innate character and intelligence of different, II 352, 353; civilised and uncivilised, mental imagery in, II 239, 240; comparison of types of, III^{A} 325, 326; finger-prints in different, III^{A} 139, 140, 143, 193, 194, III^{B} 485, 489–491; relative fertility of, III^{A} 218, 219; and stirps, II 171; composite photographs of, II 290, 294
Racial mentality, and physical characters, II 301
Radiometer, of Crookes, and 'spirit' influence, II 63
Radium, Galton's interest in, III^{B} 522
Ramsay, Sir W., on American science, III^{B} 531
Ramsay, Prof. W. M., honour conferred on, III^{B} 494
Randal (Randall or Randle!), *Sergeant*, at Galton's anthropometric laboratory, II 378, III^{B} 489; injury to, and finger-prints of, III^{A} 154
Ranking, Galton's method of, II 236; and mental imagery, II 238; of psychical characters, II 351; and the ogive curve, II 335
Ranks, theory of, II 337, applications of, II 389–391; correlation of, II 393, III^{A} 3
Rats, experiments on, II 166, 167, 176; and rat-catchers, III^{B} 532–534, 548
Rawson, Sir R. W., and International Statistical Institute, II 397
Ray, Mrs, Galton meets, III^{B} 593
Rayleigh, Lord, chairman of Roy. Soc. Committee on Colour-blindness, II 227; Karl Pearson writes to, on the trusteeship of the Weldon Medal, III^{A} 285, 286; Hon. Fellow of Trinity College, III^{A} 236, 238; remarkable head of, III^{A} 248; presidential address of, III^{A} 301; on possibility of the earth losing her atmosphere, III^{B} 470; a Copley medallist, III^{A} 431, III^{B} 614
Reaction Time, instrument for determining, II 219, 220, III^{B} 514; classification on basis of, II 229; measurement of, II 277, 278, 359
'*Reader, The*,' and Galton, II 67–69; supporters of, II 68
'*Recessive*,' discussion of term, III^{A} 337
'*Record of Family Faculties and Life History Album*,' II 363–370
Recreations of Galton, changing with age, III^{A} 353
Rede Lecture, Galton's, II 268, 270, 271, III^{B} 473
Register, national, proposed by Galton, II 120–122; general, of the insane, III^{A} 365, 366; biographical, II 355, III^{A} 264 (the 'Golden Book')
Registrators, for anthropometric data, Galton's pocket, II 340, 341
Regression, first idea of, III^{A} 6–11, 13; coefficient of, III^{A} 25, 47; and correlation, III^{A} 50; and progression, III^{A} 58; and selection, II 221, 263; in Galton's sweet-pea experiments, III^{A} 6, 7, 13; nature of, illustrated by quincunx, III^{A} 9; and heredity in stature of man, III^{A} 11–34; and the generant, III^{A} 20; establishment of breeds and, III^{A} 31; law of, II 79, III^{A} 46; and stability of types, III^{A} 46; evolution and, III^{A} 48; formula for multiple, II 54, 55; Galton's fallacious reasoning regarding, III^{A} 23, 24, 58, 61, 76, 78–80, 82–84, 86, 87, 93, 94, 170
Regression Line, Galton reaches his first, III^{A} 3, 4; determination of slope of, from contour lines of correlation table, III^{A} 13, 14, 52; of offspring on midparent for stature, III^{A} 13; earliest published, III^{A} 24
Rehoboth, Galton remembered at, I 223
Reid, Mr, photographs horses for Galton, III^{B} 506, 507. [N.B. It has now been ascertained that all the negatives of these were destroyed by him after printing]
Reid, Archdall, and inheritance of tuberculosis, III^{A} 260
Relations, average number of, III^{A} 390, 391, 396, 397
Religion, widening of Galton's views on the function of, I 207, 208; and science, II 147, 151, 152, 261; and Eugenics, II 249, 267, III^{A} 265, 272–274; and the public conscience, III^{A} 267; and celibacy, III^{A} 269; and national stability, III^{A} 88; early training and,

II 256, 257; reformation and, III^{A} 90–93, 272; three definitions of, III^{A} 89; as source of strength, III^{A} 424; three kinds of, III^{B} 564
Religions, relativity of, II 32
Religious, authority and science, II 135, 139; belief and doctrine of evolution, III^{A} 89–93; dogma, and free inquiry, II 257; faiths, Galton's attitude regarding, III^{B} 441, 442; sentiments, source of, II 82, 83; views of Galton, II 117, 119, 261, 425, III^{A} 271, 272, 424–425, II 102 Plate XII
Reminiscences, of Mrs Wheler, I 63, 64, 82; I 48 Plate XXIX
Renan, the 'Antichrist' of, III^{B} 572
Rentoul, Dr, views as to book of, III^{A} 304, 305, 312
'*Report on African Explorations*,' II 27
Reproductive Organs, Darwin on function of, II 174
Reproductive Selection, II 79, 80, III^{B} 505, 506
Reputation, a measure of ability, II 89, 91, 92
Research Fellowship in Eugenics, III^{A} 113, 221–223, 258
Resemblance, measurement of, II 329–333, III^{A} 24, III^{B} 562, 564–566, 569, 588
Retinal Picture, retention of, II 241, III^{B} 629
Reversion, coefficient of, III^{A} 9, 11; phenomena of, illustrated by quincunx, III^{A} 9
Reviewers, short-sightedness of, III^{A} 174
Rhinoceroses, shooting of, in Africa, I 237
Riano, Señora G. de, Galton sees, in Madrid, III^{B} 512
Ribot, T., book of, on heredity, reviewed by Galton, III^{B} 463; and Fechner's work, III^{B} 464
Richardson, Dr B. W., on medical life-histories, II 359
Richardson, Samuel, novels of, III^{B} 584
Richmond, painting of Darwin by, III^{A} 340 Plate XXXV
Ridges and creases of hands and feet, III^{A} 178; coarseness of, and tactile sensitivity, III^{A}, 178, 179; use of, III^{A} 178, 179
Ridges, Mrs, housekeeper at Birmingham school, III^{B} 452
Ridgeway, Prof. W., and Eugenics, III^{A} 345
Riding, Galton, aged 3¼ years, his pony unaccompanied, I 70
Riflemen, protection of, II 18, 19
Right and left sides, strength of hand and keenness of vision on, II 377
Rivet, André, biography of, II 101
Robb, Mrs, gives introductions to Galton, III^{B} 509, 513
Robbery, and robbers, I 78–80
Robertson, J. M., on forces of kakogenics, III^{A} 260; on eugenic ideals, III^{A} 263
Robinson, L., III^{B} 529; and Galton's work on finger-prints, III^{A} 176; praises Dean Inge's sermon on Eugenics, III^{B} 612
Rodwell, G. F., on statistics from Marlborough College, II 343
Roget, 'Thesaurus' of, II 269; number forms of, III^{B} 469
Roman Catholics, in the Albert Hall, III^{B} 570; Congress of, III^{B} 586
Roman Ladies, composite portrait of, II 295, II 296 Plate XLIII
Romanes, Prof., Croonian lecture of, II 191; and Galton's work, II 250; and the origin of varieties, II 272; on need for experimental zoology, III^{A} 81, 129; and use of term 'Eugenics,' III^{A} 221
Ronath, Fräulein, German opinion of Galton, III^{B} 602
Roscoe, Sir H., and 'The Reader' and 'Nature,' II 68, 69; and Eugenics, III^{A} 385; garden of, III^{A} 278; and Galton, III^{A} 400, III^{B} 546
Rosebery, Lord, visits Eugenics Laboratory, III^{A} 335, 336; speaks at Oxford on conferment of degrees, III^{B} 494; and John Morley anecdote, III^{B} 615
Rosenheim, Miss, work of, III^{A} 426
Roumania, Eugenics in, II 267
Royal Commission, on deterioration of the British race, III^{A} 251, 252, 364–366; on the poor law, III^{A} 323, 324
Royal Engineers, composite portraits of men and officers of, II 286 Plate XXIX, II 290
Royal Geographical Society, and Galton, II 2, 4, 13, 18, 24–27, 34–36, 50, 67; and Galton's African travels, I 214–216, 239; Galton elected to membership of, I 215; medal of, awarded to Galton, I 239; H. M. Stanley on, II 30
Royal Institute of Public Health, Galton at, II 345
Royal Institution, Galton becomes member of, II 88; Galton lectures at, II 243, 294, 307, 337, III^{A} 6; A. R. Wallace lectures at, III^{A} 370; Karl Pearson lectures at, III^{A} 369, 370; Galton on scope and value of, II 420, 421
Royal Society, Galton elected to Fellowship, etc., II 13; Galton's papers before, II 39, 40, 57–60, 156, 163, 164, 169, 408, III^{A} 12, 25–27, 34, 40, 50, 161–174; on kinsfolk of Fellows of, III^{A} 107–121; and large heads of Fellows, II 94; questionnaire to Fellows of, II 145, 149; Galton on Kew Committee of, II 59, 60; on Committee to investigate Colour-blindness of, II 227; on Council of, II 361; Gold Medal of, awarded to Galton, II 201, III^{B} 476; Darwin Medal of, awarded to Galton, III^{A} 236, 237; Copley Medal of, awarded to Galton, III^{A} 400, III^{B} 611, 614, 615; Committee for Measurement of Plants and Animals, or Evolution Committee of, III^{A} 126, 127, 133–135, 286–291, 311, 312, III^{B} 501; biometry and, III^{A} 241, 243, 279, 282, 283, 287; foundation of 'Biometrika' and, III^{A} 100; nominates Bateson as a referee for biometric paper, III^{A} 100, 241; referees of, III^{A} 297; and memoir on albinism, III^{A} 315; on literary standard of memoirs of, III^{A} 331, 332, 337; Galton suggests new procedure at meetings of, III^{B} 468; Karl Pearson's fellowship of, III^{A} 282–284; the Weldon memorial and, III^{A} 127, 281, 285, 286; on death of Galton, III^{B} 617; Samuel Galton and, I 44, 47
Royal Society of Literature, and low literary standard of scientific memoirs, III^{A} 329, 330, 332, 333, 336–339
Royal Society of Medicine, debate on heredity at, III^{A} 357
Royat, death of Louisa Galton at, II 281, III^{B} 447, 514
Royds, on the characters of rabbits, II 159
Rue, Warren de la, and Galton, II 59; and exhibition of scientific instruments, II 215
Rücker, Sir A., and research fellowship in Eugenics, III^{A} 222, 223, III^{B} 530; scheme for Galton Eugenics Laboratory sent to, III^{A} 305, 306; letter to, on Eugenics Record Office, III^{A} 303, 304; and Galton bequest, III^{A} 301; letter of, to Galton, accepting gift for London University, III^{A} 307; definition of Eugenics and, III^{A} 269; on lady secretaries, III^{A} 278
Rural Populations, fertility, etc. of, II 123–126
Ruskin, letter of, III^{B} 461
Russell, Count, caves of, III^{B} 559, 560; book of, III^{B} 569, 572
Russell, Dr, Galton's early tour with, I 92–97
Russia, lessons from revolution in, III^{A} 90; war of, with Japan, III^{B} 532, 534
Russians, colour associations and number forms of, II 240
Rutland Gate, No. 42, Galton's home 1857–1911, II 11, 12, II 11 Plate III; determination of latitude of, II 50; workroom at, III^{A} 238; tablet above porch of, III^{A} 311
Ryley, Kathleen V., work of, III^{A} 386, 387, 426

Sabine, Sir E., influence on Galton, II 59, 61; on Committee of Brit. Assoc., III^{B} 458

St Andrews, Dr Heron's lectures at, III^{A} 360, 361
St Quintins, the, at Keswick, I 156
Saleeby, C. W., controversial methods of, III^{A} 404, 405; attacks Eugenics Laboratory, III^{A} 408, III^{B} 601; his views on Eugenics, III^{A} 372, 428, 430, III^{B} 605; his 'Parenthood and Race Culture,' III^{B} 597; prepares abstracts of Galton's books for Harmsworth, III^{B} 601
Salisbury, Galton visits, II 130
Salisbury, Lord, never took exercise, III^{B} 569; size of head and stature of, II 150
Salvin, on Evolution Committee, III^{A} 127, 291; characteristics of, III^{A} 325
Sampler, of Lucy Barclay worked at Ury, I 46 Plate XXVII
Sandow, pupils of, III^{A} 253; physique of, III^{A} 253
Sandwich, origin of word for food, III^{B} 548
Sandwich Islands, named after Lord Sandwich, III^{B} 548
Sandys, Dr, speech on presentation of Galton for Hon. D.Sc. at Cambridge, III^{B} 495
Sanitary Administration, needs of, II 156
Savage, Elizabeth, wife of Sir Charles Sedley and ancestress of Galton, I 20
Savage Peoples, suggested tests for, II 385
Savages, character of, II 74; and domestication of animals, II 70, 71
Savants, characteristics of, II 141
Savile, Elizabeth, mother of Sir Charles Sedley, I 20
Savile, Mary, on holiday with Evelyne Biggs, III^{B} 574, 578; calls on Galton, III^{B} 608
Savile, Sir Henry, ancestor of Galton, scholarship of, I 20; portrait of, I 20 Plate XV
Savile-Sedley, ancestry of Galton, its characteristics, I 21
Sayce, Prof., in Egypt, III^{B} 517, 518, 519
Sayers, Captain, Galton meets, I 113; possible influence of, I 214
Scales, for testing hair, skin, and eye colours, II 223–226; for standardisation of colours, II 224–226
Scandinavian Hybridisation of British race, II 371
Schedules, reluctance to fill in, II 357; dealing with—age of parents and vigour of offspring, II 348, 349; persons of advanced age, II 349, 350; social stability, II 350, 351; mental fatigue, II 276, 351, 352; innate characters and intelligence of different races, II 352, 353; alleged darkening of hair in English, II 353, 354; for composite family portraits, II 356; for analysis of men of science, II 149–155; on faculty of visualisation, II 236–238; on same from Charles Darwin, II 194, 195; for phthisical patients, II 291; as four-yearly reports from 'old boys' to their schools, II 346; statistics by, II 348–356; as biographical registers, II 355; for use in pedigree work, III^{A} 105, 106; sent to Fellows of the Roy. Soc., III^{A} 108, 113–114; to form register of able families, III^{A} 121; for heredity of disease, III^{A} 71, 72
Schimmelpenninck, Mary Anne (née Galton), aunt to Francis Galton, memoirs of, II 193, III^{B} 604; her appreciation of Samuel Galton the First, I 41–43; ability of, I 53–54; Francis Galton on, III^{B} 553, 580; silhouette of, I 54 Plate XXXV, III^{B} 580
Schmidt, Dr, at Keswick, I 156, 157
Schoolboys and headmasters, I 87–88
Schoolmasters, their neglect of opportunities, III^{A} 232, 233; their opportunities for observational work, II 344–348
Schools, attended by Darwin, his views on, I 12: attended by Galton, Mrs French's, I 67, I 75 Plate XLV; at Boulogne, I 70–74; Mr Atwood's, I 77–80; King Edward's at Birmingham, I 81–90, Francis begs to be removed from, I 86–89; anthropometric data from, II 336–337, 343–346; should serve two purposes, as place of education and place of research, II 344; value of, for statistical inquiries, II 344–345; and myopia, III^{A} 356
Schuster, Sir A., number form of, II 242, III^{B} 469
Schuster, Edgar, first Galton Eugenics Research Fellow, III^{A} 258, 274, 291, 296, 300, III^{B} 536, 541, 542; resignation of fellowship, III^{A} 291, 296–298, 300; his anthropometric work at Oxford, III^{A} 328, 379; Galton's letters to, III^{B} 541–542, 554–558, 562, 577, 583; letters of, to Galton, III^{B} 561, 583–584; joint work with Galton on 'Noteworthy Families,' III^{A} 264; work on the human brain, III^{A} 315, 325, 326; memoirs by, III^{A} 258, 259, 279; on inheritance of mental characters, III^{A} 291; on promise of youth and performance of manhood, III^{A} 232; on sequestrated church properties, III^{A} 370; his opinion of Ethel M. Elderton, III^{A} 305
Schweinfurth, G., geographical discoveries of, II 31; Galton has two days in the desert with, III^{A} 240; with Galton in Cairo, III^{B} 518–519; sends Galton a photograph of d'Arnaud Bey, III^{B} 455 and Plate XLVI
Science, how sciences develop, Galton, III^{B} 463, Pearson, III^{A} 262; scope of, II 347–348; and religion, II 135, 147, 151, 152, 261; priesthood of, II 156; inheritance of—ability in, II 97, imagination in, II 98, power of analysis in, II 98; inborn taste for, II 152; need for travelling fellowships in, II 152; ways of furthering, II 153–155; as a profession and as a pursuit, II 154–155, III^{A} 331: in education, II 155; at school, Francis Galton's craving for, I 89
Scientific, memoirs, low literary standard of, III^{A} 329–332, 336–339: achievement and environmental causes, II 148–149, in relatives of scientific men, II 151, III^{A} 107–121: ability, waste of, II 153; superintendence, value of, II 153; low general culture of modern writers of memoirs, III^{A} 331, 338; research and propagandism in Eugenics, III^{A} 371, 372, 379; Societies Committee, Local, Galton Chairman of, II 362; salon of Mrs Hertz, III^{B} 464: discoverers and statesmen, II 135; tastes of Samuel Galton, I 46–48: ability of Erasmus Darwin, I 16, 49, of Sir Douglas Galton, I 53; ability, sources of Francis Galton's, I 48; imagination in Erasmus Darwin, Charles Darwin and Francis Galton, I 48: tastes, reawakening of, in Francis Galton, I 209–211, Chapter VII, I 211–242: chat, Francis Galton's delight in, III^{B} 596, 611
Scientists, fertility of, II 96: motives of, II 96, 97; distinguished, number of, in British Isles, II 151: energy of, II 151, 251: independence of character in, II 151, 155: mechanical aptitude in, II 151: religious faith of, II 152; teaching of, II 155: mediocre academic degrees of, II 155: antecedents of, II 177, 178; nature and nurture of, II 177–179: mental imagery of, II 237: phantasmagoria of, II 244; veracity of, III^{B} 478, 479: size of head and stature of, II 149, 150: English men of science, II 87, 130, 134, 142, 145: characteristics of, II 149–155, 207
Scotland, Galton's tour in, I 104–105
Scotland Yard, Galton visits, to see finger-print indexing, III^{B} 572
Scott, Dr Dukinfield, President of Linnean Society at Darwin-Wallace Celebration, III^{A} 340–341
Scott, R. H. (Secretary to Meteorological Council), III^{B} 475
Scottish Lowlanders, ability of, II 106
Seabrooke (Evelyne Biggs' maid), adventure at Biarritz, III^{B} 561; mentioned, III^{B} 583
Seal, and parchment carried by Galton in Africa, I 220; made for Galton in Egypt, III^{B} 518
Sedgwick, Adam, views on farm for experimental breeding, III^{A} 134; and Darwin Commemoration, III^{A} 369

Sedley, Catherine (Countess of Dorchester), character and ancestry of, I 19–20; portrait of, I 19 Plate XIV
Sedley, Sir Charles, character of, I 20; notices of, I 20; portrait of, I 27 Plate XXI
Sedley, Sir William, founder of Sedleian professorship, Oxford, ancestor of Francis Galton, I 21
Seeds, inheritance of weight and diameter of, in mother and daughter plants, II 392, III^{A} 3–7
Seeley, Prof. J. R., and 'The Reader,' II 68
Segregation, of the feeble-minded, III^{A} 365–367, and their happiness under it, III^{A} 373–374
Selection, and variation, II 171; Darwin's Natural, II 171; Galton's Germinal, II 171, 185, 186; Pearson's Reproductive, III^{B} 505 etc.; effect of long-continued, II 399; and man, II 72–73; artificial, II 120; of plants grown in windy and still air, III^{A} 131–132, 133; selection of germinal elements, II 172; stringent, in man, II 263–264; effect of artificial, on birth-rate, III^{A} 92; effect of continued, III^{A} 93, 94; and progress, III^{A} 94; and multiple regression, III^{A} 94; and correlation of characters in organism, III^{A} 94; of size of head in bull-dog, III^{A} 94; retrograde, III^{A} 95; and variability, III^{A} 27; natural, and new forms, III^{A} 32, 170; of size, and its limits, III^{A} 49; observations on, III^{A} 135; sexual, and finger-prints, III^{A} 168–169; natural, and finger-prints, III^{A} 169
Selector, Galton's mechanical, to pick out individuals of given character, II 305, 306, III^{A} 149
Self, multiple nature of, II 246
Self-Consciousness, in Galton, I 93, 94
Self-Control, in criminals, II 230
Seligmann, Dr, takes Galton round Vienna hospitals and museums, I 96, 97
Semlin, visited by Galton, I 133
Sensation, scale of, III^{B} 493; and its source, Fechner's Law, III^{B} 464; limit of, II 307; variation in threshold of, II 308; faint, and the imagination, II 307, 308
Senses, keenness and discrimination of, II 359, 365
Sensitivity, of men and women, II 221; and ability, II 222; to colour differences, Galton's instrument to measure, II 226
Sentiments, influence of early, II 256, 257
Sergi, G., asks Galton's aid as to anthropometric instruments, II 226; Galton meets in Italy, III^{B} 474
Serpent, Galton jumps on its tail, I 234
Services, the public, and marks for bodily efficiency, II 387, 388
Seton-Karr, hunter, III^{B} 516
Seward, Anna, and Dr Erasmus Darwin, II 192, 193
Sex, and creative genius, II 99; necessity of, in complex organisations, II 182; advantages of, II 185; and mental imagery, II 242; associated with numbers, II 253; and its origin, III^{A} 135; and touch, III^{A} 169; of offspring of aged fathers, II 210
Sextant, examination of, at Kew, II 49, 50, 59; superposition of photographs by aid of, II 285, 287
Sexual Attractiveness, and the origin of varieties, II 272, 273
Sexual Generation and Cross-fertilisation, paper by Galton, III^{A} 318
Shades, measurement of resemblance in, II 303
Shaen, Margaret, recollections of Galton and Darwin, II 196; at San Remo, III^{A} 244; at Bordighera, III^{A} 276
Sharpey, Prof., large size of head, III^{A} 248
Shaw, G. Bernard, 'Superman' of, II 109; on marriage and Eugenics, III^{A} 260, 261; lectures for Eugenics Education Society, III^{A} 427
Sheep, hermaphrodites in, III^{A} 359; blood of, at different seasons, III^{B} 564
Sheikh, incident with a, and Galton in Egypt, I 200, 205
Shell-Shock, family history of cases of, II 173; and mental overwork, II 278
Sheppard, W. F., his table of deviates, II 401; table of centiles calculated by, III^{A} 304, 312; letters of Galton to, III^{B} 486, 487; dines with Galton, III^{B} 487; his number form, III^{B} 486
Shirreff, Emily, letters to Galton, II 132, 133; meets Galton in Italy, III^{B} 474
Shooting, boar in Egypt, I 200; hippopotami in Egypt, I 202; lions in Africa, I 222, 223, 224; giraffe in Africa, I 223; rhinoceroses in Africa, I 224, 237; hyena and fowl in Africa, I 231; grouse in Scotland, I 208; hunting and, I 208–209; fox-hunting, finer sport than big game, I 234
Shorthouse, John Henry, and Abrahams family, I 39; see also Pedigree Plate D in pocket of Vol. I
Short-Lived families, Buttons, I 36, 37, 42, 43; Hubert Galtons, I 53
Shortsightedness, inheritance of, III^{A} 356. See also *Sight*
Shyness, resemblance in brothers, III^{A} 247
Siblings, and 'sibs,' origin of words, III^{A} 332; finger-prints in pairs of siblings, III^{A} 189, 190, 191
Sibship, of Francis Galton, I 63
Sicily, Weldon works on snails in, III^{A} 251; induces Galton to visit, III^{B} 525, 526
Sight, tests for, II 222, 223; maximum values for keenness of, II 374; percentile values of keenness of, II 376; values of keenness of, at each rank, II 390
Silhouette Portraits, of Elizabeth Collier (Mrs Pole and her dog), I 21, Plate XVII, I 14 Plate IV bis (as Mrs Erasmus Darwin); of Dr Erasmus Darwin and his son Erasmus at chess, I 14 Plate IV; of Dr Darwin alone, I 14 Plate IV bis; of Mrs Samuel Galton (Lucy Barclay), I 44 Plate XXVI; of Tertius Galton with his children, I 52 Plate XXXIV; of Mrs Schimmelpenninck (Mary Ann Galton), I 54 Plate XXXV; of Theodore Galton, and of Adèle Galton, I 54 Plate XXXV; of Erasmus Galton, I 69 Plate XLII; of Francis Galton, as a child, I 62 Plate XXXIX; when aged 65, II 310
Silhouettes, Galton's use of, II 304, 309; method of taking, II 309, 310; photographic, examples of, Galton's own, II 310; method of indexing, III^{A} 325, 326; Galton's researches on, III^{B} 577, 578; on a collection, including members of Galton family, III^{B} 580. See also *Profiles*
Simon [Sir R. M.], consulted as to 'medical histories' of individuals, II 359; Galton studies anatomy under, at King's College, I 105
Simpson, Mrs (née Senior), Galton visits, III^{B} 513
Simpson, second wrangler to Cayley in 1842, I 164
Sin, original, influence of doctrine of, II 85
Sires, prepotency of, in trotting horses, III^{A} 99, 100, in Basset hounds, III^{A} 44
Sisters, composite portraits of, II 288 Plate XXXII, II 288 Plate XXXIII
Sketches, of Breadsall Priory, I 74 Plate XLIII and Breadsall Church, I 74 Plate XLIV; and plan of 'The Larches,' I 75 Plate XLV; of Elston Hall, I 30 Plate XXIII; of Heydon Hall, I 246 Plate LXIV; of Claverdon, House and Church, I 48 Plate XXIX; of Loxton, of Dudson, of Whaley, of Hadzor, I 48 Plate XXIX; of Duddeston House, I 48 Plate XXX; from the Diary of Sir Francis Darwin's Boyhood, I 16 Plate V, I 18 Plate X, I 22 Plate XIX; of Francis Galton, by Evelyne Biggs, II viii, and II 425 Plate LIV, by Frank Carter, II xi, and III^{A} 432 Plate XL; of d'Arnaud Bey, III^{B} 455 Plate XLVI; water-colour, of the children of Dr Erasmus Darwin and his wife Elizabeth, I 246; of Francis Galton in the

'Fallow Years,' III^B 456 Plate XLVII; Galton's love of sketching, I 94, 95, 132, 133, 135; sketches by Galton, of Bishop's Gateway at Liége, I 94 Plate XLIX, of his rooms at Trinity College, I 150 Plate LI, III^B 453 Plate XLIII, of Ely Cathedral and of King's College Chapel, I 167 Plate LII, of last meeting of Caseo-Tostic Club, I 181 Plate LIV, of Emma Galton and Julia Hallam, I 180 Plate LIII, of corona and brushes of total eclipse in Spain, II 9–10; from Galton's Egyptian sketchbook, Bob and Ibrahim, III^B 454 Plate XLIV, of Ali from Galton's Syrian sketchbook, III^B 454 Plate XLV; from Galton's South African Diaries, 'Nangoro,' I 59 Plate XXXVIII, I 237 Plate LVIII, Ovampos and Omutchikota (Otchikoto?), I 216 Plate LVII, of lion-trap and pencil snapshots, I 215 Plate LVI, of Jonker Afrikaner walking off with Galton's code of 'laws,' I 226 Plate LVIII, of Galton's favourite hack in Damaraland, I 237 Plate LIX

Skewness, in the frequency distributions of sociological phenomena, II 228

Skin, colours of, II 224, 226; and heredity of, III^A 60

Skull, the English, III^A 253; 17th century English, III^A 257

Slaughter, Dr J. W., and Eugenics, III^A 372, 427; and Eugenics Education Society, III^B 585, 628

Slaves, monetary value of, in 1846, II 395; Galton's value as a slave, II 395; Galton's and Barclay's handling of, I 32, 39; purchase of a female slave in Damascus, III^B 454

Slavish Aptitudes in man, source of, II 72–74; slavish acceptances in man, II 257, 258

Sleeping Bag, introduced by Galton to Alpine climbers, II 6, 7

Slide Rule, Galton's use of, I 115

Smallpox, Galton's paper on, mortality and vaccination, III^B 482

Smells, tests for sense of, II 223; imaginary, and mental processes, II 275, 276; arithmetic by, II 275; do dogs think by? II 275

Smile, photograph of a, II 311, 312

Smith, Adam, on generic image of man, II 298

Smith, Mrs Archibald, her African trophies, III^B 549

Smith, Arthur H., experiments with photographing a bust 'all round,' III^B 520 and Plate LIV

Smith, Bosworth, Moslem religion better suited than the Christian to Orientals, I 207

Smith, Ethel Marshall (daughter of Sir Douglas Galton), dines with Francis Galton, III^B 532

Smith, Sir Harry, seen by Galton in Cape Town, I 219–221, 225

Smith, Dr Lyon, visits, for tea and talk, Galton just before his death, III^A 433

Smith, Prof. Robertson, pronounces Jahveh properly, is cursed by a great Rabbi and dies! III^B 608

Smith, Walter, a second wrangler, makes inquiries as to cousin-marriage, III^B 470

Smyrna, Sir Francis Darwin visits cases of plague in, I 23; visited by Francis Galton, I 138

Snails, at Biarritz, III^B 558

Snoring, Galton's discussion of, III^B 482–483

Snow, E. C., working in Eugenics Laboratory, III^A 431

Social Causes, considered by de Candolle as more important than heredity, II 146

Social Conventions, Galton's carefulness with regard to, I 93, 94

Social Duties, hampering effect of, II 154, III^A 112; yet may not be disregarded, II 246

Social Evolution, views of Benjamin Kidd on, Galton's views on, III^A 88 etc.; Prof. Haddon on, III^A 267–268

Social Influence, power of, II 91, 92

Social Phenomena, application of mathematics to III^A 1

Social Stability, permanence of elements of social strata, inquiry concerning, II 83, 350–351

Social Statistics, Galton and Florence Nightingale's letters concerning, II 416–424

Social Utility, according to Galton the primary purpose of science, I 56, 57

Société de Psychologie physiologique, circulates a questionnaire, III^B 478

Society, stability of, II 83, 350–351

Sociological Phenomena, frequency distributions of, II 227–228

Sociological Society, Galton's lecture and papers for, III^A 259, 261–267

Sociologists, their views on Eugenics, III^A 259–261

Sociology, is there yet a science of? III^A 261

Socrates, ability of, II 107; Aristides' feeling for, III^A 239

Sol, Galton's definition of a, III^B 565

Somatic Characters, Galton on, II 146, 148; transmission of, II 170; variation in, II 171; correlation with gametic characters, II 171–173; distinction between, and gametic characters, II 174

Somerville, A. A., on comparative reliability of medical and literary tests, II 388

Sons, of gifted fathers, extent of their gifts, III^A 102–103

Sorby's analysis of pigments in human hair, III^A 97; paintings of trees from these pigments, III^A 97 Plates III and IV

Sounds, association with colours, II 243; measurement of resemblance in, by least discernible differences, II 303

South Africa, travels in, I 215, 240. See also *Africa*

Späeth, J., and proportions of like and unlike twins, II 128

Spain, Galton's first travels in, II 6–7; his second visit to, III^B 507–512

Span of Arms, percentile values of, II 376; value of, at each rank, II 390

Spartan Methods, of mating, Galton on, II 110

Species, origin of, and mutations, II 84; and genera, II 171; and races, II 171. See also *Sports*

Spectacles, for divers, and vision of amphibious animals, II 34

'*Spectator, The*,' and Francis Galton's writings, II 351

Speed, of American trotting horses, II 399, 400

Speed, and accuracy of hand in men and women, II 376

Speedometer, for cycles suggested by Galton, II 60, 61

Speke, relations to Francis Galton, II 25, 27, 28; letter of, to Galton, II 26; his relations with Burton, II 25, 26, 27; death of, II 27; and African exploration, II 30, 36; Speke Obelisk, II 25; memorial to, III^B 588

Spencer, Herbert, and Galton, II 62; and 'The Reader,' II 67, 68; and agnosticism, II 102; and transmission of acquired characters, II 147, 148; stature of, II 150; and composite portraiture, II 239; goes to Derby, III^A 123; on Down as experimental breeding station, III^A 130, 134; on origin of finger-prints, III^A 142; his theories versus facts, III^A 142; Mrs Sidney Webb on, III^A 239; characterisation of, III^A 317; personal friend of Galton, III^A 434; as an investigator, III^B 614; Galton's reminiscences of, III^B 626–628; cremation of, III^B 524, 531, 614

Sphygmograph, and measure of emotional shock, II 270

Spinning Imbeciles and pigment, III^A 372

Spinoza, Pollock's book lent to Galton, III^A 313; his interest in, III^B 449

Spiritualism, Galton's interest in, II 51, 53, 62–67, 167, 169; Darwin's interest in, II 62–67, 167–168

Spoglio, at Trieste quarantine station, 1840, Galton 'makes Spoglio,' I 139
Sports, transmission of, III^A 31; in evolution, III^A 61, 62, 81, 87; and variation, III^A 79, 80, 170; breeding from, III^A 81–82; source of, in 'positions of stability,' III^A 85; place in evolution according to Galton, III^A 94, 99, 126, 170, 370; prepotency of, III^A 98–100; part in Natural Selection, III^A 170; in the human race, III^A 120, 121; are they essential to progress? III^A 221. See also *Mutations*
Spottiswoode, W., and Galton, II 11, 36; size of head and stature, II 150; large head of, III^A 248; at exhibition of scientific apparatus, II 215; letter to, when Pres. Roy. Soc. from Galton on reform of meetings, III^B 468; Galton's obituary notice of, II 245
Spring Fret, in Galton, I 197
Squeeze of Hand, maximum value of, II 374; percentile values of, II 376; values at each rank, II 390; correlation with vital capacity, II 377; 'Punch,' and the 'squeeze of 86' in a lady, II 375
Stability, or chronic variation in psychic characters, suggestion that mean and personal variation are correlated, II 103; of anthropometric constants, II 380; organic, III^A 61, 85–87; in a breed after selection, III^A 93; social, see *Social Stability*
Stability of Types, Galton's view, III^A 240
Stablemen, inferences from hats of, III^A 249
Standard Deviation, and quartiles, II 339, III^A 47; and normal curve, III^A 8. See also *Variability*
Stanley, H. M., Galton's opinion of, II 25, 27, 51; on Congo travels, II 30; methods of exploration, II 30, 31
Stanley, Lady, Galton stays with, III^B 510
Star Signals, i.e. to distant planets, II 279, 280
Starvation, Galton's experience of, I 224
Statesmanship and Eugenics, III^A 348
Statesmen, neglect the future, III^A 242; errors of ageing, III^A 365; and Eugenics, III^A 311, 312; fertility of, II 94; grading of, II 93
Statistical, fallacies, in oceanic meteorological observations, II 54; inquiries of national importance, II 156, 416–420; laboratories, schools as, II 344–345; instincts of Galton, II 149–150; results, interference by theocratic power with, II 258–259; material, difficulties in obtaining, II 276, 357, 358, 368, collection of, from schools, II 336, 337, 343–346, 370, by schedules, II 348–356, trustworthiness of, II 361, 368; methods of Galton, II 150, 268, and psychology, II 212, 236–238, and duty of Galton Research Fellow, III^A 222, of Meteorological Office, II 53, first applied to heredity, II 89, 92, 93, applied to grading of intelligence, II 89, 90, illustration of use of, by Galton, II 343–344, views of A. R. Wallace on, III^A 133, views of Darwin on, III^A 246, deficiency of de Candolle in, II 146, 147; operations, a textbook on, III^A 248; ratios, stability of, and theocratic interference, II 259; scale and the grading of the individual, II 337; Statistical Society of London and Galton, II 123
Statisticians, mechanical aptitude of, II 151; waste of effort by, II 420
Statistics, and Samuel Galton, I 48; and Samuel Tertius Galton, I 52, 57; and Erasmus Darwin, the younger, I 57; application of, to hereditary problems, III^A 1–137, to anthropology, II 334–348, III^A 57; and the Darwinian hypothesis, III^A 126; Galton's enthusiasm for, III^A 63; proposed Professorship of Applied, II 414–416, 424, Florence Nightingale on function of, II 414–415; Galton and, II 70, 129, III^B 458; sources of Galton's interest in, II 201; bearing on inheritance of ability, II 77; on twins, II 128–129; Ansell's, of families, II 128; and inheritance of stature, II 210; psychometric, II 236, 237; of population, II 265, 266; necessity of, for testing and correcting impressions and opinions, II 296, 297; and Eugenics, III^A 221, 222; modern and old schools of, II 348; origin of mathematical theory of, II 357
Statoblasts, a paper on heredity of, III^A 245
Stature, of female reduced to male equivalent, III^A 15; statistics of, III^A 54; in husband and wife, II 149; of scientists, II 150; of boys at various ages used by Galton to illustrate his statistical methods, II 343; means of racial, from small samples, II 179; French, effect of conscription on, II 191; and its inheritance (de Candolle), II 210; effect of diet on, II 210; maximum values of, II 374; percentile values of, II 376; values for each rank, II 390; correlation with vital capacity, and the isograms, II 391; inheritance of, III^A 11–34 (and regression); Forecaster of, III^A 13, 15, 17, and Fig. 5, III^A 16; assortative mating in, III^A 17; advantages of, as subject of study, III^A 18, 19; prediction of, III^A 31; correlation with cubit, III^A 51; correlation table for cubit with, III^A 52; progression in, by continuous selection, III^A 93; genometer based on Pearson's data for stature, III^A 30 Plate I
Steadfastness of Purpose, in Galton and his Quaker ancestry, I 59
Steadiness of Hand, in men and women, II 376
Stellar Characters, controversy regarding their correlation, III^A 326
Stelvio Pass, geographical model of, II 33, 34 Plate IV
Sterilisation of the Unfit, III^A 218
Sterility, and genius, II 341; in experiments with transfusion of blood, II 161; voluntary, in France, II 142
Stewart, Augusta R., second wife of Herman Galton, III^B 511
Stewart, Balfour, and Galton, II 48, 49; and 'The Reader' and 'Nature,' II 68, 69
Stewart, C. P., Galton's Cambridge friend, I 153; passes 'Little Go,' I 164; at last meeting of Caseo-Tostic Club, I 181 Plate LIV
Stewart, Dugald, on generic image of man, II 298
Stimulation, external and internal causes of, II 308
Stirp, and race, II 171; definition of, II 185–186; extinction of, II 343; and pure lines, II 171
Stirpiculture, or Eugenics, III^A 259
Stochis, on finger-prints in different races, III^A 140, 194
Stokes, Prof. Sir G. G., religious bias of, II 152; on Roy. Soc. Committee on Colour-blindness, II 227; painstaking editorial labours of, III^A 331; Galton's letters to, III^B 466–468, 471–473; handwriting of, III^B 522; remarkable head of, III^A 248
Stokes, Mr, Galton's interest in, III^B 576
Stokes, —, undergraduate at Trinity, 1842, overworks and loses his scholarship, I 171
Stout, Sir Robert, advocates Eugenics in New Zealand, III^B 612
Strachey, Miss Philippa, III^B 541
Strachey, Sir Richard, and Galton, II 44; and Kew Observatory, scores off the Roy. Soc., II 60; and maps, III^B 462
Strahan, Major-General, on use of finger-prints in India, III^A 153
Strasburger, E., Linnean Society medallist, III^A 340; death of, III^A 342
Strawberries, Linnæus' cure for gout, III^A 124
Strenuousness, national value of, III^A 401–402
Strickland, C. W., joins Galton on reading party, I 155, 159
Strikers, in Italy, III^B 540
Strutt, marriage in Galton family, I 53

Studbooks, little service for inheritance, II 321
Studdy, Lucy (Francis Galton's niece), assistance of, I viii; Galton visits, III^A 281, III^B 569, 570; stays with Galton, III^B 583, 596; letter to, from Galton, III^B 452; wins prizes for cats, III^B 515, and for needlecraft, III^B 599; leaves relics of Erasmus Darwin to Galton Laboratory, III^B 571
Subconscious Mind, II 236, 247, 307; inspiration a product of, II 412
Success, as measure of ability, III^A 111, 112; of kinsmen of Fellows of Roy. Soc., III^A 108
Sully, Prof. J., and tests on idiots, II 272
Sun Signals, Galton's, II 20–21. See also *Heliostat*
Sundial, Galton writes motto for, III^B 547; form of, in Pyrenees, III^B 573
Superman, reasoned method of producing, II 78, 79, 86
Superposition of Images, methods of attaining, II 284, 285, 287–289
Superstition, as a source of national or tribal strength, III^A 88, 423–425, III^B 594; as opposed to facts, II 260
Surgical Operations, Galton studies at King's College, London, I 121
Survival, of fittest, Galton on, II 110; of a breed menaced by regression, not a reality, III^A 94
Sutherland, Alexander, his book on 'The Origin and Growth of Moral Instinct,' reviewed by Galton, III^A 100
Sven Hedin's arrival at Simla, III^B 588
Sweet-Peas, Galton's experiments with, II 180, 181, 187, 189; III^A 3–7, 11, 13; inheritance of size in seeds of, II 392; cross or self-fertilisation of, III^A 325, 326; conclusions from experiments on, III^A 64
Switzerland, Galton's knapsack guide to, II 11; his visits to, II 144, III^B 514; Eugenics in, II 267
Sylvester, J. J., size of head and stature of, II 150; large head of, III^A 248
Symbolism, in science, III^A 330, 333, 336
Symmetry, of the two sides of the body, II 377, 379; tendency to, in finger-prints, III^B 485; attempt to show normal, in finger-print distributions, III^A 168
Syria, Galton's travels in, I 197, 198, 203–205, 207, III^B 454–455; sketchbook in, III^B 454 Plate XLV

Tables, showing mathematical tripos marks and number of candidates, II 89; 'eminence' in fathers and sons, II 105, 106; size of family in rural and urban populations, II 124; occurrence of twins among relations of twins, II 129; size of head and stature in scientists and others, II 150; sensitivity in hearing of men and women, II 221; of associated ideas and period of life at which they originated, II 235; estimate of dressed weight of living oxen, II 403; assortative mating in stature, III^A 17; correlation, for stature and cubit, III^A 52; colour inheritance in horses, III^A 96; concerning finger-prints, III^A 184–186, 189, 192, 195, 211
Taboo, influence of, on conduct, III^A 270
Tabor, proposes use of finger-prints for registration of Chinese, III^A 175
'Tactor' Machine, diagram of, II 50
Tailoring, effect of, on finger-prints, III^A 155
Talents, inheritance of, II 75–79; and characteristics of Darwin and Huxley, II 178, 179; new version of the parable of the, III^A 227–229
Tanganyika, Lake, discovery of, II 25
Tangier, Galton at, III^B 509, 510
Taylor, Henry, his couplet on advantages of unattractiveness, III^A 112
Taylor, H. M., proposal of, for the blind, III^B 584
Taylor, Sedley, and Eugenics, III^A 378
Tchapupa, and his wife from Damaraland, sketches of, I 216 Plate LVII
Tea, experiments on the making of, III^B 456–458
Teachers, replies of, to questions, III^B 478; Guild of, Galton attends a meeting of, II 276
Teaching, on effect of, II 135; men of science not made by, 155; by association with form, III^B 472
Technical Terms, in scientific memoirs, III^A 334, 336, 337
Teeth, and personal identification, III^A 188; anecdote of missing artificial, believed to be swallowed, III^B 492, 493
Teetotalism, Galton takes the pledge, I 183–185, 191
Telegony, Darwin's belief in, II 159; co-operation regarding observations on, III^A 135
Telegraph, a printing (the 'telotype'), designed by Galton, I 212, 213
Telephone, Galton's anticipation of, I 213, 214
Telescope, superposition by aid of, II 284, 285, 287
Telotype, I 212, 213, 217
Temper, measurement of, II 270–272; categories of, II 271; inheritance of, II 271, 272, III^A 70; assortative mating in, II 271, III^A 70; good and bad in English families, II 271, III^A 69
Temperament, measurement of, II 268, 269; correlation of, with mental qualities, II 229; in husband and wife, II 149
Temperature, cumulative, effect of, on vegetation, II 206–208; charts of, II 41
'Temple of Nature,' Dr Erasmus Darwin's, II 95
Tennyson, Alfred, and observation of emotional change, II 270; imagination of, II 308; Galton's appreciation of, III^A 337, 338
Terror, Galton on nature of, II 258
Theocratic Intervention, possible effect of, on statistical conclusions, II 258–260
Theologians, and Galton's work, II 258
Theory, verification of statistical, by experiment, II 405
Thermometers, standardisation of, II 59
Thompson, G., uses finger-prints to sign money drafts, III^A 175, 176
Thomson, Arthur, composite cranial photographs by, II 290, 294
Thomson, Prof. Arthur, Galton gives his anthropometric instruments to, III^A 328
Thomson, J. Arthur, his book on Heredity, III^A 335
Thomson, Sir W., aids 'The Reader' and 'Nature,' II 68, 69; at exhibition of scientific instruments, II 215
Thought, Galton's categories of, II 233; abstract, and mental imagery, II 237; mechanism of, II 256; without words, II 274, 275
Thrills, associated with roughness of surface touched by finger-tips, III^A 179
Thring, Lord, on the ablest Cabinet Ministers, III^B 568
Tichborne Trial, and need for anthropometric measurements, II 398
Tickner, Life of, II 192
Tides, big, and the earth's history, III^B 470
Tiedemann, Galton sees, in Heidelberg, I 95
Time, sense of, not absolute, III^A 354, 355
Tint, average, of a picture, method of obtaining a measure of, III^A 307, 308; instrument for testing perception of difference in, II 219
Tints, Galton's scale of, II 313, 314, II 314 Plate XLV; photograph of spinning wheel of, II 313
Titchener, E. B., finds no relation between sensitivity and ridge interval of finger-prints, III^A 168, 178
Tocher, Dr J. F., anthropometric survey of, II 380
Todd, Dr, I 179, 190; Galton studies physiology under, I 105, 106
Tönnesen, photographs Lake Otchikoto, III^B 616 Plate LIX

Tollemache, L., and Galton, III^{B} 513, 554, 557
Tones, Galton's scale of, II 313, 314, II 314 Plate XLV; painting of portrait in various, II 314 Plate XLV; mosaics of different, II 314
Tooth-Drawing, Galton's first experiment in, I 102
Topinard, scale of hair colour of, II 225; on value of anthropometric records, II 398; letter of, to Galton, II 226
Torsion Anemometer, devised by Galton, II 44
Toss-Penny, Galton's mechanical, III^{B} 545
Touch, and muscular appreciation, II 218; sensitivity of, in men and women, II 222; measurement of, II 374, and finger-prints, III^{A} 168, 169, 178, 179
Town Life, Galton on, II 118; deterioration arising from, II 124–126, 143; queries regarding effects of, II 419
Townsend, Mrs, at Rutland Gate, III^{B} 605
Toyama, K., work of, on silkworms, III^{A} 301
Trace Computer, Galton's, II 47, 48
Tracey, J., on eugenic certificates, III^{A} 296
Trades-Unionism, and racial progress, II 121
Tradition, and heredity, III^{A} 409–411
Training, early, effects of, II 96, 97
'*Transformer*,' to measure difference between individual and type, II 311–315
Transfusion, experiments begun, II 88. See also *Pangenesis*
Transition Studies, of Galton, II 1–69
Transmission, of hereditary characters, II 76, 77, 170, 182; mechanism of, II 182–186
Travel, Galton and, I 24, 25, II 2–35, 49, 53, 73, 88, 152; need of aim and preparation for, I 197; Galton's enthusiasm for, I 122; and object in, I 24, 25; and rights of natives, II 30, 31; and climate, II 36; fellowships for, II 152, 153. See also *Travels*
Travels, early, of Galton, I 92–98, 118, 120, 127–137, 178–180; in Egypt and Syria, I 199–205; exploratory in Africa, I 214–240; summary of Galton's, III^{B} 443–445; in Spain and Tangier, III^{B} 507–512; in Egypt, III^{B} 515–519; of Sir Francis S. Darwin, I 22–24; of Charles Darwin, I 24; of Admiral Sir Thomas Button, I 36
'*Treasury of Human Inheritance*,' prospectus and first appearance of, III^{A} 345, 346; suggestions regarding, III^{A} 336, 342, 343, 361, 369; Galton's introductory note to, III^{A} 361, 362; progress of, III^{A} 356, 357, 360, 371, 372, 376, 377, 379, 389, 390, 426
Trench, G., to go to Boer war, III^{B} 518
Trevelyan, Sir G., 'Life of Fox' by, III^{A} 137; hon. fellowship of, at Trinity College, Cambridge, III^{A} 236, 238
Triangulation, table for, II 23, 24
Trinity College, Cambridge, Galton's entry at, I 140; rooms at, I 149, 150; sketches of rooms at, I 150 Plate LI, III^{B} 453 Plate XLIII; Galton elected to honorary fellowship at, III^{A} 236–238, 250, III^{B} 521; Galton's portrait at, III^{A} 379, III^{B} 550–553, 572; Galton dines at, III^{B} 574
Troite, Mr A., and Galton, III^{B} 551, 552
'*Tron Kirk*,' material of, on colour of horses, III^{A} 95
Tropics, and the white races, II 33, 395, 397
Troup, C. E., and identification of criminals, III^{A} 148
Tubercular disease, course of, and physical characters, II 292. See also *Phthisis*
Tuppy, on careers of Indian Civil Servants, III^{A} 247
Turgot, and application of mathematics to social phenomena, III^{A} 1
Turner, Prof. H. H., and Galton, III^{A} 400
Turner, Sir W., and Miss Alice Lee's degree, III^{B} 514
Twins, heredity in, II 126–130, 180, 269; pedigrees of, II 129; statistics concerning, II 128, 129, 180; fertility of, II 128; Dr Ogle's case of, II 181; like and unlike, II 187, 188; hermaphrodites among, III^{A} 359; finger-prints of, III^{A} 190, 191, III^{A} 191 Plate XVIII; Mrs Jebb's account of, III^{B} 465
Tylor, and 'Nature,' II 69; and anthropology, II 334
Tyndall, J., lecture of, II 191; at exhibition of scientific instruments, II 215; a friend of Galtons, III^{A} 434
Tyndall, Mrs, at Hindhead, III^{A} 277, 323, III^{B} 584
Type, anthropometric description of, II 380; and individual, measure of difference between, II 311
Types, origin of, III^{A} 31; stability of, and regression, III^{A} 46; change of, and law of ancestral inheritance, III^{A} 48, 49; of races, comparison of, III^{A} 325, 326

Umbrella, lost and 'stolen,' III^{B} 476
Unemployed, parentage of the, III^{A} 327
Unicorns, reports of, in Africa, I 237
Unit Characters, and heredity, II 182–186, 189, 190
Universe, the, Galton on, II 262; a correlated system of variates, III^{A} 2
Universities, anthropometric laboratories in, II 337
University College, experiments at, II 162, 167
University Honours, Galton's early desire for, I 69
University of London, Galton's offer to, III^{A} 222–225, 258, III^{B} 530; chair of Eugenics in, III^{A} 224; Eugenics Record Office and, III^{A} 223, 224, 303; Galton Laboratory and, III^{A} 304–307, 386, 387, 393, 428, 429, 432; attempted foundation of a real, III^{A} 289–291
Urban Population, fertility of, etc., II 123–126
Ury, the home of the Barclays, I 28–30, I 30 Plate XXIII; Galton visits and sketches at, I 104, 105
Utopia, in dreamland of philanthropists, III^{A} 220; of Galton, II 119–122, 'Kantsaywhere,' III^{A} 234, 411, III^{B} 606–608, Galton's and others', III^{A} 411

'*Vacation Tourists*,' II 6, 7, 11
Vaccination, and smallpox (paper by Galton), III^{B} 482
Variability, regression, and correlation, III^{A} 3–11; in offspring of a given parentage, III^{A} 11; fraternal and co-fraternal, III^{A} 19, 77, 78, 221; within the family, III^{A} 26, 47; influence of selection on, III^{A} 27, 93; of mid-ancestor, III^{A} 29; interracial and intraracial, III^{A} 95; Brewster's measure of, III^{A} 95. See also *Variation*
Variation, necessity of, II 174; and selection, II 171; in human faculties, II 274; and hybridisation, II 84; heredity of, and man, II 86; measurement of, II 384, 385, III^{A} 51, 54; individual and specific, III^{A} 95; continuous, and evolution, III^{A} 126; universality of, III^{A} 314
Variations, and sports, III^{A} 79–86, 170; on observations regarding, III^{A} 135
Varieties, Romanes and Galton on origin of, II 272, 273; mode of formation of new, III^{A} 94
Vathek, legend regarding, III^{B} 573
Veins, bifurcations and interlacements of, and identification, II 306
Venables, Galton joins reading party under, I 168
Venn, Dr J., and head size of students at Cambridge, II 387, 388; his 'Logic of Chance,' III^{B} 477, 478
Veracity, in family histories, II 368, 369; in men of science, III^{B} 478, 479
Vernon-Harcourt, on Committee of Brit. Assoc., III^{B} 458
Versailles, peace terms at, and rights of small nations, II 264
Verses, of Galton, I 90, 175–178, 183, 184; of Tertius Galton on his sister Sophia, III^{B} 450; of Waller on Elizabeth Savile, I 20; on 'The Château in the Heart of the Ardennes,' III^{B} 459, 460. See also *Poem*
Vesuvius, Galton sees, III^{B} 474, 475
Victoria, Queen, Galton and his brother Darwin at coronation of, I 91
Victoria Nyanza, discovery of lakes of, II 25, 26

Vienna, Galton at, I 96, 133, 134
Viriculture, II 119. See also *Eugenics*
Virtue, characters associated with, II 197
Vision, measurement of acuity of, II 222, 223, 336; in Jewish children, III^A 426
Visionaries, Galton on, II 254
Visionary tendencies, repression of, II 244, 245
Visions and visualisation, II 244
Visits of Galton, summary of, III^B 443–445
Visualised Numerals, III^B 469. See also *Number Forms*
Visualising, faculty of, II 236, 244; in Darwin, II 194, 195; in Dr Erasmus Darwin, II 196; in scientists, II 237, 243; in artists, etc., II 243; in orators, II 252; in women, II 242; in different races, II 239, 240, 252; on starvation of, II 240; descriptions of, II 240; ideal form of, II 241; utility of, II 253; and use of words, II 274; inheritance of, II 239, 242, 253; acquired by training, II 253
Vital Capacity, maximum values of, II 374; percentile values of, II 376; growth curves for, II 377; correlation of, with strength of squeeze, II 377; values of, at each rank, II 390; and stature, isograms of, II 391
Vivacity, in pairs of brothers, III^A 247
Vogt, paper of, II 195
Voltaire, skit of, III^B 535
de Vries, H., and Galton's views, III^A 82; cited by Galton, III^A 120

Wadi Halfa, Galton reaches, I 203
Waite, H., work of, on finger-prints, III^A 140
Waldstein (Sir Charles Walston), on Herculaneum, III^B 589
Walfisch Bay, Galton at, I 221, 222, 236–238
Walker, E., at Cambridge, I 167, 190
Walker, Prof., on mimicry, III^A 370
Wallace, year books of American trotting horses, III^A 98
Wallace, Alfred Russel, and spiritualism, II 62; associated with 'The Reader' and 'Nature,' II 68, 69; on 'Hereditary Genius,' II 115; on Galton's theory of heredity, II 187; on Galton's paper on Twins, II 187; on need for experimental zoology, III^A 81, 128, 129; correspondence of, with Galton, on experimental breeding, III^A 128–133; and statistical method, III^A 133; on Darwinism, III^A 370; medallist of Linnean Society, III^A 340; Copley medallist, III^A 431, III^B 614; at Darwin's funeral, II 198; death of, III^A 342
Waller, on Elizabeth Savile, I 20
Wallington, Col., on visit to Fazakerley, with Galton, III^B 456
'*Walloping*,' and its derivation, III^B 591
Walston, Sir Charles, finds finger-prints on a Greek seal, III^A 174. See also *Waldstein*
Wanderlust, in Galton, I 24, 55, 58, 120, 122, 123, 126–128, 130, 132, 134, 135, 139, 205, II 1, etc.
Ward, Dr J., letters of Galton to, II 214, III^B 479, 493
Warming rooms, Charles Darwin's method of, III^B 591, 592
Warner, Francis, on Eugenics, III^A 259
Warren, Samuel, and a lecture at Leamington, III^B 536
Wasps, work on, III^A 312
Watson, Rev. H. W., and Galton's wave machine, II 52, 53; work of, III^B 486; on the extinction of surnames, II 341–343, III^B 461
Watson, Dr Spence, Galton's visits to, II 103, 393; and the widow of the Sheriff of Wazan, III^B 510
Watt, James, copying machine of, III^B 580
Watts, G. F., Galton's appreciation of pictures of, III^B 583
Wave Engine, Galton's invention of, II 51, 52
Wavell, Mrs W., assistance from, I viii
Wax, in the ear, Galton prescribes for, III^B 464
Wealth, evils of inherited, II 83, 118; misuse of, II 154; and social stability, II 113
Weather, charts, origin of, II 41–43; preparation of, II 42; problems concerning, II 54–58; proposals for posting up at the Meteorological Office, information regarding, III^B 481
Webb, Jonas, breeder of Southdown sheep, statue to, at Cambridge, III^B 463
Webb, Mrs Sidney, appreciation of Galton by, III^A 239, 240
Webbed Soles, for swimming, III^B 546
Weber-Fechner Law, Galton's use of, II 227; and Gaussian hypothesis, II 227; as source of error in mental impressions, II 296, 297; and tests of muscular sense, II 217, 218; model to illustrate action of, II 307
Webster, Mr, Basque scholar, III^B 559–561
Wedderburn, pamphlet of, II 181
Wedgwood, Hensleigh, and spiritualism, II 66
Wedgwood, Josiah, and Dr Priestley, I 44
Wee Ling, an albino Pekingese puppy, finds a home with Galton, III^A 391–393, 395, 397, 398, 427, III^B 600, 603, III^A 390 Plate XXXVIII; and Confucius, III^A 399
Weeds, in the garden of humanity, III^A 220
Weight, of boys in town and country schools, II 125, 126, 337; and muscular sense, II 217; maximum values of, II 374; percentile values of, II 376; values of, at each rank, II 390; of British noblemen during three generations, III^A 136, 137
Weismann, A., letter of, to Galton, on continuity of the germ-plasm, etc., III^A 340, 341; medallist of Linnean Soc., III^A 340; hon. member of German Eugenic Soc., III^A 388; death of, III^A 342
Welby, Lady, and Galton, III^A 379, 380, III^B 529, 584
Weldon, W. F. R., and Galton, II 419, III^A 239, 241, 318; and Galton's School, III^A 57; co-operation of, II 321, 322; and experiments with dice, II 405; 'Biometrika' and, III^A 100, 243–245, 250, 281, 334; on Committee for Measurement of Plants and Animals (or Evolution Committee), III^A 126, 127, 287, 288, 290, 291; early correlations determined by, III^B 483, 484; some work of, III^A 126, 127, 251, 259, 279; on farm for experimental breeding, III^A 134, 135; letters of, III^A 285, III^B 542; correspondence of, with Galton, III^B 483, 484, 525–527, 535, 540–542, 561; death of, III^A 280, 291, 292, III^B 568–570; characteristics of, III^A 282; eloge on, III^A 281, 285, 287, 289, 301; memorial to, III^A 281, 284–286, 291; portraits of, III^A 282. See also *Weldon Memorial*
Weldon, Mrs, work of, on mice, III^A 315, 360
Weldon Memorial, III^A 127, 224, 281, 284–286, 291, 297, 315, 328, 333
Wellington, army of, III^B 555
Wells, H. G., on Eugenics or stirpiculture, III^A 259; on methods of improving the human race, III^A 259, 260
Welsh, finger-prints of the, III^A 193
Wen, inheritance of a, III^A 375
West Indian Islands, excess of females in, II 337
Westcott, Bishop, Hon. Fellow of Trinity College, III^A 236, 238
Westermarck, E., chairman at Galton's lecture on marriage restrictions, III^A 266, 268, 269
Westmacott, bas-reliefs of, for Royal Exchange, I 189
Wetherby, studbook of, III^A 96
Whaley, home of Hubert Galton, sketch of, I 48 Plate XXIX
Wheatstone, Sir C., Galton on, II 208
Wheler, Elizabeth Anne, Galton's 'Sister Bessie,' on the birth and childhood of Galton, I 63–65, 82; early letters of Galton to, I 95–97, 110, 111, 116–120, 159;

on Galton's escape from drowning, I 116; on a tour with Galton in Scotland, I 104, 105; concerning Charles Darwin, I 68; early bequest of Galton to, I 69; her 'Account of the Galton Family,' I 48 Plate XXIX, and sketch of Great Barr, I 49 Plate XXXI; 'Reminiscences' of, I 51, 52; marriage of, I 193; Galton advises on education of son of, III^A 302; aged 94, III^A 238; death of, III^A 278, III^B 559–561, 570; letters of Galton to, III^B 451–453, 464, 488, 492, 506, 508, 509, 511, 512, 515–519, 527, 528, 547–550, 552, 553; portraits of, I 96 Plates L and L bis, II 332 Plate XLVI; silhouette of, I 52 Plate XXXIV
Whethams, the, interest of, in the feeble-minded, III^A 373; lecture of, III^A 427; on Eugenics, III^B 600
Whewell, Master of Trinity, his wooing, I 157–159, 162, 163
Whipple, and use of composite photography in meteorology, II 290
'*Whisky*,' occurrence of the word in the Bible, III^B 607
Whistles, Galton's, for high notes, II 215–217, 221, 222
Whistling, through the fingers, III^B 578
White, A., Galton meets in Italy, III^B 475
White, Gilbert, 'History of Selborne,' III^B 540
White Man, possible descent of, III^A 369, 370
Whiteley, M. A., her work in 'Biometrika,' III^A 256
Whiting, on the best method of cooking that fish, III^B 581
Whorls in finger-prints, types of, III^A 209, 210, III^A 213 Plate XXV. See also *Finger-Prints*
Whyte, F., and phrenology, III^B 577
Wilcox, Mr, at funeral of Mrs Galton, III^B 503
Wildbad, visit to, II 280
Wilkes, Samuel, on medical life-histories, II 359
Will, limitations of the, II 241
Willis, Galton to attend lectures of, at Cambridge, I 172
Wilmot, Rev. Darwin, and portraits of Darwin ancestry, II 192 Plates XV, XVI; assistance of, I viii
Wilmot, Emma, to have medallion of Erasmus Darwin, III^B 473
Wilson, Sir J., assists in photographing horses at Horse Show, III^B 506
Windermere, Galton on, I 155, 156
Winds, chart of, II 42, 55, 56; problems concerning, II 54–57
Winthrop, R. C., at Darwin's funeral, III^B 471
Wise Men, and foolish, III^A 310
Withering, Dr, and 'The Larches,' I 62
Wives, of eminent men, II 105, 106; of able men, III^A 102
Wolf, L., on the Jews, II 208, 209
Wombwell's Menagerie, Galton in the lions' den at, I 151
Women, strength of, II 5, 107, 374–376; and marriage, II 132–134; of Spain and England, II 7; emancipation of, and Eugenics, II 134; mental imagery in, II 242; introspection in, II 242; Galton on, II 131–134, III^A 67, 278, 317; number forms of, III^B 469; measurements on, II 374; selection of civic worth in, III^A 232; academic work of, III^A 359; noteworthiness in, III^A 117, 118; and limitation of families, III^A 322; sensitivity of, and men, II 221, 222
Woodd, S., number form of, III^B 469
Woodward, Dr, visits Galton Laboratory, III^A 385, 386
Wootton Wawen, Galtons buried at, III^B 529
Words, and associated ideas, II 235; association of, with visualised pictures, II 243; poetic and scientific use of, III^A 337, 338; proposals regarding, for types of selection, III^B 505, 506
Wordsworth, imagination of, II 308; associated by Galton with insanity, III^A 115
Work, capacity for, II 91, 92; queries regarding hours of, and value of output, II 419
Worms, Galton's and Darwin's observations on, II 196, 197
Wright, Mr, and the title page of 'Biometrika,' III^A 246
Wright, Harold, his assistance in obtaining information as to Galton's Cambridge days, I 173, 175
Wundt, W., work of, II 211, 212

X-Club, dinners of the, III^A 239

'*Yaffles*,' Galton stays at, III^A 323–326, III^B 585
Yeoman, joins reading party with Galton, I 168
Young, with Galton, on reading party, I 159
Young, Thomas, Samuel Galton and Hudson Gurney, I 47–48
Youth, promise of, as a forecast for manhood, III^A 232; on modern, II 257; and age, III^A 318
Yule, Col., number form of, II 242, III^B 469
Yule, G. U., paper of, in 'Biometrika,' III^A 251; and technical terms in science, III^A 334

Zanzibar, as centre of missionary enterprise, II 28; as travellers' starting place, II 30
Zeometer, designed by Galton, II 50
Zoological Gardens, and experimental breeding, III^A 131, 287
Zoological Society, Galton's membership of, II 88; and experimental work, III^A 129
Zoonomia, of Erasmus Darwin, Darwin's and Galton's views of, I 13
'*Zygote*,' Galton's criticism of term, III^B 515

Heredity counts for much, for more than we reckon in these matters. We breed horses and cattle with careful study of the principle; the prize bull and Derby winner are the result. With mankind we heed it little, or not at all....What is genius? None can tell. But may it not be the result in character of the conflict of violent strains of heredity, which clash like flint and steel, and produce the divine spark?

LORD ROSEBERY.

Printed in the United States
By Bookmasters